IRAQ'S
ECONOMIC
PREDICAMENT

Also available

Exeter Arab and Islamic Studies series

Water in the Arabian Peninsula
Edited by Kamil A. Mahdi
2001 • 412pp • 235 x 155 mm • Cased £35.00 • ISBN 0 86372 246 6

State and Agriculture in Iraq: Modern Development, Stagnation and the Impact of Oil
Kamil A. Mahdi
2001 • 400pp • 235 x 155 mm • Cased £35.00 • ISBN 0 86372 279 2

Available from your local bookshop; alternatively, contact our Sales Department on +44 (0)118 959 7847 or email on **orders@garnet-ithaca.co.uk** to order copies of these books.

IRAQ'S
ECONOMIC
PREDICAMENT

EDITED BY
KAMIL A. MAHDI

EXETER ARAB AND ISLAMIC STUDIES SERIES

ITHACA
PRESS

IRAQ'S ECONOMIC PREDICAMENT

Ithaca Press is an imprint of Garnet Publishing Limited

Published by
Garnet Publishing Limited
8 Southern Court
South Street
Reading
RG1 4QS
UK

First Edition

ISBN 0 86372 276 8

British Library Cataloguing-in-Publication Data
A catalogue record for this book is available from the British Library

Jacket design by Garnet Publishing
Typeset by Samantha Barden

Printed in Lebanon

This book is dedicated to the memory of
Muhammad Salman Hasan in recognition
of an intellectual debt.

Contents

Tables and Figures

Acronyms and Glossary

AAD	Arab Accounting dinar – equals 3 SDR (IMF Special Drawing Right)
AFESD	Arab Fund for Economic and Social Development
AMF	Arab Monetary Fund
Anjuman	Village councils (Iraqi Kurdistan)
APICORP	Arab Petroleum Investment Corporation
BNL	Banka Nazionale del Lavoro
CBI	Confederation of British Industry
CBIq	Central Bank of Iraq
CIRR	Commercial Interest Reference Rate
CPF	Compagnie Française de Pétrole, one of the members of the IPC consortium operating in Iraq until the 1950s
CPI	Communist Party of Iraq
DDSR	Debt and Debt-service Reduction
DFID	The Department for International Development (formerly Britain's Overseas Development Administration)
DMFAS	Debt Management and Financial Analysis System
Donum	A measurement of land. In Iraq, one *donum* (or *meshara*) = 2,500m^2 = 0.25 hectare. In some other countries in the region, a *dunum* is given a different measure of 100m^2
DRC	Domestic Resource Cost
DRS	Debt Reporting System (of the World Bank)
ECA	Economic Cooperation Agreement
ECHO	European Community Humanitarian Office
EERC	External Economic Relations Committee (of Iraq's Council of Ministers)
EURODAD	European Network on Debt and Development
FDI	Foreign Direct Investment
FSCs	Former Socialist countries
FSU	Former Soviet Union
GATT	General Agreement on Tariffs and Trade

GCC	Gulf Cooperation Council
IAEA	International Atomic Energy Authority, Vienna
ID	Iraqi dinar
IDB	Islamic Development Bank
IEA	International Energy Agency, Paris
IFIs	International financial institutions
IKBDC	Iraq–Kuwait Boundary Demarcation Commission of the UN
IMF	International Monetary Fund
INOC	Iraqi National Oil Company
IPC	Iraq Petroleum Company, a consortium of IPSA
IPSA	Iraqi pipeline through Saudi Arabia
IsD	Islamic dinar – equals 1 SDR (IMF Special Drawing Right)
KD	Kuwaiti dinar
KDP	Kurdistan Democratic Party
LIBOR	London Inter-bank Offered Rate
MEED	*Middle East Economic Digest*
mujamma'at	Relocation camps for the civilian population, particularly those set up during the 1988 Anfal operations in Iraqi Kurdistan
Mustashar	The leaders of the Kurdish irregular troops in the so-called National Defence Battalions created by the Iraqi government to protect rural areas against Kurdish guerrilla activity
MVA	Manufacturing value added
NGOs	Non-Governmental Organisations
OAPEC	Organisation of Arab Petroleum-Exporting Countries
ODA loans	Official Development Assistance loans
OECD	Organisation for Economic Cooperation and Development
OFDA	American Overseas Fund for Disaster Aid
OPEC	Organisation of Petroleum-Exporting Countries
PSA	Production-sharing agreement (in connection with foreign investment in the oil sector)
PUK	Patriotic Union of Kurdistan
RCC	(Iraq's) Revolutionary Command Council

Sirkal	Work's manager or team leader, as agent of an agricultural landlord
Swing oil	An oil market term referring to a producer (either Saudi Arabia as during the first half of the 1980s or OPEC as a whole for most of the subsequent period) adjusting supply to meet short-term demand swings, and thereby avoiding major price ratchets
UBAF	Union de Banques Arabes et Françaises
UNCC	The UN Compensation Commission created after the Gulf War and based in Vienna
UNDHA	UN Department of Humanitarian Affairs
UNGCI	UN Guards Contingent in Iraq
UNMOVIC	United Nations Monitoring, Verification and Inspection Commission
UNSCOM	UN Special Commission created by Security Council Resolution 687
US EXIM facilities	Financial credit facilities provided by the US Export-Import Bank

Acknowledgements

Most of the chapters in this book were originally written as papers and presented to a symposium on 'Frustrated Development: The Iraqi Economy in War and in Peace', held in Exeter from 9 to 11 July 1997. The contributors are all recognised authorities in their respective fields, including some with long experience of research on the Iraqi economy and intimate knowledge of the country's institutions. A number of the papers have recently been revised and updated and they represent a unique collection in their range and in the subject matter they cover.

The symposium Advisory Committee was composed of Professor Abbas Alnasrawi of the University of Vermont, Dr Sinan Al-Shabibi of UNCTAD and myself. It was a rewarding and gratifying experience for me to have cooperated closely with these two esteemed colleagues, as with all others who contributed to the book. The symposium was convened by the Centre for Arab Gulf Studies,* University of Exeter, and we received cooperation from the Iraqi Economic Forum.

It is incumbent upon me to thank all the contributors for their great cooperation in providing excellent copy and for being so willing to update their work and answer queries. This book demonstrates a twofold commitment by the contributors: to scholarship and to the people of Iraq in their present suffering, and I am confident that its publication will encourage further research on the Iraqi economy.

It remains for me to thank those who have helped in the production of the manuscript. In particular, I wish to thank Lindy Ayubi for lending her multiple skills of copy-editing, organisation and communications, as well as for ensuring consistency of style and for compiling the Acronyms and Glossary and lists. A further debt of gratitude is due to Mrs Jennifer Davies for so competently and carefully typing the papers, and at Ithaca Press my thanks go to Adel Kamal, Emma Hawker and Anna Hines for their support.

Kamil Mahdi

---

* Subsequently, Programme of Gulf and Arabian Peninsula Studies, Institute of Arab and Islamic Studies.

Introduction

Kamil Mahdi and Haris Gazdar

The two main objectives of this collection of papers are to advance our understanding of structural characteristics and behavioural features of the Iraqi economy, and to highlight the economic consequences of the momentous events of recent decades. In the latter instance, the aim is to help guide research into the new economic realities brought about by the 1991 war, by the systematic destruction of the economy by US and British bombardment, and by a decade of economic sanctions.

Although a comprehensive description and analysis of war-related actions and of their impact is not undertaken here, economic devastation is a theme that runs throughout the book. Part I deals with the general state of the economy under sanctions, and also examines the burden of war-related foreign debt that was incurred during the Iran–Iraq war. The analysis shows that the weight of the old debts increased to an unsustainable level even without taking into account the astronomical demands for compensation arising out of the occupation of Kuwait in 1990. The chapters discuss sanctions and debt as current and future restraints on the economy, and consider their consequences for its overall performance. The chapters in Part II deal, in turn, with the main commodity sectors of oil, agriculture and manufacturing industry. In each case, current post-Gulf-War conditions are discussed within an analytical framework that also encompasses the historical features of the particular sector. Both the book's objectives are clearly evident in these sectoral chapters.

Part III deals with selected aspects of Iraq's political economy. Each of its four chapters addresses a different topic and offers a different level of analysis. Chapter 6 discusses the political economy of the Iraqi state; it suggests an analytical framework for considering Iraqi etatisme through which the resilience of the political structure in the new post-1991 conditions can be understood in terms of the economic role of the state. Chapters 7 and 8 are concerned with the unresolved political situation in the aftermath of the Gulf War. Chapter 7 deals with the economic and social activities of foreign, largely Western, non-commercial agencies dealing with the humanitarian situation. These agencies entered Iraq in the spring of 1991, and have maintained an unprecedented role in the provision of some services in areas controlled by Kurdish nationalist parties. The primary concern here is to understand the contemporary economic impact of the activities of these agencies, rather than their political, humanitarian or other roles. Chapter 8 deals with the de facto division of Iraq during the 1990s. It reviews economic conditions in Iraqi Kurdistan under two rival administrations, both of which are now outside central government control. The status of the Kurdish region is connected with Iraq's current international position, and the particular experience in this region is significant for the country's economic and political prospects.

Agriculture may be expected to have a greater role, not only under sanctions but also in the early phases of post-sanctions rehabilitation. Chapter 9 examines the agrarian system and the institutional framework of agriculture that can facilitate this expanded role. Again, the chapter reflects the book's two main objectives of understanding Iraq's economic structure and the impact upon it of the 1991 war.

The final Part of this book offers three chapters that give individual perspectives on the past performance of the economy, on requirements for recovery, and on long-term consequences of war and sanctions. They point to the reforms that will be necessary, and to the extent of the changes from past economic policies and practices that will be dictated by Iraq's new economic reality, its economic predicament.

The authors of this book are aware that the two objectives outlined are not necessarily harmonious with each other. In discussing the ruptures caused by the events of 1990–1, Chaudhry (Chapter 6) casts doubt over 'the validity of the basic apparatus of conventional economic history, built as it is on the fundamental assumption of association if not continuity.' Other authors also address the dramatic transformation of Iraq's economic conditions as a result of the wars and the sanctions, and stress the fact that radical shifts in objectives, policies and institutions are required. In their different approaches and concerns, the authors discuss aspects of what we will describe as the *functioning economy* prior to August 1990, addressing macro and external aspects, structural and institutional features, key commodity sectors, and political economy issues.

[2]

A common theme emerging from several of the chapters is that although Iraq's pre-*rupture* economy was able to utilise and renew resources, to sustain production in different sectors, to deliver a wide range of services and to offer a high level of employment, it was nevertheless stagnant and structurally distorted in its use of resources and in its incentive structures. Military, security and administration activities were a heavy and unrestrained burden on the economy. The lack of economic accounting permeated through the entire structure of the economy, such that economic planning was not possible. In major areas of activity, markets were either not operational, or incomplete and biased. Huge capital and labour resources were diverted to unproductive uses, leaving existing productive capacities under-utilised, while the marginal productivity of capital was low.[1] Oil revenues, augmented by foreign credits, masked these inefficiencies, and they also obscured economic difficulties of the eight-year-long Iran–Iraq war.

Iraq's pre-rupture economy was also integrated internationally, technologically dependent in all sectors, financially reliant on external flows (both enclave sector oil revenues and foreign credit), and deeply vulnerable, due to its single commodity export and its wide multi-commodity imports. The failure to manage resources effectively, either through planning or through the market and private initiative left the economy as a series of increasingly disparate activities, each in its own way bureaucratically controlled and externally dependent. These structural weaknesses and the evident poor performance of the 1980s were also a consequence of earlier rentier growth, and of seriously flawed policies that accentuated the rentier pattern after the oil price rises of the 1970s.[2] The prolonged destruction of the Iran–Iraq war signalled the demise of development,[3] as the government placed its urgent political objectives ahead of even common-sense economics.

The economy in 1990 was thus weak and easily vulnerable to destruction by war and by the uniquely comprehensive economic sanctions imposed upon Iraq by the UN Security Council. These sanctions and the bombardment of the infrastructure together brought about the catastrophic rupture between the period of a functioning economy prior to 1990 and the post-August 1990 period when the economy suddenly confronted paralysis, fragmentation and systematic destruction.[4] Iraq's economic predicament is characterised by the simultaneous and catastrophic decline in every area of economic activity – in the destruction of human and physical resources; in the unravelling of social institutions; in the increased alienation of large sections of the population; in the disintegration of the regulatory functions of the state and the consequent collapse of communal service provisions and the loss of economic and physical security; and in the diminution of state sovereignty of a country that is having suddenly to face a proliferation of external claims against it. The collapse of an economy has such wide-ranging repercussions that it is impossible for one collection of papers to cover all its aspects. Therefore, this volume has limited itself to a somewhat narrow range of economic aspects and has steered away from many political and social aspects, as well as from issues of administrative collapse, the impact on the business culture of hyperinflation, huge uncertainties and successive shocks.

One particularly important area which has not been fully explored in these contributions concerns labour and the skills losses that arise from astronomical rates of emigration of the professional middle classes, from a surge in mortality rates, and from the evident degradation of Iraq's educational system. This matter, however, goes beyond that of labour and skills markets. In particular, the scale of emigration of professionals may have developed a qualitative dimension relating to questions of the future of Iraq's middle class, and to the role of its urban entrepreneurial culture in the acquisition of knowledge, assimilation of modern business practices, understanding of world trends and development of healthy regional and international links. All this affects the country's ability to utilise its resources effectively and its ability to benefit from opportunities of trade and technology transfer as sanctions are lifted.

Eleven years after the imposition of sanctions, Iraq's economic conditions continue to deteriorate. The country's resources are still haemorrhaging; the capabilities of its economic and policy institutions have been undermined; and the country has been isolated from the information and communications revolutions and from international economic trends and business culture. Iraq will need a positive international environment merely to reverse current trends, let alone to make up for losses or to catch up with the gains made by other countries over the past two decades. As it stands, Iraq is burdened by debt and by punitive reparations, both of which threaten any post-sanctions recovery. The economic challenges for the country are to stop the decline under sanctions, and to resume effective reconstruction and regeneration of the economy by the best means available. To this end, it will be necessary to mobilise what remains of resources, energies and ingenuity, by reordering priorities and reforming or re-establishing institutions, and by judiciously addressing the new international economic problems that Iraq now faces.

The contributions of both Chalabi and Al-Sayyid Ali to this volume note Iraq's achievement in reconstructing major parts of the physical infrastructure destroyed by US and British bombardment. Achievements such as the speedy rehabilitation of substantial electrical power generation and oil-refining capacities, the restoration of urban water supply and of communications, the reconstruction of most of the destroyed bridges, and, indeed, the prevention of widespread famine, all indicate the resilience and resourcefulness of a people working under great adversity. One of the imponderables of Iraq's present economic condition is the extent to which this resilience continues to be a factor to be mobilised under the present sanctions and after their removal. Al-Sayyid Ali (Chapter 12) notes that the costs of reconstruction in terms of hyperinflation and the destruction of the largely fixed-income professional strata were very high. It should also be remarked that since the state budget depended heavily on oil revenues, the use of incentive grain producer prices in order to supply the food rations was itself highly inflationary. All choices were therefore extremely painful, but the situation was to some extent accentuated by a number of prestige and ill-advised projects that served only the regime's self-image. As Al-Shabibi (Chapter 11) points out, inflation did not affect all strata of society uniformly and a majority were heavy losers. Income disparities widened sharply between 1988 and 1993, with the Gini Coefficient rising from 0.36 to 0.43 for the country as a whole. The disparities in the rural areas widened even more sharply from 0.37 to 0.53 over the same period.[5]

Beyond Iraq's economic predicament, further dramatic ruptures of a political and social nature have been contemplated over the past decade. Iraq, as a state and as a society, could have fractured as a consequence of extraordinarily sudden impoverishment and dislocation. War and domestic political repression, followed by United Nations economic sanctions, led many to take refuge in local subnational identities that are themselves unstable and incapable of meeting people's expectations. Government discriminatory practices and ideological representations have accentuated and encouraged this trend. The collapse of state services and the weakening of the educational system also propelled sectional identities and helped to undermine national cohesion at a time when modern civil institutions were repressed and national political life was not organised to incorporate the expression of political, cultural and religious diversity.

In bringing together this collection of papers, it is assumed that this danger of political fragmentation has now receded, and that Iraq will remain a heterogeneous society within its present borders. Economic recovery is then at least a possibility and a realistic objective in the coming period. Evidence of widespread international opposition to UN sanctions is encouraging in this regard, although ten years after Iraq's occupation of Kuwait ended, the US and British governments remain totally opposed to the lifting of economic sanctions.

Any rehabilitation of the country and regeneration of its economic activity will have to begin from the present economic reality. Many of the economic institutions are likely to be retained, at least initially, and remaining production plant, facilities, the labour force and natural resources offer some links across the ruptures of the 1990s. In fact, activities that have not been suppressed or destroyed by economic sanctions have to some extent tended to develop domestic linkages. Some businesses have managed to find local solutions while producers supply alternatives to imported products. In contrast, however, to the trend of the emergence of small-scale enterprises and new domestic linkages, short-term palliative solutions are frequently adopted, resulting in costly outcomes in terms of disruption and environmental costs.[6] Breakdowns are transmitted unevenly and magnified, while the failure of the financial system increases risks and transaction costs. Examples of costly breakdowns and distorted transmissions of their impact abound. Tractors, pumps and other equipment lie idle as their owners stop hiring machinery services for fear that mechanical failure might not be repairable under sanctions, which would then disrupt their own farming activities. Other examples of market distortions can be found in the sharp gyrations in the property markets and shifts in relative prices, as assets are switched in response to political or other factors. Risk-averting behaviour has a further stagnating effect, while agents seek safe forms of assets in conditions where production is especially risky. For many people, and for the community as a whole, survival strategies have in many cases resulted in short-term palliative solutions leading to more rapid environmental decline and to rapid capital depreciation, especially of public or communal assets. The rupture of 1990–1 is critical to understanding Iraq's present economic conditions, and this rupture has many grievous economic facets that can and should be identified through more detailed research.

Problems of Iraqi economic research

The tasks facing researchers on the Iraqi economy are considerable. Their efforts have been hampered by almost a quarter of a century of official hostility to open debate on economic issues. State budgets and public accounts have not been available to academic and professional researchers, nor have international financial statistics. Similarly, monetary and financial information, budgets of state agencies and public corporations, development programmes and expenditure have been accessible only to a small group of officials, even fewer of whom are genuine professionals who are able to appreciate the economic significance of the data and to utilise it in analytical work. Furthermore, data that is available has not been subject to sufficient critical scrutiny. With regard to national accounts, it is not clear how detailed, comprehensive and accurate is the primary statistical source material. In earlier decades and until the 1970s, Iraq had gradually built up a strong economic and social information system for use in economic planning. It established effective statistical institutions and trained competent professional staff who provided both the skills and the credibility necessary for reliable statistical information. Since then, however, not only have some of these activities been politicised, but the economy has also operated for a prolonged period with extensive sectors under administered pricing without arrangements for effective resource accounting. There is also a strong unaccountable bureaucratic and military sector, and a large semi-official private sector activity that has been subject to arbitrarily fixed charges and stamp duties, rather than to progressive income taxation necessitating careful and consistent accounting. Add to that the divergence of official from market exchange rates, numerous market distortions, and the absence of reliable price indices, and the result is that valuation and accounting for national income purposes is likely to be problematic. The picture becomes even more distorted with the massive disruption and the runaway inflation of the 1990s.

Given the problems of access and reliability of information that are faced by researchers, the chapters in this volume represent particularly significant contributions to the relevant aspects of the Iraqi economy. It is important to note here that policymakers are usually in no better position to understand economic trends. Professionals in government agencies have to do their best, but are themselves hampered by political constraints, administrative malfunction and by the same lack of reliable statistics. Moreover, under the sanctions regime, the UN committee set up by Security Council Resolution 661 of 1990 and meeting in New York, has many of the powers of a supreme central planning authority for Iraq. The powers of this committee, which is composed of political representatives from the fifteen member states of the Council, are not merely theoretical, being routinely exercised in matters of policy and contract detail. From an economic point of view, it is bizarre to have an economy that continues indefinitely under an arrangement whereby it is centrally planned without adequate knowledge, technical competence, an explicit economic objective or communication with the lower levels of the economy and without a will for the whole exercise to succeed. Economic research on Iraq has indeed been problematic, but under sanctions it is now more problematic than ever. Similarly, economic management under the sanctions regime is impossible.

Political context
No discussion of Iraqi affairs today can remain untouched by the extreme polarisation surrounding the country's political conditions and international affairs. Iraq's invasion of Kuwait in August 1990 marked, by any standard, a watershed in the conduct of international relations. The actions of the Iraqi government and armed forces in that fateful month set off a chain of events which would lead to a manifestation of a new, post-Cold-War balance of global power. The resolution of the Kuwait crisis involved nothing less than the heralding of a 'New World Order', ostensibly based on multilateral action under the leadership of the United Nations, and on the revision and reinterpretation of institutions (such as the principle of state sovereignty) that had hitherto been regarded as pillars of the post-Second-World-War system of global governance. The impact on the lives and life chances of millions of Iraqis was even more dramatic. As noted above, the crisis precipitated by Iraq's attempt at annexing Kuwait was to cause a rupture in the direction of Iraq's economic, social, and demographic development.

While rhetoric about a more just and peaceful global order was quickly overtaken by the emergence of new regional conflicts, by the increasingly arbitrary attitude of major powers to international legality, and by the relegation of the authority of the United Nations, one legacy of August 1990 has remained salient and active over the whole of the past decade. The economic sanctions imposed on Iraq in the wake of the invasion of Kuwait have remained as the formal legal basis of Iraq's interaction with the rest of the world's economy more than eleven years later. The 1990s, applauded by some as the decade of economic reform, liberalisation, the rush towards global market integration and the rise of new technology-driven forms of transactions and markets, represented, for Iraq, a period of unprecedented economic isolation, technological retrogression, cultural introversion, and institutional collapse.

The human consequences of this economic catastrophe must be placed on record from the very outset. Advancement of the human condition has, after all, to be the main point of economic growth and development. Through the 1990s, Iraq lost many of the gains it had registered in over three decades of social and economic change. According to a major United Nations Children's Fund survey, the mortality rate among under-5s in Iraq's 15 southern provinces increased from 56 deaths per 1,000 live births in 1984–9 to 131 deaths per 1,000 live births in 1994–9.[7] Estimates of excess deaths attributable largely to sanctions vary between half a million and one and half million, the majority of the dead being children.[8] The scale of such human losses was compounded by the relentless erosion of social gains in the fields of education, science and culture – gains that represented the fruit of collective action on the part of many millions of people through generations. The proportion of children attending school actually declined; diseases that had been eradicated through clinical and public health improvements returned with a vengeance; and poverty and malnutrition became the norm rather than the exception. The commitment of the United Nations to human rights, justice and economic development is facing its most severe test in the consequences of its persistent and deliberate economic measures against Iraq. If economic development is to have any purpose or meaning in Iraq, it must first be to check, and then reverse, the demographic, educational and social trends of the 1990s.

One aspect of Iraq's present reality is that international institutions and external parties have assumed for themselves extraordinary powers over economic and social policy in the country. These powers are not legitimised by any accountability for the humanitarian consequences of their exercise, nor by any undertaking to meet specified social and economic objectives. Not only have Iraqi economic activity and institutions been paralysed, but the UN's own economic and social agencies have faced insurmountable political obstacles at the Security Council. One senior foreign correspondent for the *Washington Post* has remarked that 'the [first] Bush administration purposely set its mind against grappling with the complexity of Iraqi society before, during, and after the occupation of Kuwait.'[9] Eleven years on from the Gulf War, this lack of interest in Iraqi society continues to characterise US policy. In the United States (and in Britain), debate on the social, economic and political consequences of the continuous bombing of Iraq is not encouraged. The military dimension of US policy towards Iraq appears to override all others, and impossible conditions of disarmament verification[10] are set for the return of Iraq to normal economic relations with the rest of the world. All evidence so far has pointed to the domestic impact of this policy of 'containment' being one of retrenchment into old policies and established regime practices on the one hand, and a weakening of new political forces on the other. In undermining the economy, UN sanctions have done much less to weaken the present political regime than to marginalise modern social institutions.

Conditions for recovery
Political failures and military confrontations aside, it can be argued that Iraq's economy is both historically and potentially diverse and dynamic. Before the oil era, agriculture grew very rapidly and generated resources for urban development and for a gradual expansion of infrastructure and services. Traditional urban trades proved versatile, and social organisation was flexible enough to accommodate the growth of private industry and entrepreneurial activities. Iraq traditionally had a strong public service ethos, reflected in a relatively efficient administration that was not known for open corruption. Such features seem to indicate a likelihood of recovery in the medium term, given a favourable set of conditions. These would include regional peace, domestic political stability, a sympathetic international environment, careful and successful introduction of economic reforms, development of economic policy institutions, and the pursuit of consistent, sensible and popularly supported economic

policies. The key to these conditions would be an end to perpetual conflict and the establishment of institutionalised representative power structures.

We would need to look beyond the experiences of the past two or more decades to envisage the coincidence of a combination of these conditions of economic recovery. However, as we have noted earlier, the events of 1990–1 represent a watershed for Iraq. Dramatic transformations in Iraq's political economy include the inability of the state to meet people's aspirations, the collapse of the rentier culture of earlier decades, the bankruptcy of militarism, and the loss of the appeal of the politics of ideological confrontation that has paralysed professional judgement and polarised decision making. Cold War tensions have had a major impact on domestic Iraqi politics, accentuating conflicts of class and of political aspirations and raising tensions between Iraq and its regional environment, ultimately contributing to the wars of the past two decades. As Iraq's experience shows, the present international and regional environments are by no means benign as far as Iraq is concerned. Nevertheless, Iraq's present political predicament may still be associated with the legacy of earlier conditions and with a painful process of an international and regional political transformation. There is a different emerging regional economic and political culture that has broken with the ideologically based positions that have constrained Iraq's economic policy over the past decades. It is now possible to conceive of different approaches and greater recognition of the need for rational economic calculation at state level. Iraq needs a better balance, less conflict and a clearer demarcation between the public and private sectors.

Unlike the East European transition economies, Iraq does not have a problem with the institution of property rights *per se*, and to that extent, gradual reforms might have a good chance of success. Political circumstances are also open to change, but naïve optimism is not helpful. None of this implies an improved environment; it merely suggests a relatively fluid one that allows different possibilities. What is required is sustained action that is based on an assessment of economic and political conditions. This book contributes to an understanding of economic and political economy issues with considerations of appropriate policy formulation in view. Almost all the chapters offer scenarios for recovery or present specific policy recommendations. Underlying each is a recognition that although Iraq's economic malaise has largely been the product of an acute set of political conditions, economic repercussions have to be addressed in their own right.

It is not our aim here to address the realism or otherwise of the broad conditions for recovery, nor do we intend to discuss the reasons why these conditions have been absent in Iraq's recent experience. Almost all the chapters touch upon some of the complexities of the country's political economy, and many of them address these complexities directly. In Chapter 6, for example, Chaudhry presents a historical analysis of the social basis of the Iraqi state. Her contribution draws on the country's experience prior to 1990 and raises questions regarding the kinds of adjustments that have been occurring since the imposition of sanctions in the relationship between the state on the one hand, and society's other economic and social institutions on the other.

A challenge beyond the remit of this book is to develop an analysis of how the dramatic change in Iraq's economic fortunes and in the state's role in the economy is to be reflected in the relationship between the state and social groups. In particular, the dependence of large sections of society on state provisions for their livelihood, and the disengagement of entitlement claims from social institutions of production are unsustainable under new post-rentier conditions. This dependence has been recreated through the necessity of the ration system under sanctions, but the relationship of dependence is unlikely to be stable as the economy is regenerated. With economic growth and diversification, unaccountable power that is rooted largely in the capture of the bureaucracy is unlikely to be the norm of future political patterns in Iraq. As Al-Shabibi (Chapter 11) points out, Iraq's economic conditions have been transformed in such a way that the external sector will no longer be financing the domestic economy, since it is itself projected to be in deficit. The necessary shift towards a regeneration of the domestic economy also represents a shift away from the enclave crude-oil activities towards the wider economy.

This shift in the political economy of Iraq is crucial to the process of sustained recovery and to the conditions of recovery noted above. The emergence of a developmental state as opposed to an oil rentier state would be part of the repositioning of the oil sector within the Iraqi economy, and it is not a diminution of the importance of oil. Indeed, all the chapters in this volume refer to the devastating effects of the oil embargo, and raise important expectations concerning the future ability of Iraq again to utilise its oil revenues for development purposes. The potential financial resources that can be generated by unhindered repair and maintenance of oil industry facilities and by the development of new capacity in Iraq can be substantial, as the 1999–2000 rises in price have shown. However, under the present sanctions regime, Iraq has largely been restricted to purchasing consumer items, sometimes undermining incentives for domestic production. Purchases are designed to be readily consumable and not to interact readily with the rest of the economy. Oil revenues under the food-for-oil (FFO) arrangements relieve some immediate suffering, while deepening the conditions that create misery and dependence and simultaneously inhibiting a regeneration of the economy. A successful programme to reverse the decline of the economy would have to include major reconstruction and development efforts within a stable environment and an appropriate policy and institutional framework. Although it ought to be possible to proceed from using oil revenues for emergency relief to using them in post-war reconstruction and economic regeneration, sanctions inhibit such a move, among other things through the direct disruption practised by the Sanctions Committee in New York.[11] In general, oil revenues will continue to be needed to provide the essential financial resources for reconstruction, for meeting basic needs and for investing in the social sectors through enhancing knowledge and human capability. The collapse in many sectors of the economy necessitates urgent employment and equitable distribution measures. As discussed by Gazdar and Hussain (Chapter 1), by Ahmad (Chapter 4) and Al-Shabibi (Chapter 11), Iraq also needs to institute measures to attain macroeconomic stability; an impossible challenge if the external claims on Iraq's oil revenues are not substantially reduced.

In Chapter 3, Chalabi considers the potentials and constraints of the oil sector. His chapter, and those of Jiyad (Chapter 2), Alnasrawi (Chapter 10), Al-Shabibi (Chapter 11) and Al-Sayyid Ali (Chapter 12), all emphasise the obstacles to Iraq's recovery represented by heavy compensation payments and outstanding Iraqi debt commitments. Chalabi argues that Iraq needs to recover a substantial share of OPEC production that will be commensurate with the size of the country's oil reserves. This will be a difficult task even after sanctions are lifted, because of Iraq's present financial constraints and the desperate state of its oil production facilities.[12] Difficulties will also arise from market competition with other energy producers, including those of many OPEC countries that are heavily dependent upon their oil sectors.

Compensation against Iraq is at present deducted from the proceeds of gross oil sales, with oil production running at full, and even beyond, sustainable capacity levels. Under the FFO arrangements, payments to the UN Compensation Fund and for the costs of UN operations in Iraq take up 30 and 4 per cent respectively of oil revenues.[13] Unless the buoyant oil market prevalent during 2000 persists, it would seem highly unlikely that Iraq would have an incentive to pursue an aggressive market share strategy. Such a strategy would bring the country limited financial returns once heavy investment costs had been met and after the exaggerated external claims had been paid. Attempts to unlock Iraq's oil production within the context of the new stringent financial constraints of SCR 986 (reaffirmed by SCR 1284 of December 1999), are likely to be destabilising. A new generation of Iraqis is likely to see few, if any, benefits from development in the oil sector, and political instability is therefore likely to continue.

The FFO programme, discussed below, has become the channel through which around a third of Iraq's gross oil revenues are siphoned off to meet external claims of compensation and UN monitoring.[14] Therefore it is unlikely that in their present form and scale, the FFO programme and the compensation claims can coexist with a programme for developing new oil capacity.[15] The FFO programme itself is ostensibly a humanitarian programme. However, despite its expansion in 1998 to include a wider range of needs than food and medicine alone, and despite the incorporation of allocations for infrastructure expenditure, the programme is still far from offering the prospect of economic regeneration, let alone incorporating within itself any elements of a development strategy.

Regeneration and development contain political and economic policy elements that are incompatible with the FFO programme, which has itself been bedevilled by political disagreements and administrative obstacles. An economic strategy has to be comprehensive, indivisible and coherent. It requires the authority and structures that can address detailed implementation and effective participation at every level. This means that Iraq's economic administration has to be restored, and explicit political and social objectives have to be re-established. This is tied in with the issue of Iraqi state sovereignty, which has effectively been reduced since 1990 by a series of Security Council Resolutions. An alternative UN-led economic management of Iraq is not feasible, and it also contradicts the repeated reaffirmation of Iraq's sovereignty and integrity in such SCRs.

If Iraq's ability to develop and utilise its oil revenues is inhibited by economic sanctions, the country would act as a de-stabilising factor in international oil markets. As a consequence its own economy will suffer further due to likely stop–go oil-sector policies, and due to sharper fluctuations in prices as a result of its partial exclusion. Fluctuations in oil rents, in themselves, have negative effects on overall performance.[16]

A brief background survey

A number of contributions in this volume consider the kind of economic policies and policy paradigms that are conducive to dealing with Iraq's new economic conditions. Before discussing these issues, we will briefly outline the main features of Iraq's economy and of the country's economic management before 1990.

Until the 1950s, Iraq had a predominantly agriculturally-based economy. The economy had many features of classic duality, with a dynamic urban capitalist sector that grew and absorbed rural migrant labour. The economy was also highly integrated in world capitalist markets, particularly through the financing by its agricultural exports of consumer imports for the urban sectors. Income and wealth disparities increased across the country, and this may have contributed to a rising tide of political opposition in the urban areas, especially after military-related activities during the Second World War increased economic disparities. With environmental and resource constraints, and under a debilitating agrarian structure, a period of agricultural growth was coming to an end. A subsequent period of recession ended in the 1950s with the first major oil boom. Oil companies responded to a number of major developments in the international oil industry (especially Iran's nationalisation of its oil under Musaddiq) by increasing Iraq's oil output and improving the terms of calculating its revenues.

A poor economy that was in recession and that was constrained by social conflicts was suddenly endowed with oil revenues that provided ample government finances, foreign exchange and savings. Cautious expenditure and monetary policies maintained macroeconomic stability and limited the disruptive competitive effects of imports upon the domestic productive sectors. Policy concentrated upon investment in large-scale infrastructure projects with limited attention to immediate welfare improvements or to directly productive activities. There was a combination of poor planning[17] and macroeconomic prudence. These responses were deemed to be the least threatening to the dominant commercial and land-owning interests of the time.

After 1958, policy shifted towards higher spending, raising both consumption and directly productive investment. Although military expenditure also increased (especially because the government was unable to solve the Kurdish question politically), inflation was contained due to growth in the economy and due to the government's continued adherence to its own fiscal and monetary rules. By the late 1960s, the slow growth of Iraq's oil revenues and continued growth in government consumption expenditure began to erode monetary and fiscal stability. The Ba'th government of 1968 sought foreign credits from the oil companies and from East European countries, but oil revenues soon relieved financial constraints once again.

During the two decades prior to 1973, the Iraqi government had thus mostly followed a conservative monetary and fiscal policy, and it was able to maintain macroeconomic stability. Oil revenues had financed a comprehensive development programme with detailed sectoral allocations, and Iraq was almost free of foreign debts and had a very limited domestic debt.

The implementation of land reform after 1958 was problematic in a number of ways, but it also led to more diversified agricultural production, with a dynamic middle stratum of capitalist farmers (see Mahdi, Chapter 9). On the other hand, the nationalisation of private enterprises in 1964 undermined business confidence and limited business opportunities. Private investment declined and was channelled towards real-estate speculation as well as to some sectors like transport and contracting for state construction activities. Some of these changes are briefly reviewed in this volume by Al-Khudayri (Chapter 5) and Chaudhry (Chapter 6).

As Al-Shabibi points out in Chapter 11, the new oil boom after 1973 enabled a rapid expansion of government expenditure. This was met with restrictive trade policies that held back the supply of goods, thereby creating inflationary pressures. Nevertheless, during the boom years between 1973 and 1980, agriculture declined rapidly as the shift of labour towards construction, transport and services sectors accelerated. Despite restrictions that retained labour in state employment, there was a surge in small-scale private industry catering for the domestic market. Nevertheless, institutional structures, business finance limitations, and political constraints reduced the scope for private industrial growth during that period. Overall labour shortages were quickly met by large-scale migration of workers from Arab countries, thereby stemming the rise in wages. With conscription and mobilisation of labour during the 1980–8 Iran–Iraq war, expatriate workers continued to form a substantial proportion of the labour force, even though female participation rates increased markedly. Skills shortages were being addressed with a major expansion of technical and higher education, but much public sector industrial investment was paralysed or wasted by war (see Al-Khudayri, Chapter 5).

The post-1973 oil boom therefore unleashed a massive expansion of state expenditure on an economy that was encumbered by restrictions and bureaucratic controls. Arbitrary regulations, poor management and lack of transparency ensured that many lucrative private-sector activities were speculative and parasitic. Private activity was increasingly dependent upon links to the bureaucracy and to political power. Rent seeking predominated over productive activities and, as Al-Shabibi states (Chapter 11), the economy and the institutional structure harboured strong inflationary pressures. There was abundant domestic liquidity and foreign exchange, but restrictions prevented an adequate supply of goods.

In other words, Iraq's economy during the 1970s was especially vulnerable to the rentier state/Dutch Disease effects that tended to undermine its traditional domestic tradeable-commodity sectors. The economy was also subject to arbitrary policies. As the state gained in financial and political strength, it became increasingly restrictive of the private sector, limiting the latter's range of permissible activities. The state lacked the necessary administrative capabilities and effective mechanisms for defining social priorities and practising economic restraint. However, against all this, there was a substantial programme of infrastructure development and of expansion of the education and health services, as well as a horizontal extension of administration. There was also heavy investment in the oil sector, in manufacturing, in agricultural projects, and in production and trade services.

The increases in investment and in consumption were possible through the increased oil revenues, rather than because of improved efficiency and higher productivity, but the productive potential of the economy was, in any case, greatly enhanced during this period. However, that potential was not realised, largely due to political failures. Economic institutions remained weak and, like the other oil economies, Iraq became increasingly dependent upon its oil exports for sustaining its level of economic activity and all its major economic sectors. Therefore, what followed in the 1980s came as a harsh economic shock that was somewhat cushioned by the availability of external credit. The combination of disrupted oil exports and falling prices with increased military expenditure, undermined the level of investment and curtailed consumption from 1982 onwards. The government began turning to the private sector; a development discussed by Chaudhry (Chapter 6) and Al-Khudayri (Chapter 5) below. Chaudhry also discusses the fateful short period between the end of the Iran–Iraq war and the Iraqi invasion of Kuwait. Economically, that period was dominated by military demobilisation and rising unemployment, and by the attempted readjustment of Iraq's foreign debt position and of its dependence on expatriate labour. Iraq was hampered in increasing its oil exports by a depressed market, and economic liberalisation and privatisation created political problems without offering economic solutions. At the same time, the economic burdens of the arms programmes continued. Nevertheless, Jiyad (Chapter 2) concludes that the burden of external debt was in 1990 still sustainable according to acceptable sustainability criteria.

As noted earlier, the subsequent decline of the 1990s was of an altogether different dimension. While the relentless decline of the past decade or so has been a constant feature of Iraq, it is nevertheless useful, from an economic perspective, to divide the post-August 1990 era into three sub-periods. The first period began with the imposition of an absolute economic embargo under Security Council Resolution 661 of 6 August 1990, and lasted until around April 1991 when exemptions were made in the sanctions regime to allow the import of food and medicines. This first phase was marked not only by the blockade but also, obviously, by intensive war preparations and actual armed conflict, by allied air campaigns, by the rout of the Iraqi army in Kuwait and by armed rebellions and uprisings within the country. The war which had raged on in one form or another between January and April 1991 led not only to many casualties but also to the destruction and disruption of civilian infrastructure facilities, including roads and bridges, electricity generation and transmission plants, water and sewerage treatment installations, medical facilities and food supplies. This period of around nine months was characterised by one major economic shock after another – sanctions (including a ban on food imports), war preparations, actual war, civil strife, the destruction of infrastructure, and the de facto partition of the country. It was a period of the most dramatic downturn in the economic situation.

By April 1991, following the adoption of Security Council Resolution 687 governing the ceasefire and following the first visits by UN fact-finding missions, the restrictions had been lifted on the import of food and medicine. Military confrontation between Iraq and outside powers, and within Iraq itself, did not come to an end, but did settle down to low-level tension. Air attacks were to continue and increase in subsequent years, and the military situation in northern Iraq and in the southern marshes was to remain unstable. However, from April 1991 till December 1996, there were no major shifts in Iraq's external economic relations. The relaxation of import restrictions on food and medicines was ineffectual due to lack of funds, and was not followed by any subsequent easing in the sanctions regime. There were no further significant negative shocks but there was a steady deterioration in the country's economic and social conditions as a formerly wealthy society literally ran down its stock of financial and human capital (see Gazdar and Hussain, Chapter 1).

A reconstruction drive in government-held areas began soon after the ceasefire, and continued using both public- and private-sector construction teams that had gained experience in the aftermath of the Iran–Iraq war. Work benefited from the availability of stocks of materials and equipment, and relied heavily upon cannibalising existing and partially destroyed plant. Nevertheless, the inflationary impact of the programme represented a huge burden upon fixed-income earners whose state-sector wages and salaries were no longer generally adjusted for inflation after initial rounds of partial adjustments. As a result, earnings and incentive structures broke down and public-sector services began to disintegrate. For a time, there is said to have been no single monetary authority but a number of competing centres of power, each independently printing its own supply of currency. Inflation continued, despite an effective freeze on wages leading to increasing use of the ration system as a means of tying public-sector employees to their labour contracts. As the government was facing increasing difficulties in maintaining basic services and in making ration supplies available, factional splits within the regime reached the president's closest circle. Under the conditions prevailing in 1995–6, the government agreed to implement the food-for-oil arrangement, with minor amendments offered by the UN Security Council. Until that time, Iraq had held on, hoping for the removal of economic sanctions under Article 22 of Security Council Resolution 687, or had at least attempted to decouple the payment of compensation from the emergency humanitarian measure that later became the food-for-oil arrangement. Iraq eventually agreed to implement UN Security Council Resolution 986, the so-called food-for-oil (FFO) deal, thereby consenting to the deduction of compensation from a temporary emergency humanitarian programme.

The food-for-oil arrangement
There was, thus, a major modification in the external conditions facing the Iraqi economy in December 1996, when the FFO finally began to operate. Iraq was to be allowed to export limited amounts of oil in the international market for the first time since August 1990 (the exception granted to oil sales to Jordan for its domestic consumption notwithstanding). The proceeds of these oil sales do not accrue directly to the Iraqi government, but they are placed in an *escrow* account managed by the United Nations. A part of the oil revenues was earmarked for the region of northern Iraq that was not under government control, and another part was allocated to meet the costs of UN operations in Iraq and to pay compensation claims arising out of the Iraqi action during the Gulf War. The remainder could be used to finance imports of essential humanitarian supplies, including food and medicines as well as infrastructural requirements, and to pay for oil production costs, all in the government-controlled areas.

The FFO deal represented a modification of the sanctions regime that was first instituted in Security Council Resolution 661 of August 1990, and then reiterated as part of the ceasefire resolutions (particularly Resolution 687 of April 1991). The initial FFO deal (under SCR 986) was further revised in February 1998 to increase the limits on the value of oil exports. It was subsequently amended again, in December 1999 (SCR 1284), when the ceiling on oil exports was effectively lifted altogether and when the transparency of the procedures for vetting Iraqi applications for imports was improved.

Even after these various revisions, the FFO deal retained three essential parameters set by the UN Security Council for Iraq's economic interaction with the rest of the world. First, the Iraqi economy continued to be precluded from having direct access to the country's accumulated and generated financial resources. Secondly, all trade in goods and services and all financial transactions outside the FFO programme remained proscribed (save for the very small exception of the programme of exchange with Jordan). Iraq's private sector and the possibilities of foreign investment and of any possible role for Iraqi expatriate capital were hit particularly hard. Thirdly, the extent to which the decline in the human condition and in economic activity might be reversed were constrained, not by resources available in the national economy, but by the Security Council's view of the intentions and actions of the Iraqi government.

The chapters in this volume deal in detail with developments in Iraq in the period leading up to the FFO deal. In various sections, the discussion is extended to the predicted or actual impacts of the FFO programme on the Iraqi economy, but no detailed documentation of this latest phase of the economic embargo is presented here.[18] The original papers were written at too early a time to know the full impact of the FFO deal. Unfortunately for Iraq, the impact of the FFO has not been of very great consequence to the topicality of the material presented here, and in most cases only minor revisions were deemed necessary. The FFO programme was both designed and implemented so as to ensure that the impact on Iraq's economic and humanitarian condition has been minimal.[19] There are, of course, exceptions and some of these are noted below. Regardless of the eventual quantitative impact of the FFO, the key qualitative parameters that provide insight into Iraq's economic prospects are the ones that relate to the first two phases of the sanctions period. In other words, they relate to the transition from being a 'war economy at peace' before August 1990 to a 'war economy at war' in the period from August 1990 to April 1991, and then to the establishment and maintenance of conditions of prolonged attritional decline from 1991 onwards. The post-December 1996 developments that had made it possible to maintain and somewhat supplement the basic food ration, and which may have averted widespread famine, did not represent any qualitative move away from the low-level (managed) decline that had prevailed between April 1991 and December 1996.

While the facts and analyses contained in the chapters of this book are of contemporary relevance, a brief update on some of the more recent developments can help to highlight points of continuity and change. The initial terms of the FFO agreement allowed oil sales up to the value of US$2 billion every six months, with 65 per cent of the proceeds to be made available for the importation of food, medicines and other humanitarian supplies. The US$2.6 billion dollars that were thus accessible annually were to be shared between the areas controlled by the government of Iraq and northern areas that were under the control of Kurdish parties. Government-controlled areas were to receive just over US$1 billion, or around US$58 per person; the allowance for the three northern governorates under Kurdish party control amounted to US$260 million, or around US$100 per person. Subsequent revisions to the FFO agreement led to the doubling of oil exports in February 1998, and to further increases in oil revenues and an effective lifting of the ceiling altogether in December 1999. Apart from short periods of political and military crisis, the constraint on Iraq's oil revenues has, since the introduction of the fourth six-month phase of the FFO programme in 1998, been the country's production capacity and the fluctuating oil market.

Until late into 1990, revenues available to Iraq as a result of the FFO deal amounted to only a small fraction of the country's pre-sanctions export revenues. If we take into account infrastructural damage caused by war and by years of severe economic recession, the shortfall would be far greater. The importance of debt and compensation renegotiation is highlighted also by the fact that repayments on Iraq's pre-war debt alone would, in the oil-market conditions of the 1990s, be sufficient to exhaust the country's export potential for the foreseeable future (Jiyad, Chapter 2). Therefore, the introduction of FFO, its modification and relaxation, and even the lifting of sanctions, are unlikely to lead to economic revival and humanitarian uplift unless these actions are accompanied by significant reduction of Iraq's external liabilities.

In addition to these quantitative limitations, there are qualitative aspects in the design and implementation of the FFO programme which make it a relatively ineffective tool of emergency humanitarian intervention. Oil revenues have traditionally played a dual role in the Iraqi economy. They have been used to finance imports directly, as is the intention of FFO. They have also been the means by which the state has financed its own expenditure and passed on oil revenues to ordinary Iraqis. It has managed this by using foreign exchange earnings (or foreign borrowing during the 1980s) in order to maintain the value of the Iraqi currency. While the FFO scheme allowed the government to finance specific imports, it effectively insulated private citizens' incomes from oil revenues by preventing any 'leakage' of foreign exchange earnings to the Iraqi government, the traditional conduit for the transmission of export revenues to the population. FFO made it possible, in principle, for the Iraqi government to purchase medicines from foreign companies, but not the services of Iraqi doctors for administering those medicines.[20] It is hardly surprising that the steady defection of skilled professionals, such as physicians, to the private sector, to non-related activities or to foreign lands continued under the FFO programme.

Another example of the same phenomenon is the case of the agricultural sector. As noted by Ahmad (Chapter 4) and by Gazdar and Hussain (Chapter 1), the economic collapse in the 1990s improved the relative position of the agricultural sector, which had in the past suffered years of neglect. A combination of private initiative, policy directives and better agricultural incentive structures[21] led to a large increase in crop acreage as well as to agricultural employment. Many of the requirements of Iraq's admittedly basic and inadequate ration system were satisfied somewhat precariously from domestic production prior to the introduction of the FFO programme.

However, the FFO scheme was premised on the import of food and its subsidised distribution inside Iraq. While local farmers had, in relative terms, been the beneficiaries of the urgent agricultural revival in the post-1990 period, FFO eventually reversed the incentives for farmers. The effects were particularly perverse in northern Iraq, a grain-surplus region, where the farm economy was virtually undermined by subsidised imports. In government-controlled areas, there were reports that the attention paid earlier to maintaining domestic output had given way to complacency, as a result of the exclusively import-oriented nature of the FFO programme.

There are, however, important features that also need to be noted. First, despite problems in the agricultural sector, FFO has been relatively more effective in the provision of goods in Kurdish-controlled northern Iraq than it has in government-controlled areas. One reason for this is the much larger per capita financial allocation for the former region. Another factor is that FFO has allowed some degree of local purchasing in the north, and has therefore led to some revival of general economic activity. The improvement in mortality rates in northern Iraq in the late 1990s can be seen as one tangible effect of FFO and other humanitarian interventions in that region. The longer-term institutional and political economy effects are likely to be more problematic. The FFO and the efforts of the Kurdish administrations have failed to regenerate the local economies. Low levels of investment, massive unemployment and poverty are causing a continuous flow of illegal and hazardous emigration from the region. Furthermore, overwhelming reliance on FFO could create strong economic stakes among the dominant Kurdish factions in the maintenance of some level of conflict, rather than encourage them to move towards reconciliation and towards creating conditions for a more sustainable economic development.

Another feature is the possible impact of the government's increased ability, in violation of the sanctions, to use the 'cover' of oil exports under FFO to maintain some level of exports outside the programme. In other words, FFO has provided the means for the government to bypass controls and erode the effects of sanctions. The extent of 'oil smuggling' and the extra revenues that have been available to the Iraqi government are matters for speculation. Nor is it clear to what extent Iraq's ability to circumvent the UN blockade is a result of FFO or a consequence of improved diplomatic and political relations with neighbouring countries. In any case, if – as has been suggested – some part of the foreign exchange earnings from smuggling finds its way into the Iraqi economy and provides support to the market value of the Iraqi dinar, the impact on the economy and on living standards could be positive. However, as these transactions are the least transparent, their impact is also less well known. The erosion of sanctions in this manner may also complicate longer-term prospects of Iraq's economic development. The steady erosion of sanctions, as opposed to a quick negotiated economic and political rehabilitation of Iraq in the regional and international community, is likely in the long term to lead to a skewed economic structure. It will result in the flourishing of a semi-legal private economy as well as the continued impoverishment of potentially productive sectors and of the public sector.

Economic sanctions might also be undermined in conditions of a world oil-supply shortage and of an urgent market need for the development of Iraq's oil-production capacity. This would enhance Iraq's negotiating position, which the country might variously utilise to achieve a quick removal of sanctions or to attain a reduction of external claims, or both. What the contributions to this book might demonstrate is that Iraq's economic predicament is now deeper than can be remedied with a sudden turn-on of the taps. A wide-ranging programme for recovery needs to evolve from Iraq's own professional and political milieu, working in a sympathetic international environment. A focus on the dire economic realities can only help in dealing with long-standing domestic and international conflicts from which there have been few winners.

NOTES

1 Ismail O. Hummadi, 'Al-mumkinat wa al-muskilat fi al-iqtisad al-iraqi' [Possiblitities and problems for the Iraqi economy], in *Al-iqtisad al-iraqi fi dhil al-hisar wa afaq al-mustaqbal* [The Iraqi Economy under the Blockade, and its Future Prospects], Bayt al-Hikma, Economic Studies Section Seminar, Free Table Series 32, July 1998, p. 17.
2 Abbas Alnasrawi, *The Economy of Iraq: Oil, Wars, Destruction of Development and Prospects, 1950–2010*, Westport, Conn.: Greenwood Press, 1994, pp. 72–5.
3 Ibid: pp. 79–83, 89–96.
4 The devastating economic effects of the first year of sanctions and of the 1991 war are illustrated in Jean Drèze and Haris Gazdar, 'Hunger and Poverty in Iraq, 1991', DERP Discussion Paper 32, STICERD, London School of Economics, 1991. The continuing paralysis and decline of the economy through the 1990s are discussed in Gazdar and Hussain, in this volume.
5 Hummadi, 1998, p. 20.
6 In the summer of 2000, town planning and environmental health departments announced a tightening of the implementation of industrial location regulations, after small unregistered and unregulated manufacturing and repair activities appeared to be spreading outside designated areas and in residential neighbourhoods.
7 UNICEF, Child and Maternal Mortality Survey 1999, Preliminary Report, UNICEF and Iraq Ministry of Health, Iraq, July 1999.
8 Marc Bossuy, 'The adverse consequences of economic sanctions on the enjoyment of human rights', Working Paper, Commission on Human Rights, Sub-Commission on the Promotion and Protection of Human Rights, Fifty-Second Session, Item 12 of the Provisional Agenda, United Nations Economic and Social Council, E/CN.4/Sub.2/2000/33, 21 June 2000, p. 16. According to the former UN Humanitarian Coordinator for Iraq, the Security Council 'is

guilty of intentionally sustaining a regime of killing that can only be termed genocide'. Denis Halliday, 'The deadly and illegal consequences of economic sanctions on the people of Iraq', *The Brown Journal of World Affairs*, Volume VII, Issue 1, Winter/Spring 2000, p. 230.

9   Jonathan Randall, *After Such Knowledge, What Forgiveness? My Encounters with Kurdistan*, New York: Farrar, Strauss and Giroux, 1997, p. 52; quoted in Kanan Makiya, *Republic of Fear: The Politics of Modern Iraq*, Berkeley, University of California Press, 1998, updated edition, p. xxi.

10 Scott Ritter, 'Redefining Iraq's Obligation: The Case for Qualitative Disarmament of Iraq', *Arms Control Today*, June 2000.

11 Eric Herring, 'Between Iraq and a Hard Place: A Critique of the British Government's Narrative on UN Economic Sanctions', paper presented to the International Relations Seminar, Department of Politics, University of Exeter, 26 October 1999.

12 UN, Report of The Group of Experts Established Pursuant to Paragraph 12 of Security Council Resolution 1153 (1998).

13 These percentages are deducted from gross revenues at the sea terminals. In the case of exports through Turkey, not only is the cost of oil production entirely borne by Iraq but so also are the costs of pipeline transport through Turkey and the port charges. An attempt by Iraq to charge for loading services at its Gulf terminal has faced objections from the UN Sanctions Committee in New York. The elaborate system of UN control and the ever-present possibilities of disruption on the one hand, and Iraq's inability to control and manage its blends of crude for achieving maximum revenue on the other, meant that Iraq received far less than 66 per cent of net revenues; the UN deducted more than 34 per cent. The nominal percentages deducted were reduced to 25 and 3 per cent respectively as of December 2000.

14 A British Foreign Office spokesman told *The Guardian* that the UN has no way of enforcing compensation payments if sanctions are lifted. See Brian Whittaker, *The Guardian*, 15 June 2000.

15 This appears to be recognised in proposals emanating from Russia and France for an immediate reduction in the share of compensation payments from 30 per cent to 20 per cent. See, for example, *Al-Hayat*, 22 September 2000.

16 Alan Gelb et al., *Oil Windfalls: Blessing or Curse?*, Oxford: Oxford University Press and the World Bank, 1988.

17 Cf. Alnasrawi, 1994, op. cit., p. 22.

18 The UN itself reports on its implementation of the FFO, and this implementation is discussed in Tim Niblock, *Sanctions and the 'Pariah States' in the Middle East: Iraq, Libya and Sudan*, Boulder, Colo.: Lynne Reiner, 2000.

19 The FFO programme was self-consciously designed to be a palliative and did not signal any major opening. The minimal nature of its economic and humanitarian impact was dramatically highlighted through successive resignations by senior UN officials charged with the very task of implementing FFO. So far three such officials (Denis Halliday, Hans von Sponeck and Jutta Burghardt) have resigned, each publicly calling for the lifting of non-military sanctions.

20 In practice, of course, there have been problems even in the implementation of import procedures, with complaints from Iraq, many of them substantiated by

senior UN staff, that requests for essential supplies have been frequently denied or held up on arbitrary grounds by the Sanctions Committee.

21 Ahmad (Chapter 4) notes that the immediate post-FFO impact on agricultural incentives in government-controlled areas was highly favourable due to the rapid appreciation of the dinar while official agricultural purchase prices remained unchanged. Subsequently, official price purchases were reduced and farmers have had to sell their produce at much-reduced market prices. As food became available through the FFO, the government shifted its domestic expenditure priorities to areas other than agricultural commodities. It is noteworthy that the sanctions regime bars the use of FFO funds for domestic purchases by the government.

PART I

THE ECONOMY IN RESTRAINTS: SANCTIONS AND FOREIGN DEBT

1

Crisis and Response: A Study of the Impact of Economic Sanctions in Iraq

Haris Gazdar and Athar Hussain[1]

1. Introduction

When Iraq invaded Kuwait, the UN Security Council (UNSC) responded with sanctions,[2] the intention being to force Iraq to withdraw from Kuwait by imposing on it a state of complete economic and political isolation. Following the military rout of the Iraqi army, the UNSC passed a further resolution (No. 687) on 3 April 1991 continuing the sanctions with the aim of securing Iraq's compliance with the terms of the ceasefire. Resolution 687, *inter alia*, required Iraq (a) to accept unconditionally Kuwait's territorial boundaries, as demarcated by the UN Iraq–Kuwait Boundary Demarcation Commission (IKBDC; (b) to compensate for the damage caused by the invasion of Kuwait, as determined by the UN Compensation Commission (UNCC); and (c) to submit to the dismantling of weapons of mass destruction and long-range missiles and of all programmes associated with their development or manufacture, by the UN Special Commission (UNSCOM) and the International Atomic Energy Authority (IAEA). The sanctions against Iraq, which remain in force eleven years after they were first imposed in August 1990, are unprecedented in terms of their rigour and duration.

Our aim in this chapter is to study the impact of the sanctions on the Iraqi economy and its population, focusing on the livelihood and the well-being of the civilian population in the area that has been under the control of the government; that is, the whole of Iraq except for the three governorates in Iraqi Kurdistan. Building on an earlier study carried out in 1991, when sanctions were only one year old,[3] we document changes in the Iraqi economy and the impact of sanctions on the economy between 1991 and 1996, and contrast this with the situation before 1991. Given that the sanctions are still in force, we have not attempted to update the chapter to take account of developments since 1996.

The former UN Secretary-General (Boutros Boutros-Ghali) regarded the sanctions against Iraq as 'one of the most complex and far-reaching set of decisions ever taken by the [Security] Council'.[4] The sanctions represented a landmark for the UN and international relations in three respects.

First, they were unanimously agreed by the five permanent members of the Security Council and accepted by almost all other UN member states. Some observers hailed this consensus as a portent of a new cooperative spirit in the international community in the post-Cold-War era.[5]

Second, the UN and its agencies took on unprecedented roles in implementing Resolution 687 and enforcing the sanctions. These ranged from the demarcation of the boundaries between Iraq and Kuwait, the carrying out of 'rigorous, no-notice, on-site inspection anywhere in Iraq', itemised approval of what Iraq could import, and a detailed policing of the use of revenue derived from the export of oil permitted under the subsequently concluded food-for-oil agreement. The breadth and depth of the UN involvement stands in marked contrast to its marginal role in the enforcement of internationally orchestrated economic and political sanctions against other defiant states, such as Libya, the Sudan, North Korea, Haiti and the former Yugoslavia.

Third, the sanctions have been remarkably effective in their implementation, if not in the securing of all their aims. As yet,[6] there have been few attempts by UN member states to break the sanctions. Compared to multilateral sanctions against other states, those against Iraq stand out in their scope and effectiveness. Except for food and medicines, there has been a blanket ban on all commodities unless specifically exempted by the UN Sanctions Committee.

The effectiveness of the implementation of sanctions has been aided by geographical and political conditions. There are four possible outlets for oil exports on a large scale: one is through Iraq's oil terminals off its narrow coastline on to the Gulf, bordered by Kuwait on one side and by Iran on the other; then there are three sets of pipelines, through Turkey, Syria and Saudi Arabia respectively. The shipping has been blockaded and the pipelines through Turkey and Saudia Arabia, both of which were part of the Gulf War coalition, are easy to police. The pipeline through Syria has been out of use since 1982 and has fallen into disrepair.[7] Iraq has for most of the past two decades either been at war or has had strained relations with five out of its six neighbours, while Jordan, the sixth neighbour, having refused to join in the military alliance against Iraq during the Gulf War, has since remained cautious in its relations with the Iraqi regime. The restrictions on oil exports have, as a result of their effectiveness, dealt a severe blow to the Iraqi economy. We argue in this chapter that the main effect of sanctions has come through the complete closing off of oil exports and the barring of access to the international capital market, not through the prohibitions on imports. The sharp reduction in purchasing power and the rise in the price of foreign exchange have rendered import controls largely inoperative, resulting in the 'temporary' shutdown of an economy that was once highly dependent on imports financed by oil revenues.

This shutdown produced widespread economic disruption and impoverishment. Due to limited revenues, public-sector salaries fell to as low as US$3.0 a month, and industrial production dropped by 80 per cent in many regions because of reduced purchasing power and imports. While they waited for the 'temporary' sanctions to end, millions of people took on casual work in the service sector, or shifted to food production, which remained for a time relatively profitable. A modest relaxation of the sanctions was achieved when the UN Security Council and the Government of Iraq successfully negotiated the implementation of a 'food-for-oil' (FFO) agreement at the end of 1996. This allowed Iraq to export a limited value of oil with part of the revenue available for the import of staples, medicines and other items to relieve suffering.

However, we do not consider that the FFO arrangement will be sufficient to end the extreme disruption of people's lives in Iraq. The programme is a wholly inadequate response, not least because its very design precludes the revival of economic activity in Iraq. Given the high human and economic costs of sanctions documented in this study, we believe that in order to achieve its ultimate goals in the region, the UN should seriously consider alternatives to comprehensive economic sanctions. The unfortunate reality is that over the past decade, none of the main parties to the sanctions regime seems to have been interested in examining such alternatives.

2. Context and scope of study

This study is based on a survey in 1996, and builds on research begun in 1991. The 1991 survey – made exactly one year after the imposition of sanctions, and around five months after hostilities had ended – was based on surveys of households, markets, and economic enterprises, and on interviews with a wide cross-section of individuals, including ordinary citizens, officials, factory managers, and representatives of international aid organisations and UN agencies. Researchers visited a large number of locations, including areas outside government control in Iraqi Kurdistan, and the southern marshes where the government's writ was also tentative. Both studies focused primarily on the rise of poverty, and paid particular attention to the issue of food security.

Although the format of both studies was similar, our visit in 1996 was shorter (lasting in total for one week at the beginning of May 1996), and we went to fewer places. It is important to put on record the conditions of our travel and work in Iraq in 1996. We had to submit our travel itinerary to the Iraqi authorities in advance of our travel and obtain a written permit, but the only locality that was excluded was Iraqi Kurdistan, outside government control. In and around Baghdad we visited various homes, farms, shops, and factories in a broad cross-section of districts and localities that included Al-Thawra (Saddam City), Adhamiyya, Bab-al-Shaikh, Karrada and Jamila. We also travelled to the South, to Basra where we spent one night. We made frequent stops en route for interviews in towns and small settlement in several of the southern governorates, including Kut, Amara, Nasiriyya, Najaf and Karbala. We stopped wherever we liked and interviewed whomever we chose, and the localities, families and economic enterprises that we visited were selected on the spot, with no prior warning and without any purposive criteria other than the fact that we were particularly interested in visiting poor localities. The main constraint we faced was the limited time at our disposal. Although our sample was not representative in the statistical sense, we believe that it provides a fairly accurate picture of the impact of sanctions on ordinary Iraqis.

Iraq is a heavily policed country. We came across police and military checkpoints at regular intervals, on inter-city roads as well as within towns and cities. At various points during our travels in the country, we had to show our government authorisation, but apart from one occasion we were allowed passage without let or hindrance. The exception was on the road south to the marshes, where we were stopped and prevented from proceeding until the local governor had granted permission. He allowed us to travel to the marshes on condition that we accepted the company of an armed escort, but we chose not to do so, partly because of the limited time but also because we felt that the presence of such an escort might detract from the purpose of our visit. The heavily armed presence on the roads to the marshes indicated that the area was not fully under government control, but there was regular traffic in and out of the marshes.

For most of the time an interpreter selected by the Ministry of Information accompanied us, but made no attempt to influence where we stopped, who we interviewed, or what sort of questions we asked. All interviews with household members were carried out in the privacy of their houses, and our interviewees gave no indication of reluctance or hesitation in responding to our questions. It has to be said, however, that we took care to restrict our questions to economic and welfare matters, such as wages, employment, incomes, asset ownership, prices, food consumption and the functioning of the ration system. Our respondents were generally extremely frank and hospitable. Indeed, compared with our experience of similar investigations in other countries we found that respondents were surprisingly frank on issues that concerned household earnings or profits of enterprises.[8] On a number of issues, we were able to crosscheck our respondents' replies against actual observations, both in homes and in business premises. For example, it was not difficult to gain an impression of a family's general living conditions by observing the quality of their dwelling, their ownership of large household durables, and the appearance of their children. Similarly, in factories, workshops and farms, it was quite easy to observe and crosscheck simple data such as the number of workers present and the proportion of machines running.

We conducted direct interviews with more than 20 families about their economic conditions – income, employment, food consumption, living conditions, and access to the ration system. We visited a wide range of economic enterprises, including small private-sector workshops, large factories (both private and government-owned), farms, shops, and markets, and we also visited a number of hospitals. We had meetings with the Ministry of Trade, where we met the minister, and with the Ministry of Labour and Social Welfare and the Ministry of Agriculture, where we met senior officials. We visited the institutions involved in running the food-rationing system, including the computer centre which keeps records of recipients of rations for the whole country, a flour mill, a wholesale centre that distributes ration supplies to retail outlets, and shops that deliver rations to households. This provided a way of crosschecking the responses from households about the functioning of the ration system. Although the amount of information gathered in a short time and from a limited number of interviews was not comprehensive, we believe that it is not misleading. The quantitative data is presented and utilised as a way of conveying the extensive impact of sanctions on the economy and on the welfare of the population; it also corroborates qualitative observations. These figures are particularly useful in the absence of any trustworthy and objective source of statistical data on these issues.

3. Macroeconomic overview and review of earlier findings

The results from the 1996 study showed that the macroeconomic impact of sanctions had persisted with little change since 1991. In order to understand the impact of sanctions, it is important to bear in mind the extent of the Iraqi economy's dependence on oil exports prior to the sanctions. Evaluated at the market exchange rate then prevailing, oil exports equalled 75 per cent of Iraqi GDP in 1990.[9] This overstates the importance of oil, since oil is priced in US dollars and in 1990 the Iraqi dinar was traded at a large discount in the heavily restricted foreign exchange market. However, dollar oil revenues still amounted to 15 per cent of GDP, measured at the artificially appreciated official exchange rate.[10] The actual importance of oil in the economy would have fallen somewhere between these two levels.

These oil revenues underpinned both the Iraqi economy and the government's policies, and although fiscal accounts have not been made public, it is clear that the government was highly reliant on oil for its revenues. There were few taxes in Iraq, so the bulk of the revenues came apparently from oil and from earnings transfers from non-oil state enterprises. Iraq's oil production was managed by several large state companies and these, in turn, were directly controlled by the government, the oil revenues being used to finance a range of investments, current expenditures and subsidies. The government had maintained a large public sector and military establishment, and large-scale military imports and soldiers' salaries were financed with oil revenues and borrowings throughout the 1980s.

In addition, the government built up state industry and agriculture during the 1970s and early 1980s by introducing sizeable direct and indirect subsidies. This ensured that any growth of domestic agriculture and industry was achieved with the aid of subsidies and inexpensive foreign exchange and that it was, in effect, highly dependent on oil revenues. A third use of funds was the construction of a broad-ranging social welfare system, which included consumer subsidies for a range of food items, a state safety net and a pension system, in addition to substantial investment in education and public health. By the late 1980s, Iraq had one of the best educational and health systems in the region. In short, the state was a key player in all areas of the economy: production, employment, private consumption and the provisioning of public goods. Although the main factor behind this economic structure had been the state's access to substantial revenue through the export of oil, it is interesting to note that during the war with Iran this structure was preserved through foreign borrowings.[11]

The impact of sanctions on oil revenues

The impact of sanctions caused the abrupt ending of revenues from oil exports and also cut Iraq off from other sources of finance, such as foreign borrowing. The effect on imports was dramatic. As shown in Table 1.1, total estimated Iraqi imports fell from US$10.3 billion in 1988 to just US$0.4 billion by 1991.

TABLE 1.1
Value of Iraq's foreign trade: 1987–94 (current US$ millions)

Year	Exports	Imports
1987	9,705	7,445
1988	9,687	10,286
1989	12,284	9,890
1990	10,303	6,520
1991	468	418
1992	595	621
1993	65	192
1994	307	225

Source: IMF, *Direction of Trade Statistical Yearbook 1994*, and *Direction of Trade Statistical Quarterly*, various years.

The magnitude of this shift can be illustrated using some international comparisons. Table 1.2 gives estimates of the dollar value of imports per person for various groups of countries. Controlled for factors such as the size of an economy and the diversity of its structure, per capita import values are positively correlated with per capita national income. Before the sanctions, Iraq's imports per person were close to the average of those countries ranked as 'high middle income' in the World Bank's *World Development Tables*. By 1996, they had declined to about a third of the average level for the poorest countries in the world, and an eighth of the level of the Arab countries ranked as 'low income'. Import levels in the period immediately preceding the food-for-oil (FFO) deal had fallen well below levels in countries such as Zaire (US$211 in 1991), Sudan (U$55 in 1991) and India (US$62 in 1991).[12]

TABLE 1.2
Iraq's imports per capita – some comparisons[a]

	Population (millions)	GNP per capita (US$)	Imports per capita (US$)
Iraq 1987–90	17	–	584
Iraq 1991–4	19	–	20
Low income countries[b]	3,039	380	61
Poorest ten countries[c]	167	133	37
Low income Arab countries[d]	84	650	157
Lower middle income countries[b]	1,096	1,590	304
Upper middle income countries[b]	501	4,370	781
Iraq under FFO[e]	20	–	150

Notes: [a]All comparative data for various country groups are for 1993. [b]As classified in the World Development Report, 1995. [c]Ten poorest countries for which comparable data were available: Mozambique, Tanzania, Ethiopia, Sierra Leone, Burundi, Uganda, Nepal, Malawi, Chad and Rwanda. [d]Mauritania, Egypt and Yemen were the three members of the Arab League in the low income country group for whom data is available. [e]The first phase of the 'food-for-oil' (FFO) deal allowed Iraq to import humanitarian supplies up to the value of $2.6 billion a year. By Phase V, roughly covering the second half of 1998, this figure was almost doubled, doubling again in 1999 and 2000 as oil prices rose. [Editor's note]
Source: World Bank, *World Development Report 1995*; IMF, *International Financial Statistics Yearbook 1994*.

This data can be used to examine the possible macro implications of the FFO agreement between Iraq and the UN Security Council. Under this agreement, which was signed in May 1996 (but left in abeyance until the end of November 1996), Iraq was allowed to export up to one billion dollars worth of oil every 90 days. Out of this amount, a total of US$650 million could be spent on the import of food, medicines, medical equipment, and goods specifically for humanitarian purposes. FFO implies imports of $130 per person. Added to the existing import value of US$20 per person, Iraq's total possible imports under the original terms of FFO amounted to $150 per person per annum – still well short of the figure for three of the poorest Arab states (Mauritania, Yemen and Egypt). In subsequent years, the ceiling on permitted oil exports and the import of humanitarian supplies under FFO were relaxed, thereby permitting a broadening of the FFO, including some restoration of damaged infrastructure.

It can be argued that since the figures presented above do not take possible breaches of the sanctions into account, they seriously underestimate Iraq's ability to import. In particular, if Iraq was able to export more oil than is allowed under the sanctions, its ability to pay for imports would consequently be higher. Appendix 1 examines this issue in some detail (for 1996) and argues that, apart from logistic and political difficulties in hiding oil exports, Iraq's production data, independent estimates of production by OPEC, and evidence from neighbouring countries confirm that any cheating on oil exports was probably small.[13]

Impact on government finances
The most striking evidence of the effectiveness of sanctions comes from changes in the macroeconomy. The collapse of oil revenues (and other foreign exchange flows) led to two immediate problems. First, government revenues fell, causing a fiscal crisis. Second, and most importantly, a severe shortage of foreign exchange and imports caused the relative price of tradable goods (including food and medicines) to rise.

While no data is available on the fiscal gap, it is clear that the immediate impact of lower oil revenues was to cause a large fiscal deficit. The government responded by limiting nominal expenditures, while issuing money to finance the budget deficit. With more and more money being issued, inflation rose and the dinar depreciated. Figure 1.1 shows the movement in the market exchange rate of the Iraqi dinar against the US dollar from 1990 till the middle of 1996.[14] Before Iraq's invasion of Kuwait, one dollar could be exchanged for around four dinars. By the middle of 1991, a dollar bought eight dinars, and by December 1995 it was possible to purchase 3,000 Iraqi dinars for one dollar. This rapid depreciation was driven by the government's need to print money in order to finance expenditures.

One of the important results of sanctions was that the government curtailed subsidies on producer goods more sharply than on consumer goods. Before sanctions, the import and marketing of wheat flour and some other staples was largely a state monopoly. These imports were not valued at the market exchange rate but rather at the overvalued official rate, which was thus equivalent to a subsidy to consumers. In addition, there was also a direct price subsidy on wheat flour.

FIGURE 1.1
Exchange rate trends (log)

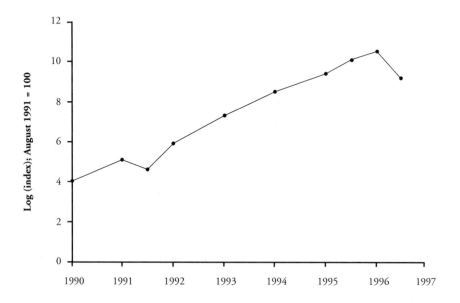

Soon after the imposition of sanctions, the government withdrew these general subsidies and replaced them with a system of subsidised rationing (see section 4 and section 6 below). The private sector began to take an active role in the import and marketing of all food items, including wheat flour. Two distinct types of price movement were noticeable. First, the initial adjustment, which involved the removal of subsidies, caused substantial *relative* price movements – in particular, the price of the basic staples relative to wages (or the price of labour) rose sharply. Second, since the government resorted to issuing money as a way of financing expenditures, the overall price level also increased and a situation of chronic high inflation prevailed. This distinction is worth bearing in mind, because the two types of price movements have different implications. The change of relative prices had a serious adverse effect on the purchasing power of Iraqi households, and was felt most dramatically in the first year of the sanctions. The second type of price change – that is, the general rise in the price level – became noticeable after the initial period of macroeconomic shock, as the government struggled to maintain its spending priorities in the absence of any significant sources of revenue.[15]

Although public finance data is not available, it can be safely assumed that the main items of expenditure were salaries of government employees and consumer subsidies, notably through the ration system.[16] It is clear that under its new and severe fiscal constraints, Iraq could no longer maintain the large public sector of the pre-sanctions period. All the evidence suggests that the government maintained its fiscal commitment to subsidise the food ration, and that it preserved employment levels in the public sector at the expense of the purchasing power of public-sector salaries, by not adjusting them with the rise in prices. Attempts were made to raise public-sector salaries (the first such pay rise happened in September 1991), but because these were financed by monetary expansion, their effects were quickly neutralised when prices rose in response. Indeed, the systematic financing of pay rises by printing more money seriously threatened to set off a spiralling cycle of inflation.

It can be argued that Iraq was on the verge of such a cycle in late 1995, when the government initiated a macroeconomic stabilisation programme incorporating reductions in nominal salaries, further cuts in government spending, and measures to raise tax revenues. This programme coincided with Iraq's announcement that it was willing to enter negotiations for a food-for-oil agreement. This announcement helped the stabilisation programme, since the market exchange rate of the Iraqi dinar appreciated and the relative price of imported goods declined. To the extent that it checked the spiralling cycle and brought down wages and prices, the programme was successful. Nevertheless, by 1996 public sector workers were commonly earning US$3–5 per month, compared to their pre-sanctions salaries of US$150–200.

The foreign exchange crisis

The most important impact of the decline in oil revenues was a severe tightening of the funds available for imports. With the reduction in supplies of imported goods, their price was bid up, relative to domestic goods and particularly to labour. This shift in the relative price of foreign exchange had striking effects on the measure of dollar GDP. The World Bank reported that, in 1989, per capita GDP was US$2,840. Our estimates for 1996 place it at roughly US$200.[17] Since real GDP had probably fallen by some 50 per cent during this period, the remaining decline would have to be explained by changes in the relative price of foreign exchange. Based on these estimates, the overall decline in dollar GDP from US$2,840 to $200 can be roughly broken down into two components: it fell by half due to loss in real production, and then fell by a factor of seven due to a sharp rise in the price of foreign exchange, itself caused by scarcity.

In our surveys of domestic markets and enterprises, we generally found that, adjusting for the exchange rate, domestic prices for imported goods were not very different from prices in Jordan. This implies that enterprises and households did not have to pay a premium (beyond transport costs) for imported goods. Whenever import restrictions were highly effective, a premium would be charged so that importers earned a reward for the risks they took in attempting to break the restrictions. Given this observation, we believe the greatest impact of sanctions came through the oil-export restrictions; with reduced purchasing power, people were simply unable to buy the goods they were used to. The importance of the exchange rate to household purchasing power is indicated by Figure 1.2, which shows changes in an index of food prices, and the exchange rate (the price in Iraqi dinars of one US dollar).[18] Except in the first year of the sanctions, the three price indexes have tended to move in tandem, corroborating our observation during the 1991 visit that the market prices of staples closely tracked the exchange rate. This shows the strong link between the open markets for staples and the foreign exchange market. It is the imported supplies at the margin that determine the price.

The scarcity of foreign exchange also had a direct impact on production. The enterprises we visited noted that purchasing power had fallen sharply, and that the cost of imported goods was prohibitively high.[19] As a result they had curtailed production, and many firms reported declining levels in the order of 70 to 90 per cent. We did not get the impression that production was restricted by shortages of inputs; rather it was constrained by the limits on market demand associated with reduced incomes. The one sector that had not declined was agriculture. Since the shortage of foreign exchange had induced a sharp relative rise in the price of food, and since food demand had been maintained even with the fall in incomes, it had become relatively profitable to be in the agricultural sector. While agricultural production appeared to fall immediately after sanctions were introduced, there was evidence in subsequent years of a substantial reallocation of resources towards agriculture.[20]

At the household level the impact of these shocks varied, depending on the sectors in which household members were employed, on whether that was in the private or the state sector, and on the opportunities for finding alternative sources of income. Broadly, while the public sector adjusted to the sanctions mainly by reducing real wages, the private sector reacted mainly by cutting employment. For the unemployed or those earning low wages, the most common means of gaining additional income was to find self-employment or casual labour.

FIGURE 1.2
Trends in logarithmic price indices

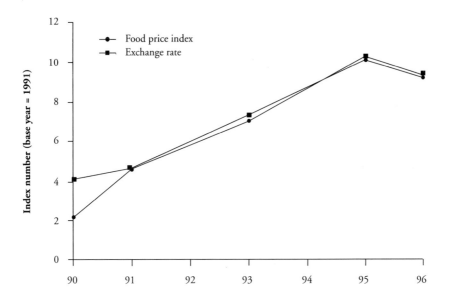

4. Livelihoods and poverty

The impact of the sanctions against Iraq is of concern ultimately because of their effect on living conditions. In this section we document and analyse changes at the micro-level in the livelihoods and the command over resources among various groups within the Iraqi population, and examine trends in wages and earnings. The price of the ration basket was revised upwards only sporadically and by a margin well below the inflation rate. For most Iraqi households, the ration system acted as the principal cushion against starvation. We also outline the details of the rationing system and its crucial role in preventing deprivation, offer some observations on the survival strategies employed by households to make ends meet, and discuss some specific aspects of vulnerability.

Trends in wages and earnings

As we have noted above, it is not easy to specify a benchmark period for 'normal' economic conditions in Iraq. Before invading Kuwait, which action triggered the sanctions, Iraq had been at war with Iran for the better part of the 1980s. The employment pattern of the labour force was heavily influenced by the war, and it is difficult to say how it would have been under conditions of peace. For what it is worth, the 1987 census showed the service sector to be by far the largest employer, accounting for nearly half of all workers.[21] The government employed about a quarter of all workers, a large proportion of whom were in the armed forces and the security services. Sectors with significant numbers of self-employed, such as agriculture and the wholesale and retail trade, accounted for smaller proportion of the labour force, at around 12 per cent and 6 per cent respectively. The 1977 census, the last to be held under conditions of peace, put the proportion of agricultural workers at around 30 per cent, and the proportion of wage-employees in the workforce at around 60 per cent.[22] It would be fair to conclude that, before the Gulf War, and even prior to the Iran–Iraq war, Iraq was a fairly urbanised society, with a relatively small proportion of its workers engaged in agriculture and a large proportion of the workforce reliant on salaried employment for their livelihoods.

In August 1991, a year after sanctions were imposed, the effects of the defeat in the Gulf War and the sanctions on the employment pattern showed clearly. First, there were obvious signs of a massive demobilisation, which implied a large reduction in the number of government employees. Second, there had been a large decline in regular formal-sector employment in the private sector because of the downturn in economic activity and closure of enterprises. This decline appeared to have been more or less compensated for by a rise in informal private-sector activities, including petty trade, casual labour, and various forms of self-employment. Third, there was no labour retrenchment in the civilian public sector, and the nominal salaries of public-sector employees had remained unchanged for a year. Fourth, a large number of foreign workers had departed from Iraq.

On the basis of these developments, the 1991 study concluded that employment had stagnated, as had nominal earnings, and that, abstracting from the subsidised rations, real incomes had declined more or less in line with the inflation rate. There had been a huge rise in prices, and according to the food-price index, real earnings had declined to between 5 to 7 per cent of their pre-sanctions levels.

It is notable that nominal earnings had remained frozen for a whole year after sanctions had been imposed. The first major pay rise was not made until September 1991, when the government made an across-the-scale increase in the salaries of public employees. This was followed by regular increases until December 1995, when the first pay cuts were announced. To understand changes in *real* wages and earnings, it is necessary to have an appropriate price deflator. One possible contender is the food-price index, of the type presented in Figure 1.2 above. Prices in Iraq at the time of our survey were, however, extremely volatile, and day-to-day fluctuations in the exchange rate were very quickly reflected in changes in food prices and in casual wage rates. Since we were interested in comparing changes in earnings between different types of activities, we used a deflator that Iraqis themselves would be likely to use as a benchmark for wages and other prices. In Table 1.3 we present all wages and earnings in terms of wheat-flour equivalents.

There was a marked decline in wages and earnings between 1991 and 1996.[23] The decline was the most pronounced for public-sector salaried employees. For example, the salary of a middle-ranking civil servant, which commanded 80 kilograms of wheat flour in 1991, was equivalent to only 16 kilograms in May 1996. Other people on government salaries, such as pensioners, workers in public-sector enterprises, and army soldiers also saw declines in the order of 60 per cent. Among public employees, professional army officers, whose real incomes declined by 45 per cent, were the best protected.

TABLE 1.3
Wages and earnings, 1991 and 1996

Wages and earnings[a]	1991	1996	Percentage decline
Self-employed, daily earnings[b]			
Petty vending	2.0	1.2	40
Unskilled labourer	4.8	3.0	38
Taxi driver	22.0	10.0	55
Average change			44
Monthly salaries			
Pensioner	22	9	60
Army recruit	34	12	65
Unskilled worker in public enterprise	52	20	62
Unskilled worker in private enterprise	70	40	43
Postman	72	16	78
Middle-ranking civil servant	80	16	80
Army officer	88	48	45
Skilled textile worker, private	120	100	17
Average change			56

Notes: [a]All wages and earnings are in terms of kilograms of wheat flour. [b]Expected earnings for a full day's work in given activity.
Source: For 1991, Jean Drèze and Haris Gazdar (1992), 'Hunger and poverty in Iraq, 1991', *World Development*, Vol. 20, no. 7, 1992, pp. 921–45; for 1996, author's survey.

Public-sector salaries were revised periodically by the government to compensate, though only partially, for inflation. Public-sector factories and other economic enterprises generally ran at a loss because the prices of many of their outputs were administratively controlled. Public-sector wages are not good indicators of the decline in economic activity in the private sector. Private-sector salaried employees were better protected against inflation but not against unemployment. The incomes of skilled textile workers, for instance, appear to have declined by 17 per cent. Similarly, unskilled factory workers in the private sector saw their real wages decline by 43 per cent compared with the 62 per cent decline in the public sector. The informal private sector saw losses of a similar magnitude. Daily wages of casual unskilled labourers as well as earnings from petty vending declined by around 40 per cent.

The group of self-employed agricultural workers were, *prima facie*, better placed than most other workers to protect themselves against price rises. Incomes in agriculture depend on the quality of the harvest and the price of the output.[24] Therefore, real incomes of farm households should not have been adversely affected by rises in food prices, if the market for agricultural produce operated without restriction. In Iraq, however, there was a state-controlled system of grain procurement, under which the government bought the entire harvest at set prices.

There was almost no change in the *relative* procurement prices of the various food-grain crops (including barley, rice and three grades of wheat) between 1990 and 1996. Here, we take the procurement price for Class 1 wheat as the benchmark. Before the sanctions, farmers received ID270 per ton. In terms of the market exchange rate, this was equivalent to around US$60. In 1991, the procurement price was ID800, or around US$100. The nominal price kept rising till 1995, but these increases did not keep up with the depreciation of the dinar. By 1994, the procurement price had been raised to ID35,000, but was equivalent to less than US$90. In 1995, the price was raised further to ID105,000 per ton. At the market exchange rate of ID2,000 per dollar in the middle of 1995 (the wheat crop is harvested in May) the real price was just over US$50. In 1996 the nominal procurement prices remained unchanged, and this implied a real gain for farmers as the exchange rate had improved considerably in favour of the dinar.[25] In dollar terms then, the 1996 price was exactly the same as the price in 1991 – that is, around US$100 per ton.[26]

Although procurement prices, and therefore agricultural incomes, fluctuated like other incomes over the six-year period under study, they were no lower in 1996 than they had been in 1991, and indeed were higher than they had been in 1990. This does not necessarily imply that farmers' real purchasing power was higher in 1996 than before the sanctions. Even if we ignore the effect of sanctions on the quality of the harvest, the purchasing power of a dollar was much greater in 1990 than in 1996 since a large number of essential consumer goods were costed at the overvalued official exchange rate, as discussed above.[27] The fact that, under sanctions, agricultural workers were somewhat better protected against price rises is of particular significance given that there was a large reported rise in the proportion of the workforce employed in agriculture during this period. According to Ministry of Labour and Social Welfare estimates, nearly 40 per cent of the workforce was engaged in agriculture in some way or another in 1996, compared with only 12 per cent in 1987.[28]

Many public sector employees supplemented their incomes with a second job or self-employment in the informal sector. For those with jobs in the public sector, recourse to secondary employment implied that they might make only token attendance at their main job. Such action by employees in particularly strategic positions within public-sector organisations – that is, individuals with professional and technical competence – would have had serious implications for the performance of their respective organisations. We also came across several cases of public-sector professionals, such as doctors or teachers, who continued to perform their professional duties only because they had other sources of income in the family. Petty trading and unskilled casual labour are, in general, 'bottom-line' activities which have virtually free entry. There appeared to have been a dramatic rise in the number of people, including young children, who were involved in these activities, and this is discussed further below. It can be quite sobering to compare the rates of remuneration in these activities with the wages of casual labourers living in regions known for chronic poverty. In 1996, an unskilled labourer in Iraq could expect to earn the equivalent of three kilograms of wheat flour for a day's work. In comparison, unskilled agricultural labourers in West Bengal, one of the poorest states of India, could expect to earn the equivalent of six kilograms of rice, the staple food in that area.[29]

Importance of the monthly ration
In the context of the dramatic collapse of private incomes in 1990–1 and their continuing decline since then, the state ration system proved to be a crucial source of sustenance. Early assessments of the humanitarian impact of sanctions identified the importance of the state ration as the main factor in preventing the onset of large-scale hunger and starvation in Iraq.[30] Five years on, in 1996, the situation remained the same and, in fact, the importance of the food ration system as a lifeline for ordinary Iraqis had increased considerably. A brief first-hand account of the ration system and some comparisons between 1991 and 1996 are provided here.

Each family was registered with a ration agent (who often owned a private grocery store) in its locality. The agent maintained a detailed record of households (the number and ages) receiving the ration. The ration bundle of a family consisted of equal entitlements for each of its members aged one year or over of various food and other items, and a fixed monthly ration of infant formula milk for each child aged up to one year.[31] There had been little change in the quantities of various items included in the ration (Table 1.4). While the ration amount for some items (wheat flour, rice, sugar and baby milk) had diminished, supplies of other goods (cooking oil, tea, soap, and washing powder) had increased.

TABLE 1.4
Changes in the state ration, 1991–6

	August 1991		May 1996	
	Quantity per person	Market price	Quantity per person	Market price
Wheat flour (kg)	8	2.4	7	240
Rice (kg)	1.5	4.1	1.25	350
Cooking oil (ltr)	0.25	10.3	0.75	580
Tea (kg)	0.05	23.7	0.1	1,270
Sugar (kg)	1.5	4.4	0.5	670
Soap (pieces)	0	–	1.5	90
Washing powder (kg)	0	–	0.2	100
Baby milk (kg)	1.8	22.2	1	3,200
	Costings for family of six persons with one infant			
Nominal price of full ration	11.1		600	
Cost at market prices	217		19,048	
Implied subsidy	206		18,448	
Implied subsidy in kilograms of wheat flour equivalent	86		77	

Source: For 1991, Jean Drèze and Haris Gazdar, 'Hunger and poverty in Iraq, 1991', *World Development*, Vol. 20, no. 7, 1992, pp. 921–45; for 1996, authors' survey.

Between 1991 and 1996, the value of the ration bundle for a family of six (including one infant) in terms of wheat-flour equivalents (that is, in terms of the market value of the bundle denominated in terms of the market price of wheat flour), declined from 90 to 79 kilograms – a reduction of around 11 per cent. The nominal cost to the family declined from around 4.6 to 2.5 kilograms of wheat flour, so that the subsidy element in the ration has declined somewhat less than the total value (Table 1.4). Given the decline in real earnings in the order of

40 per cent or so during this period, the relative value of the ration to Iraqi families had, nevertheless, increased. For an unskilled worker who supported a family of six, and who managed to find work 25 days in a month at the going wage (3 kg of wheat flour), the value of the ration was equal to an entire month's earnings. A professional army officer with a relatively high salary for the public sector earned about a third less than what the ration provided.

Compared with the levels of food consumption prevalent in Iraq before the Gulf War, the ration covered only around one-third of the total. The average monthly consumption of wheat flour, for example, was around 17 kilograms per person, compared with the ration entitlement of 8 kilos.[32] A household consumption survey from the early 1970s estimated monthly per capita wheat-flour consumption as just under 15 kilos.[33] While the differences in these consumption estimates might be due to differences in methodology and data sources, the rise in food consumption between the early 1970s and the late 1980s is consistent with the growth in per capita incomes over this period, and the extremely low food prices in Iraq before the war. It can be argued that the pre-sanctions food consumption is not a good indicator of food needs, since Iraqis were, in some sense, over-consuming.[34] Judging by a more frugal standard of consumption, the ration did indeed cover a greater proportion of needs. In 1996, for example, the adult ration supplied the equivalent of 1,290 kilocalories (kcal) a day, down from 1,417 kcal per day in 1991. It is helpful to compare these figures with a conservative benchmark for need, such as the notional calorie entitlement associated with the poverty line in India. This is set at 2,100 kcal per person a day for urban adults.[35] The Iraqi ration corresponded to 67 per cent of this benchmark in 1991, and to 61 per cent in 1996. This appears to be consistent with the subjective opinion of Iraqi households, who reported that the ration covered between a half and two-thirds of the monthly consumption of the main food items.[36]

Survival strategies

Our assessment of livelihoods, purchasing power and entitlements in Iraq was supported by a number of observations of a qualitative nature. These observations related, in particular, to strategies adopted by Iraqi families to maintain a minimum consumption level in the face of dramatic declines in their purchasing power.

It was common for workers to be engaged in several low-income activities at the same time. Although there had not been any retrenchment from the public sector, a large number of people had left voluntarily, since they found the salaries too low for subsistence. Most of those who had stayed tried to complement their meagre salaries through casual or self-employment,[37] and it was apparent that the numbers engaging in 'bottom-line' economic activities, such as casual labour and petty trading, had risen dramatically. There was free entry into these activities, so long as a person was reasonably healthy, but rates of remuneration were extremely low (see Table 1.3).

A visible phenomenon was the rise in the number of children working as petty vendors. It was not clear if the children who worked as petty traders took up these activities outside school hours, or whether they had dropped out of school altogether. In our interviews with families whose children were involved in petty vending, the parents generally claimed that their children attended school and did the petty trading in their spare time. In economic terms, these activities, such as selling from a carton of cigarettes, were extremely low in value added, and were mainly of a marginally redistributive nature. The fact that a large number of people, both adults and children, were prepared to spend long hours trying to earn a derisory margin on a few petty transactions was a measure of some desperation.

The same holds for transactions in a wide range of items that were uncommon or absent in Iraq before the war. In the period under sanctions, special markets for exchanging all sorts of used goods emerged in all major towns and cities. These markets, which were usually held in public places on specific days of the week, had hardly existed before the war, but at the time of our visit were attracting a very large number of buyers and sellers. The items bought and sold at these markets ranged from bulky household durable goods, such as cookers and refrigerators, down to items of used clothing, old door fittings, and even used transistor circuit boards. As in the case of petty vending, many transactions in these 'flea markets' also involved extremely low turnovers. Here, too, the willingness of people to spend an entire day hoping to make a narrow margin on these transactions reveals the absence of alternative opportunities for remunerative employment.

Many of the vendors were older people and pensioners, selling their assets in order to finance current consumption. The following case, observed in the main flea market in Basra, illustrates how there could be scope for sustaining such marginal activities in what had been initially a wealthy economy. A pensioner was trying to sell his cooker and hoped to raise around ID20,000 for it. His plan was then to purchase a smaller (also used) cooker for ID10,000 and to use the difference to finance current consumption. He was well aware that he might soon have to repeat this cycle with the smaller cooker that he was hoping to buy, in exchange for something of even lower value. We were able to observe another 'bottom-line' activity that is generally to be found amongst the poorest people in low-income countries – namely post-harvest gleaning.[38] En route to Basra we came across several instances of groups of women going through recently harvested fields and manually collecting the individual wheat grains that had been left behind.

Vulnerability

The economic situation in Iraq did not allow easy generalisations about patterns of vulnerability and deprivation. Certain groups, including those who had access to foreign currency (sent by relatives working abroad) or who owned productive assets or who had marketable skills, as well as those with good connections both within the country and outside, were relatively well protected. The same was true of those who had access to farmland, or other forms of self-employment in occupations that did not rely too heavily on imported inputs. Conversely, people with few assets to use or sell, elderly people with little family support, and others reliant on fixed nominal salaries were particularly vulnerable.

However, a notable aspect of the situation in Iraq was that apparently minor differences in household attributes were associated with large differences in consumption levels. As was common in Iraq, large families relied mainly on the earnings of a single worker (usually male) and could experience a dramatic turn for the worse if that sole breadwinner was unable to work through illness or some other misfortune. In this way, the effects of sanctions on incomes and livelihoods were due not only to a drop in the level of economic activities but also to the erosion of public utilities and healthcare services. Here, regional contrasts were quite striking. In many towns and cities of southern Iraq, for instance, the sewage systems had broken down, and we visited several homes that were flooded with sewage from the street. The area around Basra appeared to be far worse off than the rest, due to the fact that it had borne the brunt of the Gulf War and had, in addition, been the main battleground during the eight-year war with Iran. The region seemed to have received little investment since the beginning of the 1980s.

5. Structural adjustment

Underlying the changes in macro indicators such as price levels, exchange rates, and national income, are numerous behavioural responses to the shock on the part of the various types of economic agents. In the previous section, we discussed labour participation and the saving (or dissaving) behaviour of individuals and families in response to income declines, and documented the response of the government in first financing expenditure through monetary expansion, and then introducing a stabilisation package. This section offers some evidence as to the extent of the structural adjustment in the Iraqi economy in response to a massive reduction in its international trade, and the following section will examine some institutional responses of the public sector to the various crises that emerged in the wake of sanctions.

Agriculture

In the absence of oil revenues, agriculture, for a number of reasons, is a key sector. First, Iraq was a largely agrarian economy prior to the large-scale development of the oil sector, and its comparative advantage in terms of relatively fertile agricultural land remains. Second, the shift of the economy away from agricultural production was driven by oil revenues: directly, in the sense that much of the non-agricultural employment in the public sector was financed by oil-based public revenues, and also indirectly, in the sense that private-sector manufacturing and service activities depended heavily on state subsidies, also financed from oil revenues. Finally, from the point of view of overall food security, agriculture is obviously the key sector.

As already noted in Section 4 above, there was a reverse movement of labour towards agriculture. This makes sense, given that real incomes in this sector had declined less and, in relative terms, were likely to have improved after sanctions. In spite of the large reported increase in the agricultural workforce (Table 1.5), and increases in cropped areas of the order of 70 per cent, the estimated increase in agricultural output was relatively modest. Taking as an example the two main cereal crops

(Table 1.6), it can be seen that wheat yields decreased on average by around 20 per cent, but, with a substantial increase in the area under crop, output had been higher, by about 30 per cent on average. In the case of barley, however, although the area under crop had increased by over 70 per cent, the increase in total output had been of the order of 25 per cent, due to a large decline in yield. Moreover, as the table shows, the yields for barley in the early 1990s compare unfavourably with yields obtained in the 1950s and 1960s (see Table 1.6).

TABLE 1.5
Workforce in agriculture

Source	Year	Per cent
Population Census	1977	30
Population Census	1987	12
Ministry of Labour and Social Welfare	1996	40

Source: Population Census 1977, as reported in UN, *World Demographic Survey 1988*, United Nations Population Division; Population Census 1987, as reported in CSO, *Annual Abstract of Statistics 1990*, Central Statistical Organisation, Government of Iraq; authors' interview with Iraq Ministry of Labour and Social Welfare.

TABLE 1.6
Agricultural output and yield: wheat and barley

	Wheat			Barley		
	Area ('000 hectares)	Yield (kg/hectare)	Production ('000 tons)	Area ('000 hectares)	Yield (kg/hectare)	Production ('000 tons)
1984	493	956	471	525	917	482
1985	1,540	913	1,406	1,357	981	1,331
1986	1,240	836	1,036	1,456	718	1,046
1987	859	841	722	972	764	743
1988	1,041	892	929	1,314	1,094	1,437
Annual average, 1984–8	1,035	887	913	1,125	895	1,008
1990	1,196	1,000	1,196	1,995	929	1,854
1991	2,517	587	1,477	2,412	318	768
1992	1,677	600	1,006	2,012	750	1,509
1993	2,013	590	1,187	2,314	675	1,562
1994	1,806	743	1,342	1,535	633	971
1995	1,535	805	1,236	1,389	642	892
Annual average 1990–5	1,791	721	1,241	1,943	658	1,259
Annual average 1954–8	1,432	595	852	1,179	894	1,054
Annual average 1959–63	1,480	494	731	1,115	783	873

Source: For 1984–8, FAO *Production Yearbooks*, various years; for 1990–5, FAO (1995a); for 1954–63: Doreen Warriner, *Land Reform in Principle and Practice*, Oxford: Oxford University Press, 1969.

The main reasons cited for the decline in wheat and barley yields include shortages of farm machinery, good-quality seeds, and chemical inputs such as fertilisers and pesticides. While this might be true in the case of wheat, that of barley is not so clear-cut. The yields in the 1950s and 1960s were obtained prior to the massive farm-mechanisation programme and before the introduction of other new import-intensive technologies had been affected by the sanctions regime. The fact that reported yields were much lower in the mid-1990s than they were in the 1950s and 1960s cannot be explained entirely or even substantially with reference to the shortage of these imported or manufactured inputs. It is possible that official figures for agricultural output are downward biased or that cultivation has expanded into considerably poorer soils.[39]

Supply and demand constraints

As noted above, the Iraqi economy was highly dependent on oil revenues for the financing of imports, and various sectors, such as manufacturing, had relied heavily on imported inputs such as raw materials, machinery and spare parts. The economic blockade affected various industries and enterprises differently. For larger enterprises with specialised or bulky input requirements, quantitative restrictions on imports were sufficient to cause a substantial reduction in capacity. However, for numerous small-scale consumer industries, inputs could be, and were, purchased on the black market and smuggled through neighbouring countries. The main problem that they faced was the sharp rise in the relative price of their inputs. Whereas in the past they had rationed access to foreign exchange at the overvalued official exchange rate, they had now to achieve profitability at the discounted exchange rate. In 1991, their main difficulty had been in making the transition to this change. It is reasonable to expect that the surviving economic enterprises in Iraq would be those that were subsequently able to make an adjustment to the new situation. In other words, for industries that did not face very serious quantitative constraints in obtaining imported supplies, only those firms that remained competitive at world prices would have survived.

Our interviews with small-scale industrial units revealed that such an adjustment had indeed occurred. The surviving firms reported that in their expansion of output and employment or in their utilisation of capacity, they were constrained largely by the declining demand for their products, resulting from the overall fall in incomes. However, with a pick-up in incomes and consumer demand, many industries would be able within a short period to expand output and employment. We did find evidence, therefore, that the private sector had adjusted to the new economic situation in accordance with expectation. An unintended effect of the sanctions had been the revival of the private sector and the weeding out of activities that flourished under the protective regime of the dual exchange rate.

6. The rationing system and other public interventions

Here we examine the rationale for the government's responses to the economic crisis precipitated by the UN sanctions. First, as already discussed above, the government has been the dominant force in the allocation of resources as a result of the past availability of oil revenues. Secondly, in the post-sanctions period, specific state interventions have been crucial in avoiding large-scale hunger and starvation, notably through the monthly ration scheme. Finally, any government's response to major crises reveals a great deal about both its underlying political priorities as well as its administrative and managerial abilities. In this section we focus mainly on Iraq's state ration system, and also suggest some comparisons with other public services and government interventions.

Coverage

A central feature of the state ration system is that it is comprehensive and non-targeted: everyone is entitled to the ration regardless of means, and the ration and its price are uniform across the country. This was the principal finding of the 1991 study, and was based upon surveys of households and markets in urban and rural regions from all government-controlled governorates at the time. Up until that time, there had been a great deal of speculation outside Iraq that the ration system was selectively used by the Iraqi regime as a reward and punishment device against various sections of the population. The 1991 visit discovered no evidence for this. We found, on the contrary, that even people who privately expressed intense opposition to the regime were satisfied with the functioning of the ration system. During our 1996 visit, we covered fewer locations and interviewed a smaller number of households than in the 1991 visit, but found that, apart from a few details, the ration system operated exactly in the same way as had been observed in 1991.

Section 4 above gave details of the quantities of the various commodities that were supplied as part of the monthly ration. We did not find a single family that reported even a small difference between the officially announced quantities and the quantities it actually received. Furthermore, the ration agent was usually a local grocer, and people faced little difficulty in obtaining the ration on time. It is fair to assume, therefore, that the ration system did function quite effectively.

The fact that the quantities were uniform all over the country, and that these quantities were regularly broadcast in the public media, was an important factor in ensuring that people had good information about their entitlement. However, although the basic ration entitlement was uniform across the country, we did find some minor variations, two of which are worth mentioning. First, the quality of the goods supplied, particularly the quality of the wheat flour, varied from month to month, and also from location to location, though we did not observe any systematic variation between locations in different regions of the country. Secondly, supplies of items that were not normally part of the monthly ration were made available from time to time. In particular, all respondents in the Baghdad area told us that they had received small quantities of poultry meat during the Muslim holy month of Ramadan in the previous year. Some families, also in the Baghdad area, reported receiving occasional supplements of lentils and chickpeas. Outside Baghdad, particularly over most southern towns and villages, we did not come across any household that had received supplements to the usual ration. While there is some discrimination in the dispensation of public resources, including the ration system, this discrimination is limited to extra items that are *outside* the normal ration entitlement. The basic ration is, from all evidence, comprehensive and equitable.

The system of delivery

All families were registered with a local ration agent, and received coupons corresponding to the number of members of the family. Ration agents were usually local grocers for whom the disbursement of the ration was a side activity. The agents received supplies from government-controlled wholesale stores and silos and flourmills on presentation of the coupons which they had collected from the families in return for the delivery of the previous ration. These ration coupons were the currency within the ration system. Though they were not transferable, the goods supplied on the ration were obviously marketable. As each agent had only about 40 or so local families registered with him, he was in a position to monitor their situation. Equally, his clients could monitor his performance just as effectively.

There was potential scope for mutually advantageous collusion between the agent and his clients at the expense of the government. Some instances of collusion are bound to occur, but the possibilities appeared to be few, given the severe punishments meted out as a deterrent. These included revocation of the licence to act as a ration agent (which was a guaranteed source of income), and in extreme cases, the closure of the business. Furthermore, the whole rationing system was computerised with special built-in checks to weed out false claims such as exaggerated household numbers and registration with more than one agent.

To obtain ration tickets, household heads had to register with the neighbourhood ration office by presenting the identity cards that all citizens and foreign residents are obliged to carry. We visited one such office in Baghdad. The data collected by the neighbourhood offices was collated and centrally stored and cross-checked against the data collected at the time identity cards were issued. Ration tickets, which carried the name of the head of the household and the month for which they were valid, were colour-coded by governorate and were distributed to households by the ration office. It was interesting that the checks against abuse of the system focused on preventing false claims rather than on the reselling of rations on the open market. Certainly there seemed to be no attempt to brand the goods supplied on ration, but given that the ration was designed solely to prevent starvation, there was little incentive for households to sell their ration. However, we were told that poor families often sold infant formula milk and fed their infants ordinary food, occasionally with fatal consequences for the infant.

The effective functioning of the rationing scheme relied heavily on the elaborate system of policing and surveillance that reaches down to the grassroots, and it is interesting to note the dual use of this system for both coercion and for social welfare. Another important feature of the rationing system was the symbiosis of the government and the private sector. Shops distributing the rations were privately owned. Generally, they were neighbourhood grocery stores that supplied rations as just one of their lines of business, and they also sold the same items that were supplied on ration at market prices. Many of the mills that supplied wheat flour for ration were also privately owned. There appeared to be sufficient economic incentive for the private sector to take part in the operating of the ration system.

We also visited a centre for the distribution of sugar, tea and cooking oil to ration agents. On presentation of the tickets collected from households for delivering the previous ration, agents were issued with the entitlement to collect the next set of rations. We visited all the links in the rationing system chain, all of which were in the Baghdad area, and all seemed to be efficiently run. Certainly the Iraqi rationing system operates with considerable efficiency, and is apparently successful not only in preventing false claims but also in making it relatively easy to collect ration tickets and rations. This is in marked contrast to the operation of similar rationing system in a number of developing economies, such as India and Pakistan, where abuses of the system were extremely widespread.[40]

Availability and entitlement

The ration system in post-1990 Iraq can be justifiably regarded as an example of a successful food-security programme in a time of economic crisis. The government's response to the economic shock can be viewed at two levels.

In the first instance, it can be examined as an emergency food-distribution system that prevented the onset of famine at a time when purchasing power collapsed dramatically. There is considerable evidence from international experience to indicate that famines are likely to occur not simply because of inadequate food *availability* in aggregate terms, but because the entitlements of particular sections of the population have collapsed.[41] In the case of Iraq it is also clear that in the absence of public intervention, famine conditions might well have prevailed. The basic food entitlement of all sections of the Iraqi population was protected by the ration system from falling below the point where mass starvation could have occurred.

The ration system not only played a key role in preventing famine at a time of economic shock, but emerged subsequently as an effective mechanism for converting the country's aggregate food availability into a basic year-on-year universal entitlement. While there is now a great deal of literature on the experience of various countries in dealing with dramatically declining food entitlements at a time of emergency, there are few examples of similar interventions that have been sustained over such a long period.[42] The Iraqi state's ration scheme stood out in the sense that it combined an efficient system of food distribution with a sustainable system of procurement.

In our view, the procurement system was sustainable because it was largely based on Iraq's domestic grain production and therefore relied only minimally on the availability of foreign exchange. In this regard, it represented an important departure from the way that most sectors of the Iraqi economy had been organised prior to the sanctions. Even if we accept the official data on the output of food grains,[43] Iraq's aggregate output was sufficient to cover the needs of the ration system in 1995 (Table 1.7). Admittedly, for this to have happened, the 'wheat flour' supplied in the ration would have consisted of a good measure of grains apart from wheat, such as barley. Reports on the quality of the flour that was supplied through the ration system suggest that such dilution did take place.

TABLE 1.7
Wheat-flour requirement and availability: various estimates

Requirement for 1996[a] $7 * (18,000,000) * 0.95 * 12 =$	1.4 million metric tonnes
Estimated availability as per cent of 1996 'need'	
1. 1995 wheat harvest	71
2. 1995 wheat + 20% barley supplement	88
3. 1995 wheat + barley harvest	122
4. 1995 wheat + 29% barley supplement	100

Note: 'Total requirement is calculated as follows: per capita requirement per month = 7 kg; estimated population in government-controlled areas = 18 million; estimated proportion of the population above the age of one year = 95%; number of months = 12.
Source: FAO (1995a) for crop production and wastage ratios, authors' survey for grain requirement of the ration system.

The fact that a country is able to produce enough grain to cover its basic needs does not imply that it does in fact do so. The recent history of famines around the world amply demonstrates that potential or actual output does not translate freely into food security for the population. In the absence of a well-designed and effectively-managed procurement system, production targets may not be realised and whatever is produced may not actually reach the ration system. If, for example, the procurement price is much lower than the market price, farmers would have an incentive to conceal their output from the procurement agency. If the gap between the procurement price and the market price were large enough and if concealment were difficult, people might actually withdraw from grain production altogether.

In the Iraqi procurement system, the procurement agency was a local committee of the Ministry of Agriculture, and farmers were required to supply the agency with the entire harvest. A farmer was also obliged to supply grain up to a minimum threshold. This was calculated by multiplying the cropped area with the designated yield of the land in question; this was, in turn, calculated by taking into account various production conditions and was based on some notion of the average yield of that land. The penalty for any shortfall below this minimum threshold was the market value of that shortfall – that is, a farmer who failed to meet the minimum target set for him had to purchase the shortfall from the market and supply it to the procurement agency. Farmers could apply for consideration if the harvest turned out, for some specific reason, to be much lower than expected. In such cases, a local committee of the Ministry of Agriculture with knowledge of local production conditions adjudicated.

Procurement systems of this type are common in economies where there is a government food subsidy to consumers. Part of the subsidy is raised by imposing a tax on farmers, who are required to sell some part of their output (in the case of Iraq, the entire harvest) to the government at lower than market prices. The tax is the gap between the market price and the procurement price. Like any system of taxation, procurement systems also attract evasive behaviour on the part of the taxed population. As with the case of the food ration, where good system design is required for effective operation, the issues of design and implementation are of paramount importance if a procurement system is to be sustainable. A number of features of the working of the procurement system in Iraq indicate that the authorities were attentive to such considerations.

First, the implicit tax on farmers should not be onerous enough for them to engage in large-scale evasion, or worse, withdrawal from grain production altogether. International experience has shown that attempts by the procurement (or tax) authorities to shift much of the burden of the consumer subsidy to producers of the subsidised commodity can backfire badly, thereby resulting in a depletion of output. Given the paucity of data on a number of relevant variables, there are no reliable estimates of the implicit tax on agriculture in Iraq. On the basis of known differences between procurement and market prices of wheat, we have attempted to arrive at some very rough benchmark numbers, and our rough estimates of this implied rate of taxation for 1996 range from 20 to 35 per cent of the value of produce.[44] Considering the fact that the ration system involved a subsidy of over 95 per cent to consumers, the part of the burden borne directly by farmers was relatively modest.

Secondly, because the minimum target for farmers was set according to the area of land they had under crop, they might have found a potential form of evasion in under-reporting the cropped area. The incentive for farmers in Iraq to do this was reduced by linking the supply of subsidised inputs, such as seeds and fertilisers, with the reported cropped area. The grain-procurement agency also acted as the supplier of these subsidised inputs, which were rationed according to reported cropped area. Finally, evasion at the time of harvest was kept in check by the procurement agency exercising some direct physical control over the harvesting operations. In the major grain-growing regions of the country, farmers could hire mechanical harvesters either directly from a local branch of the Ministry of Agriculture (which also happened to be in charge of procurement) or from private contractors, who in turn also reported to the procurement agency. At the time when the harvesters went into operation, an official of the agency was usually present and made a record of the condition of the harvest.

In sum, the Iraqi procurement system appeared to be well protected against various forms of evasion and avoidance that could have crippled its functioning. Most importantly, while the procurement price did indeed imply a net tax on farmers, this tax did not appear high enough substantially to dampen incentives. In fact, as we have argued above, the proportion of the workforce in agriculture increased substantially and cropped area also rose significantly during the sanctions period. The procurement and the ration systems, which worked in tandem, had therefore been successful in turning Iraq's agricultural potential into higher domestic food availability, and transforming this aggregate availability into basic food entitlements for the population as a whole.[45]

Repression and welfare

It is clear from the foregoing discussion that the Iraqi government invested considerable financial, administrative and political resources to maintain the ration system in good order. This may appear paradoxical to some, given the regime's notorious record in terms of brutality and oppression within the country.[46] It might be argued that the government's active role in preventing the onset of mass starvation served the existing leadership's narrow and cynical aim of prolonging its rule by averting political strife. Such calculations clearly do have a place in explaining the behaviour of governments. The actual calculation is likely to be somewhat more sophisticated than a simple 'bread-for-loyalty' trade-off. By responding in an effective manner to a threat of food insecurity instituted by 'outsiders' (that is, economic sanctions imposed by the UN Security Council), the leadership would have sought to enhance its claim to political legitimacy within Iraq.

Even if the Iraqi regime's commitment to the ration system was motivated primarily by narrow political calculations, it needs to be stressed that this was not a new-found commitment to welfare and that it has to be viewed in the historical context of welfarist interventions by successive governments in Iraq. These interventions, which included governmental action on a variety of social and welfare issues such as education (particularly of girls), public healthcare, development of infrastructure and, indeed, the development of radical land reforms, have been consistent and substantial features of public policy, at least since the late 1950s. These welfarist concerns have coexisted for most of this period with political repression and the widespread suppression of civil liberties.[47]

Both these aspects of the Iraqi state – that is, political repression and social welfare – have been documented over time. The fact that the Iraqi state and the present government have little respect for civil and political liberties is well established, nor is there much doubt that the sanctions regime has done little to alter this situation. Our investigation into the government's response to economic sanctions has also highlighted the continuity in that other, less widely recognised, aspect of the Iraqi state – namely, its commitment to social welfare. This point is important for developing an understanding of what might happen in Iraq if and when the sanctions regime is relaxed or modified. There has been a tendency to equate political absolutism with the absolute appropriation of public resources for private ends. It has, for instance, been argued that the overriding and exclusive financial priorities of the Iraqi leadership are to amass private wealth and to bolster the coercive and military apparatus of the state.[48] While there is no doubt that private appropriation and military expenditure are important priorities for the Iraqi leadership, commitment to social welfare expenditure is also an important government priority in its own right. Any considered discussion about the future of the sanctions regime needs to keep this point in mind.

7. UN sanctions: impact, objectives and perpetuation

The main aim of this chapter has been to establish the impact of sanctions on Iraq's economy and, in particular, on the economic conditions of the civilian population. Most of the discussion thus far has been focused on changes in economic variables, on the responses of the private sector and the state to these changes, and on the livelihoods and living conditions of the people. This section provides a brief summary of our main findings, and by shifting the focus, draws attention to issues that concern the objectives of the sanctions regime.

As it was instituted, the sanctions regime did not include any time limit: the sanctions could continue indefinitely into the future unless specifically terminated by the consent of a majority of Security Council members and of all the permanent members. It therefore offered tremendous leverage to individual permanent Security Council members, since any one of them could veto a proposal for terminating the sanctions or modifying their terms. Given the extraordinary leverage enjoyed by the permanent members, it is clear that the political and strategic aims of these powers played a crucial role in the continuation of sanctions.[49] Here, we restrict our attention to the formally stated goals and objectives of the United Nations, and examine whether, or to what extent, these goals and objectives were in fact realised.

Finally, this section examines the modifications in the sanctions regime that were initiated in late 1996 in the shape of the so-called 'food-for-oil' (FFO) agreement between the UN Security Council and the government of Iraq. Since our first-hand exposure to humanitarian conditions inside Iraq did not extend into the period when the FFO agreement was in operation, we will not comment directly on its actual impact. There is now substantial independent evidence to support the contention that FFO should be seen as a relatively modest change in the sanctions regime.[50] FFO was never supposed to be a programme of economic revival, and while it did provide welcome relief it did not obviate the need to seek a quick and just solution to the crisis that has blighted the lives of millions in Iraq. Moreover, certain important contradictions in positions taken by the main protagonists were exposed by the food-for-oil deal, and here we do offer some comments.

Summary of main findings

The scale of economic sanctions against Iraq has been unprecedented in recent history. The imposition of sanctions can be regarded as a macro-economic shock of massive proportions. Although the sanctions regime allowed for an easing of restrictions on the importation of foodstuffs and medical supplies, the availability of these essential supplies continued, among other things, to fall far short of the requirements of the civilian population.

The main source of this shortfall lay in Iraq's inability to pay for imports. The Iraqi economy was denied the revenue from petroleum – its most valuable asset. The *effect* of the ban on Iraqi exports was the same as the effect of a quantitative restriction on the importation of food or medicines, and in this sense, the allowances in the sanctions regime for 'humanitarian' imports were somewhat illusory. In the past, Iraq had relied heavily on oil revenues and its economy had developed around the surpluses generated by oil exports. This being the case, Iraq was particularly vulnerable to an oil embargo. The mechanisms through which the oil embargo affected the economy were two-fold: the revenues available to the government and the value of the Iraqi currency both suffered a dramatic reduction.

The impact on earnings and livelihoods was disastrous. Real earnings fell by around 90 per cent in the first year of the sanctions, and then fell by around 40 per cent more between 1991 and 1996. There was a steady shift of people into casual labour. In this form of employment, wages in Iraq were much lower than were wages for casual workers in some of the poorest parts of the world. Qualitative observations on the type of survival strategies that Iraqis were resorting to confirmed the impression that in many ways Iraq was very much like some of the poorest countries in the world, whereas before the sanctions it could have been placed on a par with the upper-middle income countries.

The government responded in a number of ways to the economic crisis brought on by the sanctions. There was a policy of generating some revenues through seignorage, or by way of printing money,[51] but, as economic theory has warned, this process has its limitations, and in Iraq these appear to have been reached in late 1995. However, under the circumstances the management of the macroeconomy overall was effective. Arguably, the most significant achievement of the government in response to the sanctions was the establishment of a ration system, which appeared to be efficient and equitable, and it would be fair to say that the ration system, in tandem with the grain-procurement system, was instrumental in preventing the onset of mass starvation and famine in Iraq.

The enormous human and economic cost of the sanctions has largely been borne by Iraq's civilian population, and there is no question of seeking justification for *policy-induced* human suffering of this magnitude. Nevertheless, it is imperative to determine what the ultimate objectives of this policy were, and whether, and to what extent, those objectives have been attained.

Objectives of sanctions

The United Nations Organisation, and in particular its Security Council, can properly be regarded as the body charged with representing the will of the international community in the resolution of the Gulf conflict. Punitive sanctions against Iraq ought, therefore, to be seen within the wider context of actions taken by the international community in the pursuance of its objectives. It is worth recalling at length the words of Dr Boutros Boutros-Ghali, the former Secretary-General of the United Nations, who oversaw many issues arising from the Gulf crisis during his tenure:

> The breadth of actions taken by the United Nations in more than five years of work – from the immediate, intense worldwide diplomatic activity aimed at ensuring universal support for the restoration of Kuwaiti sovereignty, to the ongoing challenge of building long-term peace and stability in the region – have confirmed the extraordinary relevance of the United Nations in addressing the most complex political issues facing the international community . . . A principal concern for the Organization throughout its involvement has been to alleviate the hardships the conflict has visited upon the Iraqi civilian

population . . . By the use of sanctions and other enforcement measures, through efforts to institute a programme of disarmament and weapons control, in the provision of humanitarian assistance, and with programmes in areas such as boundary demarcation, compensation for damages caused by the invasion and the promotion and protection of human rights, the United Nations has broken new ground as peacemaker, peace-keeper and peace-builder.[52]

These reflections on the United Nations and the Gulf conflict highlight two important aspects of how the organisation itself perceived its own role. First, while the restoration of Kuwaiti sovereignty and the cessation of armed conflict were regarded as immediate tasks, the ultimate objective was to build 'long-term peace and stability in the region'. Secondly, the organisation regarded the alleviation of hardships faced by Iraqi civilians as a 'principal concern'.

It is pertinent to ask how far, and to what extent, the objective of building long-term regional peace and the concern with the conditions of Iraqi civilians has been reflected in the resolution of the Gulf crisis. In addressing these questions it is useful to examine the mechanism adopted by the Security Council in pursuing its objectives, and in this regard, the ceasefire resolutions, and particularly resolution 687 are of fundamental importance. According to Dr Boutros-Ghali,

> The primary basis for the wide range of actions that the United Nations has taken to accomplish this goal [that of establishing regional peace], is Security Council resolution 687 (1991), adopted on 3 April 1991, which imposes numerous obligations on Iraq and, in doing so, constitutes one of the most comprehensive sets of decisions ever taken by the Council.[53]

Resolution 687 does, indeed, impose a comprehensive set of conditions on the government of Iraq. It also calls for unprecedented coordination between various UN agencies and international organisations for its implementation – particularly in the area of weapons control since, apart from demands for compensation, the acceptance of international boundaries and the repatriation of prisoners, resolution 687 is overwhelmingly concerned with military matters. Besides its scope and detail, there are three aspects of resolution 687 that deserve mention, especially in the light of the broad objectives mentioned by Dr Boutros-Ghali.

First, the only reference to humanitarian considerations contained in resolution 687 is that it allows for the relaxation of the sanctions regime for the import of food and medicines.[54] This relaxation, however, is conditional 'on the policies and practices of the Government of Iraq, including the implementation of all relevant resolutions of the Council'.[55] Since these 'relevant resolutions' and conditions are largely concerned with weapons control, there appears to be a clear hierarchy in the goals of the Security Council in favour of military objectives over and above humanitarian concerns. Although, to our knowledge, no specific threat has been made since resolution 687 was adopted to revoke the relaxation of imports of food and medicines, it is remarkable that the conditionality exists at all. The declaration that the alleviation of the suffering of Iraqi civilians was a primary concern also rings hollow, in the light of admission by Dr Boutros-Ghali that 'donor response to the five United Nations inter-agency humanitarian appeals issued since January 1992 has . . . fallen well short of the funding goals'.[56]

Secondly, resolution 687 shows clearly that the Security Council failed to make any meaningful distinction between mechanisms designed for the achievement of immediate and urgent objectives (ending Iraq's occupation of Kuwait), and 'peace-building' measures (such as the demarcation of boundaries and the limitation of Iraq's offensive potential). The regime of comprehensive economic sanctions was imposed under resolution 661 within a few days of Iraq's invasion of Kuwait in August 1990. Quite justifiably, the objectives of this resolution were extremely short-term ones, since at that time it was imperative to secure a speedy withdrawal by Iraq from Kuwait and the restoration of Kuwait's sovereignty. However, the aims of the ceasefire conditions as set out in resolution 687 are, by definition, over a longer time horizon. They include, among other things, the identification and destruction of Iraq's capacity for the production of chemical, biological and nuclear weapons and their delivery systems, as well as the institution of an ongoing programme to monitor possible breaches of weapons control in the future. No time period is specified for the completion of these tasks.

When resolution 687 was adopted in April 1991, the political and military situation and the demands of regional peace and security were vastly different from what they had been in August 1990. It is striking, then, that instead of considering any alternative instrument for ensuring Iraq's compliance with the ceasefire conditions, the Security Council simply reiterated the maximalist position of comprehensive sanctions as put forward in resolution 661. This was in spite of the fact that the UN did recognise the distinction between its peacemaking, peace-keeping and peace-building tasks.

Thirdly, resolution 687 advances a purely legalistic and military settlement to the Gulf conflict. The resolution contained detailed prescriptions for limiting Iraq's military potential and concerning its legal obligations under international law. However, neither resolution 687 nor any of the other resolutions or negotiations included proposals or considerations for long-term political or diplomatic initiatives for confidence building and conflict resolution in the region. This has been a serious omission, because although adherence on Iraq's part to its international legal obligations and to the limiting of its military potential might be necessary conditions for regional peace and stability, they are not sufficient conditions. Given the country's importance in terms of its population, resources and political history, it is difficult to envisage a long-term settlement in the region that excludes Iraq.

In sum, resolution 687, which provides the basis for the continuation of the sanctions regime, is very inadequate in reflecting the international community's twin objectives of building long-term peace and alleviating the suffering of Iraqi civilians. Even its wording fails to live up to the claim that the suffering of civilians is a primary concern. This lack of success in making a distinction between the very different contexts of peacemaking and peace-building – in other words, between the imperatives of immediate aims such as the restoration of Kuwait and longer aims such as the monitoring of Iraq's armaments programme – has set the scene for prolonged and continuing economic hardship in Iraq.

It can be argued that the sanctions regime under resolution 687 has been successful in securing Iraq's compliance with its international legal obligations, and in implementing one of the most comprehensive internationally monitored disarmament systems in the world. Indeed, this has been the main argument put forward by those who advocate the continuation of sanctions. Progressive compliance by Iraq with the demands of resolution 687 is undeniable, though the claim that comprehensive economic sanctions are the *only* – or even the most effective or least costly – instrument for ensuring compliance is open to question.

This is particularly the case if we take a longer view of the conditions required for peace-building in the region. Implicit in the prescriptions contained in resolution 687 is the perception that Iraq's disregard for international law and its advantage in military capability over other states were the main causes of regional instability and aggression. Such an analysis ignores other political factors, such as the potency of Arab nationalism and the sense of historical injustice in Iraq among both the rulers and the people.

The present regime, as well as almost the entire spectrum of political opinion in Iraq, is strongly influenced by these ideas, and any political leadership would be required to affirm Iraq's role as a regional power in order to establish its domestic credentials. The issue of long-term peace and stability in the region cannot therefore be properly addressed without due recognition of these realities. While the sanctions regime might have been effective for the time being in reducing Iraq's military potential and in securing its compliance with international law by inflicting a heavy cost on Iraqi civilians, it adds further fuel to the sense of historical injustice felt by ordinary Iraqis. In addition, given that it is widely perceived inside Iraq as a vengeful measure, it also gives rise to further cynicism amongst ordinary Iraqis with regard to international institutions.[57] In the medium term, the sanctions and other punitive measures against Iraq might enforce an uneasy peace, but their long-term effects are likely to be counterproductive and unpredictable.

The 'food-for-oil' agreement

As noted earlier, the 'food-for-oil' deal was first proposed in August 1991, and was approved in May 1996 when a Memorandum of Understanding was agreed between Iraq and the Security Council for implementing UNSC resolution 986 of April 1995. It was eventually implemented at the end of 1996, having remained in abeyance for over six months due to the intervention of the Iraqi army in factional fighting in Iraqi Kurdistan. The terms of resolution 986 were initially valid for six months and were renewable for further six-month periods. The agreement was revocable and did not supersede any previous UN resolution.

Background to FFO

As noted, resolution 687, the key Security Council document governing the sanctions regime, contained a proviso that empowered the sanctions committee to allow limited and ad hoc exports from Iraq in order to finance the importation of humanitarian supplies. Between April and

July 1991, the government of Iraq made several representations to the sanctions committee to exercise this power. This period also saw several fact-finding missions sent to Iraq by the Secretary-General in order to ascertain humanitarian conditions. The reports of these missions, particularly that of the inter-agency mission also known as the Aga Khan mission, painted a bleak picture of conditions inside Iraq.

The Security Council's response was to adopt resolution 706 in August 1991, which allowed the export of US$1.6 billion worth of Iraqi oil for a period of six months.[58] A part of the proceeds was earmarked for a compensation fund and to pay for various UN activities in Iraq. The remainder was to be placed in an escrow account and used for the import of essential supplies approved by the sanctions committee. The resolution also proposed strict monitoring of Iraqi finances, including imports and the distribution of food and other supplies. The Iraqi government rejected these terms since they were overly intrusive and compromised its sovereignty. Although negotiations on the implementation of resolution 706 continued sporadically, it was clear in early 1992 that the Iraqi government would not accept its terms. This diplomatic impasse allowed the Security Council to wash its hands of any responsibility for the condition of the civilian population of Iraq. It stated that the government of Iraq

> by acting in this way [that is, discontinuing negotiations on FFO], is forgoing the possibility of meeting the essential needs of its civilian population and therefore bears the full responsibility for their humanitarian problems.[59]

After this, there were no significant initiatives for over three years. In April 1995, a new resolution (No. 986) was adopted and formed the basis of the food-for-oil agreement. Nor was there any significant attempt by the UN to monitor humanitarian conditions in Iraq during this period. According to the former Secretary-General, resolution 986 'took into account some of Iraq's concerns' over resolutions 706 (1991) and 712 (1991) by reaffirming 'the commitment of all Member States to the sovereignty and territorial integrity of Iraq'; he also described the new exercise as 'temporary'.[60] The implicit admission that Iraq's objections to the resolutions could be (and were) accommodated, somewhat contradicts the Security Council's position during the three-year period of stalemated negotiations that Iraq bore "full responsibility" for the suffering of its civilian population.

Impact of the agreement
The food-for-oil agreement that was agreed at the end of 1996 allowed Iraq to export US$2 billion-worth of oil every six months, excluding that delivered to Jordan for its own consumption. The proceeds were to flow into an escrow account in an international bank chosen by the UN Secretary General, and were to be used for three purposes in the following proportions:

(a) 53 per cent to be used by the Iraqi government for procuring humanitarian supplies for the whole of Iraq except for the three Kurdish governorates outside Iraqi control;
(b) 13 per cent to be used by the UN for humanitarian supplies for the three Kurdish governorates of Arbil, Duhak and Sulaimaniyya;
(c) 30 per cent to be paid into the UN-controlled Kuwait Compensation Fund;
(d) 4 per cent for UN administration expenses.

The amount of over one billion dollars (every six months) that was at the disposal of the Iraqi government for humanitarian supplies had to be spent as specified in the categorised procurement plan approved by the UN Sanctions Committee.

The agreement was expected to bring about a substantial reduction in the suffering of the Iraqi population by two means: first, through the appreciation of the Iraqi dinar relative to the US dollar; and secondly, through increased resources for importing food, medical supplies, and inputs for restoring the infrastructure, including sewage works and water purification plants. The effect of the first was immediate and, as noted above, the market prices of food and medicines closely track movements in the exchange rate. The second, especially as far as the restoration of infrastructure was concerned, was expected to take effect after a certain time lag. As remarked in Section 2 above, the food-for-oil agreement was a contributory factor in the relative success of the macroeconomic stabilisation programme initiated by the Iraqi government in late 1995.

However, it must be noted that any positive impact of FFO is actually an indication of the depth of the economic recession in Iraq.

The agreement fell far short of any measure of the country's basic needs. In 1989, Iraq's import bill for food and medicines was US$2.36 billion, and expenditure on water sanitation and purification was around US$100 million. Assuming 4 per cent inflation, these two categories of import would have cost around US$3.2 billion at 1997 prices. This excludes the repair of facilities and the replacement of medical equipment. Even allowing for cuts in non-essential food and medicines, US$2 billion per year would fail by a significant margin to cover the bill for essential food and for preventive and curative healthcare.

The amounts stipulated under the FFO also contrasted markedly with the estimates of needs prepared by various UN agencies. They appeared to be based upon the six-monthly limit on oil exports of US$1.6 billion as first proposed in resolution 706 in August 1991, adjusted for price increases. The origin of the US$1.6 billion figure is itself somewhat obscure. According to the report of the Aga Khan mission, the last authoritative inter-agency mission, Iraq required in 1991 US$6.8 billion over a one-year period, even for a 'greatly reduced level of services'. The breakdown of this amount was US$180 million for water and sanitation, US$500 million for health services, US$53 million for supplemental feeding, and US$1.62 billion for general food. Import requirements of agricultural inputs were estimated at US$300 million, and those for the restoration of the oil and the power sectors were put at US$2.0 billion and US$2.2 billion respectively.

The discrepancy between the estimates of the Aga Khan mission and the limit set under resolution 706 was recorded by the UN Secretary-General in a report in September 1991, which stated that 'the actual sum from the sale of Iraqi petroleum and petroleum products will have to be determined by the Security Council after its consideration of the present report'. The report estimated the six-monthly essential requirements at US$1.8 billion.[61] At 1996 prices, this would have been equivalent to over US$2.2 billion; that is, twice as high as the allowance set by the food-for-oil deal.

The modalities of this agreement, which was in essence the only concession by the UN to humanitarian conditions inside Iraq, raised two important (and uncomfortable) questions. First, what was the rationale of restricting oil exports to an amount that falls well short of meeting humanitarian needs? Second, what were the prospects for the resolution of the sanctions crisis?

The limitation on the export of oil to US$2 billion per six months did not lend itself to justification in terms of the misuse of additional revenue. The agreement provided for stringent monitoring by the UN Sanctions Committee of the proceeds set aside for humanitarian needs. It was difficult to comprehend why the monitoring mechanism would not have been be able to cope with an import bill exceeding US$1 billion per six months.

In subsequent modifications of the FFO regime, Iraq was allowed, more or less, to export as much oil as it could. Just as the initiation of FFO had amounted to an implicit backtracking by the Security Council on its early claims about the minimal humanitarian impact of sanctions, the eventual relaxation of FFO export limitations vividly demonstrated the illogicality of the Council's earlier insistence upon a strict ceiling. In fact, an abiding feature of negotiations between Iraq and the Security Council since 1991 has been the absence of any sense of urgency with regard to humanitarian concerns. Each small relaxation of the economic blockade that the UN has hailed as a major humanitarian triumph has, on closer examination, turned out to be no more than a grudging concession. Many (along with the Iraq government) claim that the FFO deal and its modifications (such as UNSC Resolution 1284 of December 1999) are simply the means by which members of the Security Council can prolong economic sanctions while maintaining the appearance of humanitarian concern. With each passing day, this charge becomes more difficult to answer.

The sanctions may have succeeded in forcing the Iraq government to relinquish a substantial proportion of its stock of long-range rockets and nuclear and chemical materials as well as the facilities for their manufacture. The continuation of sanctions may help to unearth and destroy what has until now been concealed; but sanctions have not changed the Iraqi leadership or the character of the regime, and it is doubtful that they will succeed in doing so in the future. Sanctions prolong the suffering of the Iraqi population and contain Iraq through a stranglehold on its economy. These conditions are likely to hinder rather than help the cause of meeting 'the ongoing challenge of building long-term peace and stability in the region'.

APPENDIX 1

Leakages under the sanctions regime

Under the sanctions resolutions, the only exports that were exempt were oil exports to Jordan for its domestic consumption. Before the Gulf War, Jordan's total annual imports of crude oil averaged around 17 million barrels, with an annual import bill of around US$320 million.[62] It can be safely assumed that nearly all of Jordan's oil imports originated from Iraq, and IMF sources in fact reported the total value of Jordan's imports from Iraq at around US$300 million. After the Gulf War and the sanctions, this pattern remained unchanged, except for one year (1993) when Jordan's imports from Iraq were recorded as nil. In subsequent years, the pattern of trade appears to have been revived. It is likely that Jordan's domestic oil consumption increased over the years because of economic growth.

Even allowing for growth, Jordan's domestic consumption would still be a tiny fraction of Iraq's pre-sanctions estimated oil output of over one billion barrels per annum. Current industry estimates put Iraq's output at under a fifth of that level. For example, the Organisation of Petroleum Exporting Countries (OPEC) reported in 1994 that Iraq's output stood at 570,000 barrels a day (or over 200 million barrels a year). Even at these low levels, Iraq's output far exceeded its exports to Jordan.

The difference is, however, consistent with Iraq's pre-sanctions patterns of domestic consumption. Between 1987 and 1990, Iraq consumed, on average, 173 million barrels of oil per annum. Whether domestic consumption had declined substantially is far from clear. Industrial recession would certainly have led to lower demand for power, and thus for fuel oil. Similarly, with massive declines in income, the demand for oil by motorists and other consumers ought also to have reduced. However, the prices of petroleum products had declined dramatically, so that petrol was virtually free in Iraq. In 1996, petrol cost ID1.5 per litre, or the equivalent of 0.15 cents at the market exchange rate. This price reduction probably counteracted somewhat the effects of low incomes and low demand from industry. It is, of course, possible that actual output was substantially higher than OPEC estimates, and it is interesting to note that Iraqi government figures for oil production in 1993 and 1994 were over 30 per cent higher than OPEC estimates.

This would not be consistent with a government that was trying to conceal the level of output.

It is useful to consider the implications of the most extreme case of leakage from the sanctions regime. If we take the higher figure supplied by the government of Iraq, the country's total oil output in 1994 would have been around 270 million barrels compared to the OPEC estimates of 208 million barrels (using secondary industry sources). Even under the extreme assumption that this *entire* amount was exported, at the improbably high return to Iraq of US$10 per barrel, total export revenue in 1994 would have been US$2.7 billion. The term 'improbably high' is used advisedly, since the average price of crude oil in 1994 was US$15.5 per barrel.[63] It is highly unlikely that smuggled oil would have fetched anything close to its market value. Iraq's return per barrel, after subtracting the cost of extraction, is likely to have been lower than US$10. In per capita terms, this would have been the equivalent of US$150 per person. This is still lower than the value of imports per person of the three poorest Arab countries (see Table 1.2).

APPENDIX 2

Rough estimate of Iraq's 1996 GNP

Our household surveys suggest that workers earned US$25–30 per month in casual labour. This was generally more than the wages paid in the state sector. At US$30 per month, total annual wages per worker were US$360. Cross-country comparisons show that profits fall within the range of 0.25 to 0.75 times the wage bill. Assuming that they were 0.5 times the wage bill, and assuming a 50 per cent participation rate in the labour force, then GNP per capita would have been US$270 (before subsidies, rents or indirect taxes). Since these calculations are very rough, the bias could be substantial and in either direction, but they show that dollar GNP had fallen sharply from the levels above US$2,000 recorded in the late 1980s.

An alternative measure comes from the National Accounts data published by the Iraqi authorities. According to these, Iraqi GDP was ID128.1 billion in 1993. The average exchange rate during the year was 260 ID/dollar; hence GDP per capita would equal US$26. Since inflation and controlled prices for rations would skew the value of GDP, this will also be a very rough estimate.

APPENDIX 3

Rough estimate of implied tax rate on farmers in 1996

The following estimates are worked out for wheat. The benchmark procurement price of wheat grain is taken as ID100 per kilogram. The benchmark market price of wheat flour in May 1996 was ID250 per kilogram. The market price is determined on the margin by the import price. The implied tax on the margin is calculated by working out the difference between the price a farmer could expect to obtain on the market, and the procurement price.

Starting from the market price of wheat flour (ID250), it is necessary to work out the market price of wheat. Part of the difference between the wheat-flour price and the procurement price is accounted for by the costs of converting wheat into wheat flour, and other transportation and transaction costs. Allowing a margin of 33 per cent for the costs of milling, transport and other expenses will give the implicit market price of wheat flour as ID167.

However, the procurement price of ID100 does not take into account the implicit subsidy to the farmers for seeds and fertiliser. If we assume that the implied rate of subsidy is as high as one third (that is, ID33) of the procurement price, the net tax paid by a farmer per kilogram of wheat sold to the procurement agency would have been around ID34. If, on the other hand, the implied subsidy was one tenth (ID10), the net tax paid would have been ID57. The tax rate, on the margin, would have been between 20 to 35 per cent, depending on the value of the subsidy.

NOTES

1 This contribution is based on an earlier report for the Center for Economic and Social Rights in New York that formed part of a larger study on the impact of sanctions on the well-being of the civilian population of Iraq. The earlier report was prepared by H. Gazdar (now in Karachi), and A. Hussain at the Asia Research Centre, London School of Economics, and Peter Boone at LSE's Centre for Economic Performance. The authors are grateful for their comments and suggestions to: Ghassan Abu-Sitta, Jean Drèze, Kamil Mahdi, Abdullah Mutawi, Roger Normand, Sarah Zaidi, and participants at the conference on 'Frustrated Development: the Iraqi Economy in War and Peace' held in July 1997 at the University of Exeter.

2 The UNSC passed its resolution 661 on 6 August 1990, within a week of Iraq's invasion of Kuwait. This resolution required all member states to prevent any trade or financial dealing with Iraq and occupied Kuwait.

3 For detailed reports of the earlier study and related surveys see Jean Drèze and Haris Gazdar, 'Hunger and Poverty in Iraq, 1991', DERP Discussion Paper 32, STICERD, London School of Economics, 1991; and Jean Drèze, and Haris Gazdar, 'Hunger and Poverty in Iraq, 1991', *World Development*, Vol. 20, no. 7, 1992, pp. 921–45. See also, International Study Team, 'Health and Welfare in Iraq After the Gulf War', Mimeo, Medical Education Trust, London. 1991.

4 See, for example, United Nations, *The United Nations and the Iraq–Kuwait Conflict, 1990–1996*, New York, UN Department of Public Information, 1996.

5 United Nations, 1996, p. 111.

6 The situation began to change following the rise in oil prices in 1999 and the expansion of Iraq's trade under the FFO, and to a lesser extent, outside it. Free Trade agreements have been signed between Iraq and each of Egypt, Syria and Tunisia, and similar agreements have been negotiated with Jordan and with Yemen. There has also been a limited, almost token, resumption of international civilian flights to Baghdad during 2000 and 2001, without the explicit prior approval of the UN Sanctions Committee. In March 2000, the Committee adopted new procedures permitting the import of some basic supplies with the approval of the UN's Office of the Iraq Programme, and without prior circulation to the Sanctions Committee. The list of items eligible for this "fast track procedure" was subsequently widened, but it remained highly specific, leaving wide scope for the Sanctions Committee to exercise its control over Iraq's finances to a great disruptive effect. Thus, sanctions remained strict and fully effective in preventing the regeneration of the Iraqi economy. Iraq's ability to circumvent UN control over its oil exports and revenues has been limited. At the time of writing this note, it is not clear whether the US and Britain are planning to resume a failed attempt to obtain UN support for new mandatory border controls under the so-called "smart sanctions" proposals in order to further tighten the effect of sanctions. [Editor's note]

7 However, there was agreement between Iraq and Syria in 1999 for the restoration of this pipeline. Oil industry sources believe that the pipeline has been pumping an average of 150,000 barrels of oil per day since October 2000 which Syria has been receiving under a heavily discounted price without Sanctions Committee approval.

8 One reason for such openness with income data in Iraq compared to many other developing (and indeed developed) countries is that direct taxation is practically unheard of. In many other countries people are worried about disclosing their earnings lest the information gets passed on to tax authorities.

9 Economist Intelligence Unit, *Iraq: Country Profile 1995*, London: EIU, 1995.

10 Government of Iraq, *Statistical Yearbook 1994*, Baghdad, 1994.

11 It is worth noting that issues relating to external debt formed the backdrop in 1990 to events leading up to Iraq's invasion of Kuwait. For detailed analysis of the evolution of Iraq's external debt and its sustainability, see Ahmed Jiyad's contribution on the development of Iraq's foreign debt in this volume. On the historic welfarist aspects of the Iraqi state, see Kiren Aziz Chaudhry's contribution, also in this volume.

12 World Bank, *World Development Report 1993*, New York: Oxford University Press for the World Bank, 1993.

13 Relations between Iraq and its neighbours have witnessed some improvement, particularly since around 1998. It stands to reason that Iraq's ability to circumvent the export embargo will have been enhanced as a result of this improvement. Paradoxically, the very stringency of the sanctions regime might have led to its erosion as trading partners refuse to participate in policing a regime which is increasingly regarded as unjust and morally repugnant by popular opinion in the region.

14 The official exchange rate remained at US$3.2 per Iraqi dinar throughout this period.

15 About the only form of effective taxation that existed in Iraq during this period was an implicit tax on agricultural production. See Sections 5 and 6 for discussion.

16 Even if we assume that a large proportion of the government's resources was taken up by expenditure on the military or on conspicuous consumption by the ruling elite, in the absence of significant opportunities to import merchandise, much of this expenditures is also likely to have been in the form of wage and salary payments.

17 See Appendix 2 for details.

18 The food price index is based on a basket of food items with consumption weights corresponding to *pre-sanctions* consumption patterns. For details of the consumption weights used, see Jean Drèze and Haris Gazdar, 1991.

19 Section 5 gives further details.

20 This is discussed in greater detail in Section 5.

21 Drèze and Gazdar, 1991, based on Government of Iraq, *Annual Abstract of Statistics*, Baghdad, 1990.

22 United Nations Population Division, *World Demographic Survey 1988*, UN, 1988.

23 It must be emphasised that the comparisons in Table 1.3 are for the period between 1991 and 1996 – that is, in the period *following* the initial decline in incomes from their 1990 levels.

24 On the question of the quality of the harvest see Section 5.

25 Sources for procurement prices in the various years are: Food and Agricultural Organisation (FAO), *Evaluation of Food and Nutrition Situation in Iraq; Terminal Statement Prepared for the Government of Iraq*, FAO, Rome, 1995a, for prices up to 1995; and authors' survey for 1996 prices.

26 World export prices during these years were US$129 in 1991, and US$151 in 1994. See FAO, *Annual Report* 1995, Rome, 1995b.
27 See discussion of exchange rates and prices in Section 3 above.
28 For further discussion of grain procurement and changes in Iraqi agriculture, see Sections 5 and 6.
29 See, for example, Sunil Sengupta and Haris Gazdar, 'Agrarian Politics and Rural Development in West Bengal', in Jean Drèze and Amartya Sen (eds), *Indian Development: Selected Regional Perspectives*, Delhi: Oxford University Press, 1997.
30 Drèze and Gazdar, 1991.
31 Details on the functioning of the ration system are discussed in Section 6.
32 Estimated by dividing the total wheat consumption in the country in 1989 (3.3 million metric tons) by the estimated population over the age of one year in 1989 (14 million).
33 FAO, *Review of Food Consumption Surveys*, Rome, 1977. This review was based on a nationwide survey of food consumption in 1971 carried out by the Ministry of Planning, Government of Iraq.
34 This view is corroborated by the pattern of disease in Iraq before the Gulf War. According to the World Health Organisation (WHO), for example, the main child nutritional disorders were related to obesity. See also, World Health Organisation, *The Health Conditions of the Population in Iraq Since the Gulf Crisis*, WHO Report, 1996.
35 EPW Research Foundation, 'Poverty levels in India: norms, estimates and trends', *Economic and Political Weekly*, 13 April 1993.
36 There were, of course, food items such as vegetables that were not on the ration at all. This qualification implies that the food consumption levels of Iraqi families were higher – though not by a very large margin – than those of families in India who had incomes close to that country's poverty line.
37 Indeed, we came across several cases of army conscripts who worked as casual labourers or in family enterprises on the days that they were on home leave.
38 For observations from eastern India see, for example, Tony Beck, *The Experience of Poverty: Fighting for Respect and Resources in Village India*, London: Intermediate Technology Publications, 1994.
39 See the contribution by Kamil Mahdi in Chapter 9 of this volume.
40 For an account of the disfunctioning of the ration system in Pakistan, see, for example, Harold Alderman, 'The twilight of flour rationing in Pakistan', *Food Policy*, August 1988.
41 For a seminal contribution see Amartya Sen, *Poverty and Famines: An Essay on Entitlement and Deprivation*, Oxford: Clarendon Press, 1981; for evidence from various countries see also Jean Drèze, Athar Hussain and Amartya Sen (eds), *The Political Economy of Hunger*, Oxford: Clarendon Press, 1995.
42 Notable amongst examples of sustained food security programmes under conditions of severe economic and military strain is the case of Britain during the Second World War.
43 See Section 5 for a discussion of why the official data might be downwardly biased.
44 For details, see Appendix 3.
45 Mahmood Ahmad in Chapter 4 of this volume argues that this has been done at a higher financial cost than was probably necessary and suggests a targeted subsidised ration. After the implementation of the FFO, free market prices for

food declined, but the procurement system and price support were maintained with modification. Procurement prices became higher than market prices, and farmers were allowed to sell their produce in the domestic market, but not for directly for trade across international borders, where crops and livestock tended to command higher prices due to the depressed dinar exchange rate.

46 Samir al-Khalil's book, first published in 1989, provides a harrowing account of the systematic use of terror and violence as instruments of control in Iraq. He also comments on the development achievements of the current and past Iraqi governments: 'A regime of terror actually presided over an across-the-board increase in the standard of living in Iraq, and it significantly improved the lot of the most destitute layers, furthering the levelling of income differentials that began after 1958.' See Samir Al-Khalil, *Republic of Fear: Saddam's Iraq*, (paperback reprint 1991), London: Hutchinson Radius, 1991, p. 93.

47 This combination of political repression and social welfare is, of course, not unique to Iraq.

48 For instance, while arguing against the relaxation of sanctions in April 1996, the US Secretary of State asserted that Iraq had spent over US$2 billion on building palaces for President Saddam Hussain (CBS, *60 Minutes* television programme). This estimate was found to have been based on highly dubious methodology, which included satellite-based estimates of the total covered area of these palaces, and the application of the estimated per-square-metre cost of construction in Iraq, converted at the grossly overvalued official exchange rate.

49 For example, Geoff Simons documents statements by leaders of various permanent members linking sanctions against Iraq with political and strategic objectives such as the overthrow of Saddam Hussain or the 'containment' of Iraq. See Geoff Simons, *Iraq: From Sumer to Saddam*, (2nd edn), London: Macmillan, 1996.

50 These observations are restricted to government-controlled parts of the country. In Kurdish-administered northern Iraq, the impact of FFO has, arguably, been more substantial.

51 In its original meaning (1444), a duty levied on the coining of money for the purpose of covering the expenses of minting, and as a source of revenue to the crown, claimed by the sovereign by virtue of his prerogative; now, by extension, an income-raising strategy for governments.

52 United Nations, *The United Nations and the Iraq–Kuwait Conflict, 1990–1996*, New York: UN Department of Public Information, 1996, p. 3.

53 United Nations, 1996, p. 4.

54 Mindful of the fact that allowing the import of food and medicines was meaningless in the absence of Iraq's ability to pay for such imports, the resolution also empowered the sanctions committee 'to approve [Iraqi exports], when required to assure adequate financial resources' for humanitarian imports. This formed the basis for the food-for-oil deal, which is discussed further in section 7. See Resolution 687, section F, paragraph 23 in United Nations, 1996, p. 197.

55 See extract from paragraph F-21 of resolution 687, in United Nations, ibid., 1996, p. 197.

56 United Nations, 1996, p. 113.

57 Even the Iraqi opposition parties that have suffered extreme brutality at the hands of the present rulers remain far from unequivocal in their support for sanctions.

58 The basis for these figures is somewhat obscure. See further discussion in Section 7.

59 Statement by the President of the Security Council, 5 February 1992, in United Nations, op. cit., 1996, p. 391.
60 United Nations, 1996, p. 103.
61 Ibid., pp. 299–300.
62 United Nations, *UNCTAD Commodity Yearbook, 1992*, Geneva, 1992.
63 International Monetary Fund (IMF), *International Financial Statistics Yearbook 1995*, Washington, DC.

2

The Development of Iraq's Foreign Debt: From Liquidity to Unsustainability

Ahmed M. Jiyad

1. Introduction

During the Iran–Iraq war, and especially after the second half of 1982, Iraq relied primarily on international economic diplomacy to gain the support it needed to overcome its economic and financial difficulties. This support was reasonably successful in enabling Iraq to cope with such difficulties, although the eventual effect was to push the country into the trap of indebtedness.

Up to the invasion of Kuwait, Iraq's accumulated debts were, in this author's view, manageable. From a longer-term perspective, they did not represent a severe strain on the economy, although the country was encumbered with somewhat serious short-term difficulties. However, with the devastating consequences of the Gulf War and its long-term and structural impacts on the economy, Iraq's external debts have become an increasingly serious developmental problem that bears strongly on national resources and will therefore have detrimental impacts on the socio-economic well-being of the country and its people.

In addressing the subject of Iraq's debt problem this chapter is organised in three parts. The first part reviews the formulation and development of Iraqi debt policy by identifying components and phases of implementation and their outcomes, in relation to a broad multi-functional debt management system. The second analyses and assesses debt structure, debt ratios, profile of interest, estimation of net flows, net transfers, and present value of debt stock, and also assesses debt sustainability. Two sets of data are used: a detailed time series based on OECD statistics, and official Iraqi data. The third part explores some of the possible available alternatives or policy initiatives that might constitute a feasible debt strategy for post-sanctions Iraq.

There are certain major constraints to bear in mind: the first part of the chapter draws largely on the author's close involvement in Iraq's financial and debt negotiations worldwide during the period from 1981 to 1988, but lapse of time and changed personal circumstances may have influenced his recollection of major events or important details. The second part is affected by non-availability of suitable data. While OECD data is detailed and covers the whole of the 1980s, it does not cover all Iraqi debt. Iraqi data, on the other hand, covers all the debt, but it is highly aggregated. Finally most, if not all, of the policy initiatives discussed in Section 4 are largely hostage to exogenous factors and are thus externally determined.

2. Debt policy and management

Within a broad framework and under normal circumstances, sound debt management is multifunctional. This involves five basic functions:[1] the *policy function* determines the country's sustainable level of external borrowing; the *regulatory function* provides a well coordinated administration of external borrowing; and the *operational function* makes proper choice of markets, instruments, currencies, maturities, sources of funding, and so on. The *accounting function* decides what constitutes an external debt; and the *statistical function* offers a comprehensive reporting system and analysis of the main variables – such as interest rates, exchange rates, terms of trade, and so on – and their impacts on debt service.

The decision maker may be unaware of the scope and requirements of such management and the interdependencies of its main functions. Furthermore, and even if aware of multifunctional debt management, the decision maker might, in a crisis situation (for example, involving severe short-term liquidity or a prolonged lack of adequate foreign reserves), lose sight of the overall picture and concentrate instead on partial solutions. Within this broad framework, the formulation, development and accomplishments of Iraq's debt management and policy during the 1980s will be examined.

During the 1970s, Iraq used some of its oil revenues for domestic development, to accumulate foreign exchange reserves, for a modest expansion of bilateral and multilateral foreign investments, to enhance its international economic relations, and to provide developmental assistance. In international trade and among business communities as

[86]

well as among export credit agencies, Iraq was regarded as a 'cash country', with minimum economic and commercial risks to its trading partners, or to the international companies and contractors implementing various projects within the country.

Investment had increased from ID1.14 billion for the 1970/74 Plan to ID14.2 billion for the 1976/80 plan.[2] According to the IMF, trade balances had enjoyed surpluses that increased annually from US$3.2 billion in 1975 to a record high of US$14.4 billion in 1980. Over this period and until the beginning of the 1980s, Iraq was a donor and a creditor country,[3] with an estimated external reserve of US$36 billion in 1980. Its foreign investments, though mostly in Arab joint ventures, were around US$2.5 billion.[4]

By September 1980, and the start of the Iran–Iraq war, the 1976/80 Development Plan was in its final months, and the 1981/85 Plan, which included ID27.6 billion of investment and business opportunities, was in its final stage of approval. Preparations for awarding project contracts under the new plan had been under way and would be confirmed once the plan was approved. Major and very costly military projects such as the 'navy contract' with Italian companies were also under way.

The decision to go ahead with the 1981/85 Development Plan, despite the serious setback suffered by oil exports following the closure of the Mina al-Bakr and Khor al-'Amaya terminals (with a total capacity of 4.3 million barrels per day (mbpd)) on the Arabian Gulf, marked the beginning of Iraq's debt problem.

The core of Iraq's international diplomacy on debt was based on the formulation that the country could benefit in two ways from more international financial involvement. First, it would enhance Iraq's future leverage in dealing with its developing financial crises. Second, it would increase the foreign 'financial assets' and, hence, increase the commitment of the creditors to take the necessary measures to 'protect' their assets. Politically, this meant they would be expected to take a stand to support Iraq in case the war shifted in favour of Iran. In domestic and party politics, the regime interpreted this to its followers and supporters as an indication of international consensus in support of Iraq. Another key element in formulating this policy was to diversify the sources of supplies as much as possible. It was thought that the best way to ensure such diversification was through multi-purpose financial agreements with as many trade partners as was feasible.

The stages through which Iraq pursued its debt policy to secure some of the necessary external financing and the outcomes of that policy will now be examined.

Phase one

This phase, which covers the period from early 1981 to mid-1985, witnessed both the emergence of the debt problem and the laying down of the operational foundations for the institutional arrangements of debt policy.

Deferred payment of contractual dues

Since the decision was taken to proceed with the 1981/85 plan, all ministries took the necessary steps to implement their respective plans. The ministries were in fact subjected to pre-war norms of plan implementation in terms of budgeting and spending, as was clearly demonstrated by the level of economic activities – investment, concluded contracts and imports – for 1981.

The value of non-military projects where contracts were concluded was US$24.3 billion for foreign corporations alone.[5] Such a level of spending represented an increase of 64.2 per cent over the more normal level of 1980; it constituted 27 per cent of the resources of the whole plan for the period 1981/85; and, finally, it did not include military contracts or contracts concluded with local enterprises. Total non-military imports increased from US$13.5 billion in 1980 to US$20.7 billion in 1981 when total exports were down from a record high of US$27.9 billion to US$10.6 billion during the respective years.[6] Similar trends continued in 1982; exports went down further to US$10 billion, non-military imports went up further to a record high of US$21.5 billion, and military imports accelerated to US$6.4 billion.[7]

However, it became rather obvious to the decision makers that the 'business-as-usual' thinking could no longer be maintained when, in April 1982, Syria closed the oil pipeline going through its territory and immediately caused a loss of an export outlet for 700,000 barrels per day (bpd). By then, many of the Plan projects were under construction and large investment funding had already gone into them. If these projects were stopped and their contracts terminated, then Iraq, the international companies and their governments were likely to face some rather awkward consequences.

For Iraq, the development process would suffer a serious setback that would undermine the economic premises of the regime's domestic policy. For the international companies and contractors there would be the dilemma of lengthy, complicated and unpredictable arbitration processes since most, if not all, of these contracts were governed by Iraqi laws. This would, of course, result in serious cash-flow problems, especially as many of these companies already had significant fixed assets in the country. For the governments of the trading partners, themselves already facing severe budgetary constraints, the termination of so many large contracts would mean significant levels of financial compensation by their specialised export credit and guarantees agencies. Politically, Iraq was considered by most of these countries as the 'lesser of two evils', thus strengthening the justification for the financial support given to it.

It was not very difficult to see that, in the prevailing circumstances, the convenient option available for all three parties was to provide the necessary financing for these projects by deferment of the contractually due payments.

Later on, new directives and guidelines were issued for non-military projects, both new and those where contracts had not been finalised. The ministries and state entities were permitted to go ahead with implementation if certain conditions were met. Such a project would, on completion, have to enhance the country's foreign currency reserves through export promotion and/or import substitution;[8] it would also have to support the military and war efforts. External financing was to be available to cover all or most of the foreign currency portion of the contract.

Putting rhetoric aside, the availability of external financing and/or using oil for total or partial payments was the deciding factor.

Market sensing via Arab links
Iraq tried to penetrate the international financial market in 1982 through two Arab financial links in which it is a shareholder. In mid-1982, a 'frame' agreement was concluded between Rafidain Bank, the state bank and at the time the only commercial bank, and UBAF (Union de Banques Arabes et Françaises). The agreement was to provide the Rafidain Bank with a US$500 million Euro-loan for general purposes. In December 1982 a second 'frame' agreement was reached between the Iraqi National Oil Company (INOC) and the Arab Petroleum Investment Corporation

(APICORP).[9] This was intended to provide US$120 million to finance the first expansion of the pipeline through Turkey, and increase its capacity from 0.6 mbpd to between 0.9 and 1.0 mbpd.

Though these two 'frame' agreements were signed in 1982, the 'final' agreements were only concluded in 1983. In the view of the top decision makers (ministers, the Cabinet Office, the Presidency Office, the Governor of the Central Bank, and so on), the outlook for this course of action was not very encouraging, especially since borrowing from the Eurocurrency market was structurally complicated and demanding, and took quite a long time to finalise. Nevertheless, the experience proved to be very useful for the Iraqi negotiators since at that stage they had only a limited knowledge of the functioning of this market in particular and of borrowing in general.

The economic breakthrough with the USA

In December 1982, and after many months of contacts and negotiations, Iraq received the first tranche of some US$480 million in agricultural credits offered by the US government.[10] This proved to be a very significant breakthrough, not only in terms of economic relations between the two countries but also in terms of Iraq's economic and financial relations with many other countries. The US government continued to provide Iraq with this credit facility in increasing amounts so that it reached an estimated cumulative total of US$1.8 billion by the end of the fiscal year 1985. This credit was increased significantly to US$1.2 billion for the fiscal year 1987/88 and US$1.05 billion for 1988/89.[11] Consequently, the US share in total Iraqi imports increased from 4.3 per cent in 1982 (rank 4) to 10.7 per cent in 1987 (rank 2).[12] The resumption of full diplomatic relations between the two countries in November 1984 was also a significant added factor in helping to realise Iraq's economic diplomacy on debt.

Following the American move, other Western governments for the first time provided Iraq with new official financial facilities. During the first half of 1983, Iraq established diplomatic contacts with Austria and with the UK, which led to the conclusion of financial agreements/protocols in October of that year. The British facilities continued, and by the time of the Financial Protocol of 1988 amounted to US$1.7 billion. These protocols helped push UK exports to Iraq to US$700 million in 1988. This was the highest level since 1982 and put Iraq into fourth place among the UK's biggest Middle East markets in 1988.[13]

Compared with the period that had followed the outbreak of the war, Iraqi–Soviet relations continued to improve from mid-1982 onwards. Significantly, Iraq's improved economic relations with the USA induced the Soviets to offer new financial facilities. In 1983, for example, the two countries had concluded an agreement to develop Iraq's West Qurna oil field, based on financial credits. In 1985, the two sides concluded a 'general' economic cooperation agreement and in 1986 'commercial' cooperation agreements were agreed, these too being based on financial facilities.

In other words, the Iraqi–US economic breakthrough in 1982 had initiated a worldwide process of providing Iraq with financial facilities, which was to continue until the invasion of Kuwait in August 1990.

Domestic organisational arrangements

It was imperative that certain domestic arrangements were put in place as a necessary prerequisite for the successful and orderly implementation of economic diplomacy and for utilisation of externally available financial resources and credit facilities. Various mechanisms and practices were enforced, the most important of which were the following.

CENTRAL 'POLITICAL' MANAGEMENT

All matters related to the formulation and implementation of economic relations were put under the domain of the First Deputy Prime Minister, Taha Yasin Ramadan, who was also Chairman of the then influential External Economic Relations Committee (EERC) of the Council of Ministers.[14]

The EERC had a crucial role as an advisory as well as a follow-up body. The sectoral ministries, as well as the Central Bank of Iraq, had to report to and receive directives from the EERC on all matters related to external economic relations in general and to financial arrangements in particular. From the viewpoint of the external partners, this arrangement worked rather well, since the Deputy Prime Minister had 'executive' as well as 'legislative' authority through his membership of the Revolutionary Command Council (RCC). The RCC was the highest and only political body capable of issuing decrees which have the power of law.

LEGALISATION

Before 1983, only the Ministry of Finance was able to borrow externally on behalf of the government. Discussions in 1982 to deal with and address this matter centred on two alternatives: to continue using the option of the Ministry of Finance, or to empower all state entities to borrow from external sources when necessary. Influenced on the one hand by the UBAF and INOC experiences mentioned above, and on the other by the need for more flexibility to deal with the expected high and diversified level of borrowing in the future (since most of the financing needed was project- and trade-related), evaluation of these alternatives gave preference to the second option.

Based on this assessment the RCC issued a decree in the first half of 1983, authorising all entities to borrow from external sources, subject to the approval of the First Deputy Prime Minister. This effectively consolidated the central, political and legal aspects of Iraq's economic diplomacy and, accordingly, the economic role of Mr Ramadan and his External Economic Relations Committee.

THE MINISTERIAL JOINT COMMITTEES

During the 1970s, Iraq had concluded many economic and trade cooperation agreements which involved establishing Ministerial Joint Committees. These commissions were used by Iraq in its economic diplomacy as the main agencies for securing external financing and for rescheduling its debt. When joint commission meetings took place outside Iraq, the Iraqi delegation was, in the majority of cases, headed by Deputy Prime Minister Ramadan. This was intended to give political as well as economic dimensions to such events, and thereby to strengthen economic diplomacy. In fact, such visits provided more opportunities for conducting high-level diplomacy than did the usual ministerial joint commission meetings.

In addition to the joint commission meetings, many ministerial and high-level delegations were formulated to discuss financial and other matters related to economic cooperation with the rest of the world. Those delegations proved to be very flexible mechanisms and were useful devices for the performance of economic diplomacy.

SPECIALISED NEGOTIATION AND FINANCIAL COMMITTEES

Many committees were formed to negotiate such financial matters as new protocols, deferred payment arrangements, new financial facilities, debt rescheduling, and so on. Some were based on a bilateral/country basis, others on a project basis, and some even on a company basis. They also negotiated various civilian and military projects and supplies. Most of the committees were usually headed by ministers. The External Economic Relations Committee was the coordinating and supervising body.

Phase two

This phase, which lasted from mid-1985 and mid-1987, was operationally rather an easy one for Iraq, since the main foundations of its international economic diplomacy to facilitate the acquisition of the necessary external financing had already been laid down in phase one.

During this second phase, various types, cover, conditions, magnitude and sources of external financing continued to be available for Iraq. However, the financial situation from mid-1985 onwards brought new problems, concerned mainly with debt rescheduling.

There were many signs of optimism in Iraq that instalments due for payment in 1985 from deferred payment arrangements of 1983 would be met on time. Much of this optimism could be attributed to three factors. It was felt that debt service of US$2.1 billion in 1985 was relatively manageable. It was also thought that the US dollar, the main currency of Iraq's foreign exchange earnings, would continue in its strong position against other major currencies. This, by itself, reduced the real value of the non-dollar debts in terms of barrels of oil, as was the case with the deutschmark and French franc instalments paid then. Indeed, Iraq had even renegotiated the 'navy contracts' with the Italian companies and had gained a marginal discount on the grounds of a strong dollar against the Italian lira. Furthermore, up to half a million barrels per day of new export capacity would become available at the Saudi Red Sea port of Yanbu' by the end of September 1985, following the construction IPSA-1, the first stage of the pipeline which linked the southern Iraqi oil fields with the East/West crude-oil pipeline in Saudi Arabia.

However, such optimism began to fade when the value of the US dollar started to decline towards the end of the first quarter of 1985, and when this decline was finally institutionalised by the G5 Plaza Accord in September of that year.[15] In addition, oil prices continued their downward trend, causing further and serious deterioration in Iraq's terms of trade and deepening its financial difficulties. In 1986 the financial situation deteriorated even further, and financially speaking, this year proved to be a disastrous one due to the collapse of oil prices. Though Iraq's oil production increased by nearly 18 per cent in 1986, compared with 1985 when daily production had increased from 1,433 to 1,690 thousand barrels, oil export earnings nevertheless declined by 27.2 per cent, dropping from US$10.3 billion in 1985 to US$7.5 billion in the following year.[16]

Faced with these conditions of declining purchasing power caused by the combined effects of collapsed oil prices and a weakened dollar, Iraq intensified its efforts for more bilateral financial facilities, in terms of new credit and guarantees on the one hand and debt rescheduling on the other.

It was not difficult for Iraq to receive more credit and guarantees to finance its imports and projects, since the mechanism for such facilities had been worked out during phase one. In addition to the availability of these facilities, this second phase witnessed the emergence of debt rescheduling. Most of the credit facilities and related deferred payment arrangements that had been concluded in 1983 and 1984 with the West Europeans were of relatively similar structure; 85 to 90 per cent of the foreign currency payments due in each year would be deferred for two years, with repayment by four equal semi-annually consecutive instalments. The remaining 15 to 10 per cent would either be paid in cash or financed through commercial credits. Though Iraq had requested these conditions, they were in fact agreed upon by the OECD Export Credit Group within the Berne Union, in 1982/3.[17]

In short, this phase is characterised by debt rescheduling. Because of this two-year debt structure and the volume of deferred repayments involved, it was expected that the weight of these repayments would be spread between mid-1985 and mid-1987, with the peak in 1986.

Phase three

The third phase, which lasted from mid-1987 to the invasion of Kuwait in August 1990, witnessed some negative as well as positive developments.

Central debt management suffered a serious blow when the External Economic Relations Committee was abolished in July 1987. EERC negotiators and debt expertise were transferred to other ministries and the Ministry of Finance resumed its central role. A period of disintegration, inter-ministerial rivalry and chaos prevailed, and poor performance in 1988 in terms of repayment, disbursements, net flows and net transfers all led to the sacking of the Minister of Finance, Dr Hikmat al-Hadithi, in October 1989.[18] Nevertheless, the country did obtain new financial facilities and made some payments on its debt which marginally improved its creditworthiness. In February 1988 West Germany, for the first time during the war period, made DM300 million available in new financial credit, guaranteed by the official export credit agency, Hermes. A similar facility was made in 1989.[19]

In April 1989, Iraq's Composite Risk Rating was 36.0, a marginal improvement over its rating of 32.0 a year earlier. This raised Iraq's April 1988 ranking from 125 out of 129 and to 120 out of 129 a year later.[20] However, Iraq's repayment of its debt was not uniform across the country's creditors. That was mainly due to the non-availability of sufficient funds to serve all due debts on the one hand, and to the fact that some of the debts (such as the US EXIM bank facilities), were of a revolving nature, on the other hand. This prompted some of the creditors to complain about the way they were being treated by Iraq. As one Western diplomat reportedly commented, 'They've been robbing Peter to pay Paul'.[21]

One interesting development that occurred during this phase, concurrently with Iraq's privatisation process, was the extension to the private sector of external financial credit that had previously been available only for state and mixed-sector enterprises. In this connection, a group of West German banks had extended a line of credit of DM200 million to finance the sale of machinery and equipment to private sector businesses.[22] This facility was the outcome of an agreement reached in Baghdad in mid-June 1989 between a team of more than 20 West German companies and the Iraqi Federation of Chambers of Commerce and Industry (FCCI). It followed an announcement in March by the trade minister that about 26 per cent of the import finance allocations would be for private sector purchases.[23] Before this new arrangement with West Germany, the first financing of contracts placed by the Iraqi private sector was through the 1988 Financial Protocol between Iraq and the UK.[24]

The outcome: agreements and financial facilities

The outcome of Iraq's international efforts during the three phases discussed above was a long and diversified list of agreements, protocols, agreed minutes and arrangements. Practically all of Iraq's trade partners, from the poorest (Bangladesh) to the richest (Norway), were included on this list in one way or another; for reasons of debt rescheduling, credit and financial facilities, deferred payments, or any combination thereof.

The OECD countries
The new credit and financial facilities from West European countries followed a more or less uniform pattern, based on the OECD's system of export credit financing, under the known principles of the so-called 'Consensus Arrangement'. The basic guidelines of the Arrangement consisted of the following:

(a)   A minimum cash down payment of at least 15 per cent of the foreign currency portion of the value of the contract. In most of the deals with Iraq, this down payment was arranged through commercial credit.
(b)   Repayment terms of two to five, five to eight and a half, or eight and a half to ten years, depending on the nature of the project, by equal semi-annual instalments starting from a certain point. Facilities granted to Iraq were included in the two- to five-year terms.
(c)   A matrix of minimum interest rates, established for maturities and for three categories of countries: relatively rich, intermediate and relatively poor. Iraq was included in the second category. (See section below on interest rates and payments.)

The above principles were applicable for projects, capital goods, equipment and services. Military equipment, raw materials and agricultural commodities were excluded from the 'Consensus' but were provided under separate bilateral agreements. Most of these facilities were granted to Iraq on a regular basis, more or less annually, through a joint commission, a financial committee or specialised delegation meetings.[25]

The implication of the conditions of these facilities is reflected in the structure of Iraq's imports from the EEC. In the first nine months of 1988 nearly 81 per cent of such imports consisted of manufactured goods, chemicals, machines and transport equipment, 15.3 per cent of food, and 0.5 per cent of raw materials. Additionally, and because

France, Ireland and the Netherlands had offered Iraq convenient financial facilities for foodstuffs, 68.5 per cent of these imports, amounting to 173.1 million ECUs, had come from these three countries during the same period.[26]

US credits

Two types of US financial credits were offered to Iraq: agricultural credit and a short-term Export–Import (EXIM) Bank programme. The agricultural credit, which had been in place from 1982 until it was terminated due to the invasion of Kuwait, was an annual programme; up to the end of the fiscal year 1988/89 its accumulated total value was US$4.05 billion. The facilities granted under this credit in November 1988 were divided into US$1.0 billion in three-year loans and $50 million in seven-year loans. The EXIM Bank guarantee programme, which was introduced in mid-1987 was modest both in its value, at US$200 million, and in its maturity, as a short-term insurance cover. Both credits were operating well and were actively used since Iraq was up-to-date on them and the repayments were made on time (but remember the Peter and Paul story!).

As to be expected, Iraqi imports from the USA grew between 1985 and 1988 from around US$437 million to more than US$1,156 million, or by an increase of approximately 171 per cent.[27] Not surprisingly, a very high percentage of these imports consisted of agricultural raw materials and supplies of bulk consumer items. This, as noted above, was to substitute for or to complement similar imports from the EEC and other sources of supplies.

The Former Socialist Countries (FSCs)

Most of the financial facilities from the Socialist countries before 1984 were arranged on an annual basis or on a case-by-case basis through joint meetings. However, from 1985 onwards a new model emerged to govern and regulate these facilities: this was the Economic Cooperation Agreement (ECA).

Though there were no uniform export credit arrangements in the former socialist countries similar to those of the OECD Consensus mentioned above, these ECAs were, nevertheless, based more or less on the following principles:

(a)   The rate of interest was fixed for the entire period of the credit and
      these rates were relatively low (see below for further details).
(b)   The duration and validity of the agreement were for four years,
      renewable.
(c)   Some of the agreements had a credit ceiling, though others did not.
(d)   In the majority of cases, financing could reach 100 per cent of all
      the foreign currency portion of the contract concerned.
(e)   The repayment period for projects could be as much as ten
      years, but was much shorter for supplies. The starting point for
      each repayment period was defined in each contract/agreement, and
      payments could be made by equal annual instalments.
(f)   ECAs were inter-governmental agreements and most of them
      required mutual ratification.

   In terms of cost of financing, time availability of the facility, automatic
renewability, and the length-of-payment terms, these agreements proved
to be more convenient than those concluded with the Western countries.
Nevertheless, and despite the obvious advantages of these ECAs, the levels
of Iraqi imports from all the socialist countries, excluding Cuba, remained
modest. The percentage of imports from all the socialist countries to total
Iraqi imports was less than 9 per cent for 1983 and 1988 respectively.[28]

The developing countries
Due to the economic circumstances of the developing countries and
their inability to offer attractive and generous financial facilities, many
of the arrangements concluded with them were different from those of
either of the other groups of countries.
   Arrangements between Iraq and many of its trading partners
among the developing countries were different in magnitude, terms and
conditions; even so, common denominators can be observed based on
the following general principles:

(a)   Rates of interest were close to the commercial rates in the inter-
      national financial markets. In some of these facilities, the London
      Inter-Bank Offered Rate (LIBOR) was taken as a reference point.
(b)   A higher percentage of the amount concerned was paid in cash,
      while the rest was deferred and repaid in short and medium periods
      by quarterly and/or semi annual instalments.

(c) Oil was actively used in the repayment of dues to the major trade partners.

(d) Most of the arrangements with the developing countries, with the exception of those with Brazil, Turkey, former Yugoslavia and, to a certain extent, India, were not regulated, either by regular annual meetings or through long-term agreements.

More than 90 per cent of total Iraqi imports from non-Arab developing countries was generated from four countries during 1983 to 1986. These were, in order of magnitude, Turkey, Brazil, South Korea and India. Two observations can be made: first, the share of these four countries in total Iraqi imports had doubled from 10 per cent in 1983 to 22 per cent in 1988. Second, in 1988 Iraq imported from Turkey alone much more than it imported from the whole socialist bloc.[29]

Interest rates and payments

Interest on long-term credit provided by the various creditors discussed above was as follows.

The long-term credits offered by OECD countries and banks under the official export credit and guarantee systems carried interest rates according to the matrix of the 'Consensus'. This matrix was revised twice a year. Interest rates in the matrix were dependent on the category of the country and on the duration of the credit; that is the maximum repayments in years. However, once a particular credit agreement or contract was finalised, the interest rate remained fixed for the entire duration of the credit. Accrued interest was payable semi-annually.

Iraq was classified under 'category II' as an intermediate country, and all its long-term credit had a maximum repayment period ranging between two and five years. For category II and such payment durations, the interest rate was 7.5 per cent between May 1978 and October 1981. During the period from November 1981 to December 1985, it began to fluctuate between 10.35 per cent and 11.55 per cent. From then until June 1989 rates ranged between 8.25 per cent and 9.65 per cent.

Long-term credit offered by the financial markets carried an interest rate that was usually based on the LIBOR, plus a 'spread' margin. Unlike interest rates under the 'Consensus', LIBOR was not fixed for the entire period of the credit but during the interest period only. If interest is paid annually, semi-annually, or quarterly,[30] LIBOR is fixed during the agreed period of interest payment. The magnitude of 'spread' or the margin over LIBOR usually reflects the borrower's creditworthiness, its exposure with the creditors, and the position of the supplier or contractor concerned.[31] For this category of long-term credit, I am not aware of any that carried more than one percentage point over LIBOR.

Long-term credits provided by the FSCs had much lower rates of interest compared with those provided by the OECD countries and the commercial banks. For this group of countries, interest rates were between 2 per cent (as was the case with the former Soviet Union for project-related financing), and 6 per cent (for credits from former Czechoslovakia). In addition to being low, interest rates were fixed for the entire maturity of the credit, and were payable annually.

Most of the multilateral long-term credit to Iraq came from regional institutions. The non-concessional facilities, such as those provided by APICORP, carried interest rates similar to those of the long-term credits that were provided by commercial banks and based on LIBOR. Concessional credits, such as those offered by the Arab Monetary Fund (AMF), the Arab Fund for Economic and Social Development (AFESD) and the Islamic Development Bank (IDB), had low and fixed interest rates or loan fees. AMF loans,[32] for example, had interest rates at 4.15 per cent, while AFESD loans[33] carried interest at 4.5 per cent.

In spite of the outcomes mentioned above, debt policy and management in Iraq was far from being the multifunctional integrated system mentioned at the beginning of this section of the chapter. There were several reasons for this. First there was the 'newness' factor. As already noted, Iraqi financial institutions had had little experience in external borrowings before 1980, and they were overwhelmed by the scale, diversity and requirements of new borrowing during the decade. This undermined their ability and readiness to perform the necessary accounting and statistical functions.

Secondly, under the pretext of national security considerations, debt data was not made available without the prior approval of some high level authority. Moreover, financial institutions and state entities had developed a tendency not to reveal such information unless they were requested to do so, in order to secure participation in debt negotiations, especially when such activities took place abroad. This contributed to the further weakening of the accounting and statistical functions.

Thirdly, the country's financial position and ability to face its obligations was highly sensitive to a multiplicity of factors, such as war requirements, export revenues and import needs. With such a constraint, the operational function had a limited menu to choose from. Furthermore, debt decisions had been largely of a reactive nature. Since availability of external funding determined projects implementation, and availability of foreign exchange determined payments, very little policy function was thus possible.

Finally, by the time the impact of the debt problem on the economy began to become obvious, the process of debt management, though modest, had disintegrated, and its status and level of authority had fallen, leaving the door wide open for inter-ministerial rivalry that seriously hindered and undermined the regulatory function. (For an explanation of why the Ministry of Finance was not able to conduct effective debt management, see *Phase three* above, and endnote 18.)

3. Debt structure, ratios, present value and sustainability

The efforts made during the 1980s had accelerated the pace of debt accumulation to a level and scale unprecedented in the entire recent history of the Iraqi economy. The following analysis is made on the basis of detailed time series data on the 'identified' Iraqi debt according to the OECD. This data, *which does not cover all Iraqi debt*, is utilised here for analytical purposes on composition, trends, structure and sources of debt.[34] Another set of aggregated official Iraqi data will be used later in this chapter to estimate the present value of debt and to assess debt sustainability under different assumptions.

Debt stock and debt structure

Debt stock, which is often defined and reported as debt outstanding at the end of a given period of time, normally a year, measures a country's total debt liabilities.

At the beginning of the 1970s, Iraqi debt was almost negligible. It was less than US$2.5 billion in 1980 before it began to increase gradually and continuously to reach US$22.8 billion in 1990. In fact, such increases took place on a yearly basis, except in 1988 – the year the Iraq–Iran war ended – when total identified debt decreased slightly by around US$300 million due to a decline in the long-term non-concessional and short-term debts. It should be noted that debt accumulation had accelerated from 1984 onwards. This was a natural outcome of the diplomatic efforts of 1982/3 after the US breakthrough and the signing, for the first time, of financial protocols with the UK, Austria and many other European countries.

The maturity structure of the debt is divided into long-term and short-term debt. In general, long-term credit, which has a maturity of longer than one year, is related mostly to project financing and major trade financing. Short-term credit, which has a maturity of one year or less, is mostly related to non-major trade supplies.

Iraq's long-term debt had increased annually from less than US$2.5 billion to US$16.3 billion between 1980 and 1989, before it fell to US$14.3 billion in 1990. Short-term debt, on the other hand, displayed a fluctuating pattern. After its slight decline in 1983 to US$2.2 billion, it began to increase annually to reach US$5.8 billion in 1987, then fell again to US$5.2 billion a year later, before picking up to a record level of US$8.5 billion in 1990.

By looking at the data and composition in Figure 2.1, the following can be observed.

The proportion of short-term debt to total debt had, as shown above, fluctuated in both absolute and relative terms. The share of short-term debt in the total had decreased from 40 per cent in 1982 to 28 per cent in 1984, then up to 34 per cent in the following year. Then it decreased continuously to 26 per cent in 1988, before it began to increase again to reach 28.6 per cent and 37.4 per cent in 1989 and 1990 respectively. A comparison between the changes of general trends in total debt and in short-term debt indicates that total debt was not sensitive to such fluctuation in the short-term debt.

Since short-term debts are, by definition, non-cumulative on a yearly basis, then the higher its share and the faster it grows, the more of a liquidity problem it represents.[35] In addition to its increasing proportion to total debt, referred to in the previous paragraph, short-term debt had registered an extremely high annual rate of increase in 1989 (25 per cent) and in 1990 (31 per cent). In fact such annual rates for short-term debt had outstripped those of long-term debt, both concessional and non-concessional, and of total debt during 1988–90.

FIGURE 2.1
Structure of *identified** Iraqi debt 1980–90 (billion US$)

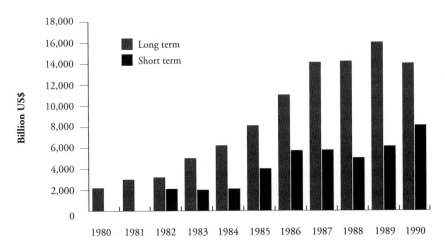

Note: *The part of total Iraqi debt identified in OECD statistics.
Source: OECD, *Financing and External Debt of Developing Countries*, Paris, various annual surveys.

Composition of long-term debt

Long-term debt comprises concessional and non-concessional debt. While the contribution of concessional debt to total long-term debt during the period prior to 1984 was not significant, ranging between 22 per cent in 1980 and 3 per cent in 1983, the gap between the two types of debt began to narrow from 1984 onwards. In 1984, 63 per cent of long-term debt was non-concessional and this had been reduced further to 52 per cent in 1990. This represented an improvement in Iraq's favour, since concessional loans, such as those extended by the OECD, included a grant element ranging from at least 25 per cent for Official Development Assistance (ODA) loans to at least 35 per cent for loans under the OECD 'Consensus'.[36] However, as we shall demonstrate later, the continuous decline of the interest rate in the financial market seriously undermined the concessionality of OECD loans.

COMPOSITION OF CREDITORS

The identified long-term credit was generated from three groups; (a) OECD countries and capital markets; (b) multilaterals; and (c) non-OECD countries. Among the three sources of credit, the multilaterals

[103]

stood as the lowest, with a range of debt between US$79 million and US$368 million in 1981 and 1990 respectively, the major part of which was non-concessional. Obviously such a low contribution and low concessionality on the part of the multilateral institutions reflect the attitude of these institutions (such as the World Bank and IMF) towards Iraq because of the war situation with Iran on the one hand, and the inability of Iraq to engage these institutions on the other. Furthermore, most of this category of debt is under-reported and belongs to regional multilateral institutions.[37]

FIGURE 2.2
Creditors' composition of *identified** long-term debt
1980–90 (billion US$)

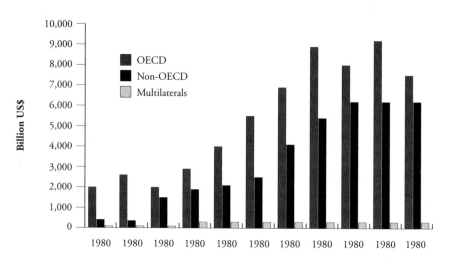

Notes: In this chart all identified Iraqi long-term debts are divided between three main creditors' groups; OECD, non-OECD and Multilaterals. *The part of total Iraqi debt identified in OECD statistics.
Source: OECD, *Financing and External Debt of Developing Countries*, Paris, various annual surveys.

Iraq's long-term debts to OECD countries and banks peaked at US$9.4 billion in 1989 (Figure 2.2), representing 58 per cent of Iraq's total long-term debt in that year. The following observations may be made on the credit offered by this group:

(a)   The majority of this debt was generated by official, officially supported, or guaranteed export credit.

(b)   Commercial bank credits had shown a remarkable jump during the period from 1988 to 1990, compared with the previous period.

(c)   The Official Development Assistance contribution was very low indeed: it reached its highest level of US$408 million in 1987, and thereafter gradually went down.

Non-OECD creditor countries come next to credits from the OECD area. Credits from this group of countries had shown a continuous increase during the whole period from 1981 to 1990. Almost all of this credit came from the former socialist countries. As indicated above, two significant upward movements took place: the first was in 1982, due to what I have called the 'Iraqi–US economic breakthrough effect'; and the second was in 1986, after the visit by Saddam Hussain to the then Soviet Union and his meeting with Gorbachev.

Debt service

Debt service statistics usually include payment of principal debt (i.e., amortisation of long-term debts) and payments of accrued interest on both long-term and short-term debt,[38] and do not include payment of the short-term debt.

On the basis of identified debt, Iraq had serviced its debt during the period 1980 to 1990 to the tune of US$24.2 billion. The major part of this, amounting to US$20.2 billion (or 95.6 per cent), was made to the OECD countries and capital markets. The remaining part was paid to the multilaterals (US$438 million representing only 2.1 per cent), and to non-OECD countries (US$501 million representing 2.3 per cent).

These figures might indicate that favourable/preferential treatment was given to OECD creditors, since their share in Iraq's debt service was much greater than their share in Iraq's total identified debt. They might also indicate substantial differences between the terms of maturity that were made available by OECD and non-OECD countries. As has already been noted, most OECD credit which was offered to Iraq had a maturity of two to five years, while that from non-OECD countries, mainly the FSCs, had a maturity of ten years; thus the US$501 million paid by Iraq to this group of countries could represent interest payments only. Rescheduling of non-OECD debt beyond 1990 might be another explanation. Finally, these figures might indicate that while the non-OECD countries made their credit to Iraq identifiable, they did not fully report the servicing of their debt by Iraq.

Total debt service payments of US$24.2 billion were divided between amortisation of the long-term debt amounting to US$17.5 billion, interest on the long-term debt of US$3.6 billion, and interest on the short-term debt of US$3.1 billion.[39] Amortisation of the long-term debt had followed a systematic fluctuating pattern. It increased from a low level of US$1.1 billion in 1980 to US$1.9 billion in 1982, then fell gradually to US$965 million in 1984 before increasing again to reach US$2.5 billion in 1986. During the next two years, it fell gradually to US$1.2 billion in 1988, increased to US$2.5 billion in 1989, before again falling sharply to US$601 million during the year of the invasion.

This almost-repeated pattern of a low amortisation for two years followed by a high one during the third year represents a 'bunching' phenomenon that can be explained by the following.

Since OECD creditors had received around 96 per cent of what Iraq had paid to service its debt, it can reasonably be assumed that the practices of this group of creditors shaped the bunching pattern. The first such practice concerned deferment of Iraq's annual dues by OECD members of the Berne Union (as explained above); the second was related to a repayments pattern of two to five years for new credits offered under the 'Consensus', where most of the utilised credits had the lower instead of higher band of maturity period.

The attitude of the Iraqi decision makers was characterised by bilateral and case-by-case approaches. In spite of their advantages, such approaches make smooth repayment of debt very difficult indeed. The difficulties are compounded when the three main groups of debt overlap: deferment of dues on an annual basis, rescheduling of arrears, and fresh credit, both long- and short-term.

It is rather disquieting to note that while short-term debt was less than half of the stock of long-term debt in any given year, except 1982 and 1990, the difference between interest payments on the two types of debt was only US$500 million. This clearly indicates that the interest rate on short-term debt was much higher than that of long-term debt. Such a possibility does exist, especially when one considers that almost all short-term credit was made available by the commercial banks not only at the interest rates then prevailing in the financial markets but also outside official export credit and guarantees facilities.

Net flows and net transfers
These two concepts describe the net effect of borrowing and repayments on the flow of financial resources into and out of the debtor country. Net flows, defined as disbursements minus principal repayments, measure whether new financing exceeds repaid debt. While positive net flows of long-term debt increase debt stock, this reflects an appreciated credit-worthiness of the borrowing country, especially when such positive net transfers persist over a reasonable period of time.

Net transfers refer to disbursements minus the sum of interest payments for both short- and long-term debts and principal repayments. Negative net transfers imply that total debt service payments exceed gross inflows, that net real resources are being transferred from the economy, and that a trade surplus is thus required.

Since OECD data does not include net flows, net transfers, or disbursements, the writer has estimated these three variables (Appendix One explains the method of estimation). The results are presented in Figure 2.3.

From these estimates it seems that long-term lending had been favourable to Iraq; a trend that was interrupted by Iraq's invasion of Kuwait. All three variables – disbursements, net flows and net transfers – indicate the positive net effect of borrowing and repayments on the flow of financial resources to and from Iraq during the years prior to the invasion. The estimates also show that 1988, the year the Iran–Iraq war ended, was a downward point in the above positive direction. Disbursements or new money declined sharply from US$5,143 million in 1987 to US$1,533 million a year later; net flows in 1988 were less than one tenth their level a year earlier; and net transfers in the same year registered the only negative value. While a significant recovery of all three variables took place in 1989, the invasion of Kuwait caused them to slide deep into negative territory.

By way of comparison, the World Bank estimated that in 1987 the Middle East and North Africa region had a positive net flow of US$3 billion and a negative net transfer of US$12 billion.[40] During the same year, Iraq had a positive net flow and a net transfer of US$3.2 billion and US$2.3 billion respectively.

FIGURE 2.3
Parameters of long-term lending to Iraq 1981–1990 (million US$)

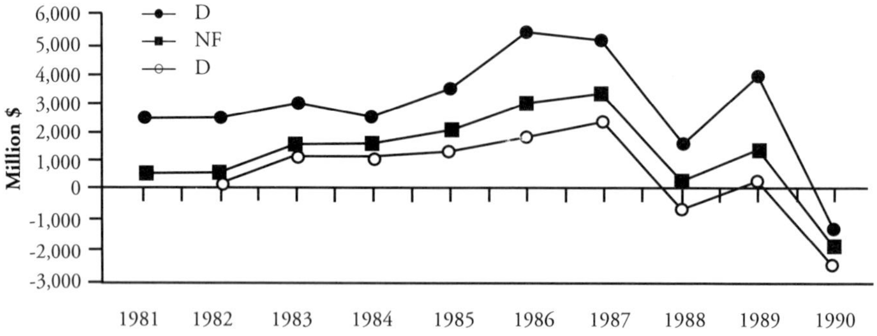

Note: The estimates of debt net flows and net transfers are made according to the methodology mentioned in Appendix 1 and are shown in this graph.
Source: See text and Appendix 1.

Debt ratios

In the literature on debt, one can find different and various types of indicators and ratios which are used to measure the magnitude, scale and severity of the debt problem or burden and to classify the level of indebtedness of countries. Some of these measures emphasise the absolute volume, such as the stock of outstanding debt or debt service, while others are of a relative and proportional nature.

The absolute measures do not say much, since debt magnitude by itself is not the problem. The real problem is the ability of the national economy to service its debt in a sustainable manner. Therefore, it is more meaningful to look at debt in relation to the capacity of the borrowers to fulfil their contractual obligations vis-à-vis the creditors. Though debt comprises both a stock, which is total debt, and a flow, which is debt service, it is the flow more than the stock that constitutes the problem; and this flow is better understood in relation to other flows generated by the borrower's economy. Relative measures could be related to economic variables such as GDP or GNP, or to financial variables such as export revenues, or to fiscal variables such as government budget or expenditure, and so on.

Debt to GDP (D/GDP) ratio

One of the widely used measures of indebtedness is total debt (D) in relation to the gross domestic product (GDP) or the gross national product (GNP) of the borrowing country.

The World Bank uses this ratio to determine the level of indebtedness.[41] If the *present value* of future debt service (current debt stock plus interest, and taking account of concessionality of debt stock) exceeds the *critical* value of 80 per cent of GNP, the country is considered as severely indebted. If the critical value is not exceeded, but the ratio is 60 per cent or more of the critical value, the country is classified as moderately indebted. Finally, if the ratio is less than 60 per cent of the critical value the country is less indebted. Others have used the nominal value of debt stock instead of the present value of future debt service.[42]

By using the annual nominal value of debt stock, the D/GDP ratio was calculated during the 1980s. The ratio had increased continuously from 5.2 per cent in 1980 to reach its highest level of 35.4 per cent in 1987, then declined to 30.5 per cent in 1990. Obviously, when this ratio was on the rise it implied that growth rates of external debt were higher than those of GDP, which registered negative growth rates in some years during that period.

As calculated, this ratio was much less than the critical threshold devised by the World Bank. However, this calculation was based on most, but not all, Iraqi debt and was in nominal current values, not the present value of future debt service. If all debts were identified, future payments were known, and the discount factors[43] were chosen, it would then be possible to estimate this ratio according to the Bank's procedure, and one would be in a better position to determine the level of this ratio for Iraq.

Debt/Export (D/E) ratio
A more reasonable way to look at debt is through export revenues (or total inflows including Foreign Direct Investments (FDIs), expatriate remittances, grants and aid, and transit revenues, in addition to exports), since debt has to be satisfied and paid for by foreign currency. This ratio and the debt service/export ratio are, as we shall see, widely used as indicators of sustainability problems or solvency.

Based on the same partial figures, Iraq's debt/export ratio had been over the 200 per cent level during the second half of the 1980s, except 1989 when it was 185 per cent, due to the marked improvements in export revenues.

Debt Service/Export (DS/E) ratio
Comparing debt service to export revenues can be considered as a suitable measure of the country's ability to service its debt: a lower DS/E ratio implies solvency and a higher ratio implies stress, liquidity or cash-flow problems. According to *The Economist*, any country that spends more than 40 per cent of its annual export revenues on servicing its debt is likely to face big problems.[44] UNCTAD uses a lower threshold range of 20 to 25 per cent,[45] and the World Bank and IMF use 15 per cent.

DS/E ratio registered its highest level of 48.6 in 1986, when the first bunching of debt service coincided with the lowest level of export revenues. The second highest DS/E ratio of 30 per cent accrued in 1989 at the second bunching in debt service, but this time it was counterbalanced by the highest export revenues. As with the previously calculated indicators, these figures are based on partial accounting of Iraq's debt.

DS/GDP ratio
From a wider perspective, and considering the potential of the country, the DS/GDP ratio can provide a reasonable measure of long-term indebtedness, especially in a country endowed with an internationally tradable commodity that can generate sufficient foreign exchange, such as oil. However, this ratio is more meaningful for strategic rather for than short- and medium-term decision making on the parts of the creditors.

Iraq's DS/GDP ratio (with the same proviso as above) had been less than 5.5 per cent during the entire 1980s, except in 1986 where it registered its highest level of 7.5 per cent.

Needless to say, any improvement in oil export revenues could have significant effects on the values of some of the above-mentioned ratios. However, as noted, these ratios do not reflect the true picture of Iraqi debt, especially during the last three years of the decade, since they are based on OECD statistics, which cover part of the debt. Actually, total Iraqi debt identified by the OECD in 1991 at US$21.2 billion, was little over half of what the Iraqi government had admitted, as will be shown.

Nevertheless, Iraq's financial difficulties prior to the Gulf War were a reflection of many factors: an ambitious industrialisation – especially military – programme,[46] military requirements during the Iran–Iraq war,[47] the debt-rescheduling structure, a relatively high level of short-term debt and, of course, export revenues. The implication was that Iraq's debt problems and the financial difficulties it was facing were manageable in relation to the country's resources and capabilities in a long-term perspective and under conditions of normality. This is why many observers had concluded in early 1989 that the years 1991–2 were to be crucial.[48]

Present value of Iraqi debt

On 18 May 1991, the Iraqi representative at the UN presented a memorandum to the Security Council which included, for the first time, official Iraqi data on the country's indebtedness. The reported stock of external debt was US$42.097 billion, excluding funds from Gulf countries, which the memorandum considered as 'grants'. The repayment of these debts was scheduled as follows:[49] US$23.468 billion[50] in 1991 (due but not paid up to the end of 1991), US$4.922 billion in 1992, US$4.824 billion in 1993, US$4.612 billion in 1994, US$2.950 billion in 1995, and US$1.321 billion after 1995.

In our assessment, the Iraqi figures represent the totality of the country's military and civilian debt excluding, as stated by the memo, interest and GCC funds. Since no new credits were extended to Iraq after August 1990, it must be assumed that the OECD figure of US$22.846 billion represents a very high percentage of non-military debts to the OECD countries at the time of the Kuwait invasion. The remaining $19.251 billion covers military-related debts to the OECD countries, as well as all debts to developing countries, to non-GCC Arab countries, to multilateral institutions, and to the FSCs. The figure of US$42.097 billion represents the outstanding stock of Iraqi debt at the end of 1991. On average, Iraq reckoned it paid 8 per cent per annum as interest on the outstanding balance of the said debt.

The present value of total debt is here calculated under three scenarios (see Table 2.1): scenario zero assumes that Iraq will not pay any interest from the end of 1990 until the end of 1997 and thus the present value equals (calculations were made at the end of 1997) the nominal value of debt stock of US$42.1 billion. Scenario one assumes Iraq pays interest, at 8 per cent per annum, on its outstanding debt, and this interest is not capitalised. Therefore, the present value of debt (end of 1997) is US$65.7 billion (US$42.1 billion principal debt and US$23.6 billion accumulated interest). Finally, scenario two is similar to scenario one but here interest is capitalised. Thus, the present value of Iraqi debt at the end of 1997 is US$72.2 billion, including US$30.1 billion accumulated compounded interest. For all scenarios, no default penalties or surcharges were considered, and the exchange-rate effects on both principal debt and accrued interest were not taken into consideration.

Sustainability analysis and assessment
As mentioned earlier, debt sustainability means the ability fully to meet current and future debt obligations without resorting to debt relief, rescheduling or accumulating arrears, and without compromising economic growth. Many key indicators with specific values or thresholds were suggested and used to assess debt sustainability. Though analysts[51] accept the two most common indicators – the present value of debt to export ratio and debt service to export ratio, chosen by the IMF and the World Bank – they do, however, raise three concerns. First, there should be more indicators; second, there is a degree of arbitrariness in the choice of value or threshold; and, finally, there is the matter of the time horizon – how long should the indicator's value be over the specific threshold to indicate sustainability?

As noted earlier, the IMF and World Bank use a 15 per cent threshold for debt service to export ratio, and UNCTAD uses threshold ranges of 20 and 25 per cent. For the present value of total debt[52] to exports, the IMF and World Bank use a 200 to 220 per cent threshold and UNCTAD uses a threshold range of 200 to 250 per cent.

Appendix Two outlines the methodology and procedure used to estimate present value and sustainability assessments of Iraqi debt, and the results are presented in Table 2.1.

Export needs to satisfy sustainability requirements
How much *should* Iraq export in order to be able to pay for its debt service within each of the three sustainability thresholds? Naturally, one must look not to the maximum but to the minimum export levels needed to satisfy these two conditions. As the figures indicate, the lowest export levels correspond to the upper band of the sustainability threshold. The question now is how attainable are these minimum export levels? The answer, based on export performance during 1981–90, is not at all.

The conclusion, then, is that Iraq is incapable of sufficient generation of export revenues to cover even the minimum requirement of debt sustainability.

TABLE 2.1
Sustainability assessment for Iraqi debt

	10 years	15 years	20 years	25 years
Annual debt service (in US$ billion)				
Scenario zero	6.27	4.92	4.29	3.94
Scenario one	9.79	7.68	6.69	6.15
Scenario two	10.76	8.44	7.35	6.76
Total debt obligation (in US$ billion)				
Scenario zero	62.74	73.78	85.76	98.60
Scenario one	97.91	115.14	133.83	153.87
Scenario two	107.60	126.53	147.07	169.09
Sustainability requirement (level of export revenues) at:				
a) (DS/E) threshold 15 per cent (in US$ billion)				
Scenario zero	41.83	32.79	28.59	26.29
Scenario one	65.27	51.17	44.61	41.03
Scenario two	71.73	56.23	49.02	45.09
b) (DS/E) threshold 20 per cent (in US$ billion)				
Scenario zero	31.37	24.59	21.44	19.72
Scenario one	48.96	38.38	33.46	30.77
Scenario two	53.80	42.18	36.77	33.82
c) (DS/E) threshold 25 per cent (in US$ billion)				
Scenario zero	25.10	19.67	17.15	15.78
Scenario one	39.16	30.70	26.77	24.62
Scenario two	43.04	33.74	29.41	27.05
DS/E ratio (assuming historic export levels) (per cent)				
a) DS/E (Av. 1981–90: US$9.8 billion)				
Scenario zero	64.02	50.19	43.75	40.24
Scenario one	99.91	78.32	68.28	62.80
Scenario two	109.79	86.07	75.04	69.02
b) DS/E (Mx. 1981–90: US$12.3 billion)				
Scenario zero	51.01	39.99	34.86	32.06
Scenario one	79.60	62.40	54.40	50.04
Scenario two	87.48	68.58	59.79	54.99

Debt service and actual export performance
If one looks at the issue from another angle by comparing annual debt service requirements with what had been exported during the 1980s, and then tries to find any of the three sustainability thresholds (15 per cent, 20 per cent, and 25 per cent), there is none to be found. The lowest DS/E ratio was 32.1 per cent.

Total debt and actual export performance
The present value of debt to export is the second sustainability indicator, and as mentioned above it should have a threshold of less than 250 per cent. By considering only present values of debt at the end of 1997 (the

three scenarios) and relating each one of them to the maximum export levels attained during the 1980s, we find the lowest ratio was 342 per cent.

Two conclusions can be drawn from the above analysis. First, by using any sustainability indicator, Iraq's debt problem is severely unsustainable. Second, even with the deliberate use of the optimistic scenario zero (interest forgiveness during 1991–7), the debt problem is unsustainable. The implication is that a significant debt reduction in both principal debt and future interest payments, together with convenient rescheduling, will be needed to bring back control over Iraq's indebtedness.

4. Options and possible solutions

The previous section of this chapter demonstrated the scale, severity, and complexity of Iraq's debt problem. If debt ratios and indicators are calculated on the bases of export revenues and GDP during the period of most restrictive sanctions from 1990 until 1998, then Iraq could be the most indebted country in the world. Ironically, some well-informed Arab[53] and European[54] institutions still consider Iraq as a creditor and donor country. Iraq has no debt-reporting system (DRS) arrangement with the World Bank,[55] but it is classified by the Bank as being among the severely indebted middle-income countries.[56]

Even though Iraq is a resource-rich country, wealth in the ground means nothing if it is not tapped to serve the national development objectives and requirements, as well as to meet the country's financial obligations. Despite this obvious and major constraint, we will try, in this section, to explore some of the viable options that might be available to Iraq in addressing the debt problem.

Debt forgiveness, reduction or freeze

This option aims at reducing the total amount of debt through a reduction in the principal debt and/or accumulated interest. The emphasis here is placed on official debt.

The available G7 possibilities

Over the years the OECD, and more specifically the G7 countries, have adopted many measures to tackle debt problems and reduce their impact on the low-income developing countries. These measures have include the 1988 'Toronto' terms, the 1990 'Houston' terms, the 1991 enhanced

'Toronto' terms, Naples 1994, and so on, all of which envisaged debt reduction for the said group of countries.

According to the new Naples terms, for example, the eligibility criteria to qualify for a 67 per cent reduction is that the country must have either a per capita GDP of less than US$500 or a ratio of the present value of debt to export exceeding 350 per cent. Iraq qualifies on both counts for these eligibility criteria, the debt ratio since 1990 having far exceeded this 350 per cent threshold.

Under the lowest scenario of non-capitalisation of interest, the ratio of total debt at the end of 1991 was US$45.465 billion (US$42.097 billion principal plus US$3.368 billion annual interest), plus an additional amount of $3.368 million per annum as interest from 1992 until the time of writing. Though Iraq is eligible for debt reduction according to the above-mentioned criteria, the possibility of its being considered is constrained by the following factors:

(a) It could be argued that Iraq is a resource-rich country, though it currently has a very low income.
(b) A significant portion of Iraqi debt is owed to non-OECD countries and, considering the 'burden-sharing' principle, unless these creditors agree to consider Iraq for debt reduction, the OECD countries will not have the country on their minds or their lists.
(c) Lack of precedent or the 'bandwagon factor': to benefit from a debt-reduction facility a country should have been included in the previous facilities. For example, countries eligible for Naples-1994 terms are those that have received Toronto and enhanced Toronto. Iraq did not have Paris Club official debt rescheduling nor did it receive any of these sorts of terms. Political considerations play significant roles in considering a country for debt reduction or forgiveness. Examples are Egypt[57] – which after the Gulf War had a 50 per cent debt reduction and which was exempted from the rules of the Houston terms[58] – and Poland.

Accrued interest
As was estimated earlier, the amount of accrued interest at the end of 1997 could range between US$23.6 to US$30.1 billion, depending on whether interest was capitalised or not. This accumulated and unpaid interest as from August 1990 could be reduced in the following manner:

(a) *Interest forgiveness* The argument for interest forgiveness for the above-mentioned period could be made on the principle of *force majeure* since Iraq was prevented from earning foreign currencies to enable it to service its debt. The possible counter-argument could be that there was no *force majeure*, since Iraq was prevented by the UN (that is, by international law) from exporting oil in order to enforce international legitimacy.

(b) *Non-capitalisation of interest* Many national laws, such as those of Germany, do not allow the capitalisation of accrued interest. Furthermore, there are many examples during the 1980s of debt negotiations when interest was not capitalised, based on legal and/or *force majeure* grounds.

A possibility exists for non-capitalisation of interest, especially with the FSCs, since all credit matters are governed by economic agreements that contain the principle of 'amicable solutions'. It is possible that interest capitalisation issues could be discussed together with other issues pertinent to these economic agreements, such as the possibility of extracting concessions from Iraq, probably by Russia.[59]

Certainly, as of the end of 1997, a total interest forgiveness could have reduced Iraq's obligations by US$23.6 billion and a total non-capitalisation of interest would have saved the country US$6.5 billion.

Debt freeze
As a sub-option, debt freeze, or keeping debt outstanding for a certain period of time, can be visualised as an intermediate debt-relief arrangement. Depending on other conditions, debt can be reduced or forgiven at the end of the freeze period.

Debt restructuring
The specific target of this option is to reduce the rate of interest on the officially-supported credit and financial facilities concluded by Iraq with many OECD countries during the 1980s. As we have seen, interest on export credit offered by this group of creditors ranged between 10.35 per cent and 11.55 per cent for credit concluded during the period from November 1981 to December 1985, and between 8.25 per cent and 9.65 per cent for credit concluded during the period from January 1986 to July 1989.

These fixed interest rates of the 'Consensus' were, and are, much higher than the prevailing commercial interest rates. Since 1989, interest rates for credits donated in US dollars have fallen sharply. During 1992 and up to the first quarter of 1996, the LIBOR (the main US dollar interest rate for international loans), on period averages in per cent per annum, was 3.9, 3.41, 5.07, 6.1 and 5.33.[60] Furthermore, the World Bank has projected LIBOR to average 5.8 per cent for the period from 1994 to 2003.[61]

In addition, when the commercial interest rate is lower than rates under export credit for a considerable period of time, the interest payment (or its accumulation due to default and non-payment of scheduled instalments) of export credit-related debt then converts the intended concessionality of such credits into a significant financial burden. A comparison between the LIBOR and the 'Consensus' interest rates mentioned above implies a staggering burden on Iraq for its debt to the OECD countries.

Some of the project-related credit agreements contain a non-automatic acceleration clause as an option for Iraq to make total, whole credit or partial-instalment prepayments in order to avoid high interest rates, or when economic conditions and oil revenues permit it to do so. Needless to say, Iraq will not be able to use such an option for the foreseeable future. However, restructuring this debt on the basis of LIBOR should be pursued by Iraq using two time references: the first and most desirable is from 1990 onwards in order to avoid accumulation of the 'Consensus' interest, whether capitalised or not; then the second will at least be from the commencement of negotiations on structuring such debt when it happens in the future.

Debt rescheduling: bilateral or multilateral approaches

Debt negotiations during the 1980s were concluded on bilateral levels through government-to-government or state organisations-to-company approaches, and this was the option followed for deferment of dues, for rescheduling of debt and for fresh credit. The other option was the multilateral level. Multilateral debt negotiation, which was then something new for the Iraqis, required Iraq to negotiate on at least 'two fronts', comprising all the 'official' creditors via the Paris Club, and, possibly, all the 'commercial' creditors via the London Club as well.[62]

Obviously Iraq did not pursue the multilateral-debt-negotiation approach during the 1980s, partly for the following reasons.

One of the fundamental principles for rescheduling through the Paris Club is conditionality. This implies that as a precondition to the Paris Club negotiations, the debtor country should conclude a standby arrangement with the International Monetary Fund, the main purpose of which is to look for the causes of the country's debt problem and specify the measures needed to eliminate these causes. Since Iraq's financial problems were the direct result of a war situation, the causes of the problem in hand were therefore known, and the IMF was unable to eliminate them. In addition, the Fund is probably not able, on legal grounds to conclude a standby arrangement with a country engaged in such a large-scale and prolonged war.

There was no genuine pressure on Iraq by any of its official creditors to go for the Paris Club option. Probably many of them had discovered that it was in their interest if Iraq was kept away from the Paris Club, since greater advantage accrued to them through bilateral arrangements than through the multilateral approach of the Club. Nor was the international banking community keen to see the Paris Club forcing the rescheduling of Iraqi debts, even after the Iran–Iraq war.[63]

The Paris Club itself seemed to have been unprepared for the debt crisis of the 1980s, and its response may have been rather slow. The numbers of agreements and the amounts rescheduled by the Club both indicate that it might have been overwhelmed by what needed to be done. From 1980 to 1987 the numbers of agreements concluded during the relevant consecutive years were 2, 7, 6, 15, 12, 17, 15 and 17 respectively. The corresponding rescheduled amounts were US$72 million, US$1,074m, US$662m, US$7,823m, US$2,813m, US$5,340m, US$13,122m, and US$24,705m.[64] This rapid increase in Paris Club commitment could also have been one of the reasons why Iraq was not pressurised by its creditors to go to the Paris Club.

Is it possible, and preferable, for Iraq to pursue bilateral negotiations, as it did in the past, or should it choose the multilateral approach? The choice of either option can be considered in the light of the following factors: (a) the principle of precedence or the 'bandwagon factor' (that is, the provisions of the agreements concluded before the Gulf War and whether they work for the benefit of Iraq); (b) the new post-sanctions economic and business opportunities which could be made available to the international companies from the major creditor countries, thus influencing the preference of the major creditors for either option; and (c) the availability of a sufficient number of qualified Iraqi specialists and negotiators in debt and financial credit arrangements.

In view of the above, the writer inclines towards multilateral debt negotiation for the following reasons:

(a)   The scale and complexity of the debt problem require an 'exit strategy', with a relatively long repayment period and suitable grace period, which can only be achieved through multilateral negotiations. Moreover, even if sanctions were lifted today, the country's economic situation cannot cope with either quick repayment or bunching of payments.

(b)   Although a multilateral approach entails concluding some arrangements with the IMF, which in turn imposes some tough economic policies under what is known as conditionality, no matter how harsh and difficult they may be, the impact of these policies on the economy and people of Iraq would be easier to bear than the prevailing conditions or the conditions of quick repayment of all debt. Furthermore, the latest forms of conditionality imposed by the international financial institutions (IFIs)[65] – including good governance, accountability, transparency, and the rule of law – would impose economic discipline and maintain a close scrutiny of Iraq's economic policy and help the country to design a recovery programme.

(c)   The country had already undertaken some degree of economic liberalisation and structural adjustments, especially privatisation, which would make it easier to digest some of the difficulties usually associated with these programmes.

(d)   The bilateral approach, given Iraq's lack of specialised professionals and debt expertise, would not generate sufficient and concerted efforts to address a problem of such magnitude in a convenient time frame.

(e)   The recent development in the relationship between Russia and the G7 might motivate Russia to pursue the Paris Club option in order to collect her debts from Third World countries, including Iraq (see editor's note 57 below).

Having said this, the writer has reason to believe that Iraq's major official creditors – France, Russia, China, Italy – prefer the bilateral approach: first because most of their debt is military-related and they may prefer not to reveal the terms of such deals; and second because they are already in a process of business-related rapprochement with Iraq, and debt matters may already have been discussed. Other major creditors from the developing countries, such as Turkey, Brazil, India, and South Korea, are even less keen to use the multilateral forum and prefer to discuss their credits with Iraq individually.

Debt-'X' swap

Here 'X' could be an economic, a political or a geopolitical variable.

For 'X' as an economic variable, one thinks of operations similar to debt–equity swaps, where some of Iraq's debts could be converted into equity in domestic enterprises. Debt–equity swap is possible, given the economic liberalisation and privatisation that took place prior to the Gulf War. The oil sector might stand as the most attractive option for foreign investors, especially from Europe, Japan, South Korea, China and Russia.

There is reasonable international experience with this type of operation, and enough information is available on the legal, procedural, and operational aspects for it to be accomplished.

Theoretically, Iraq can pursue this sort of operation in order to reduce its commercial or even official debt. In reality, however, there is a long way to go before such operations can be performed. First, foreign ownership of national assets is still a very sensitive issue, not only in Iraq but in most of the Arab countries, as the recent experience of privatisation has demonstrated. Second, the mere availability of a profitable business opportunity is not enough to attract foreign investors. Other conditions related to guarantees, transferability and liquidity are also important but not yet developed in Iraq. Finally, regionally speaking, no debt–equity conversions seem to have taken place in the Middle East and North Africa region during the period from 1984 to 1992.[66]

Having said this, Iraqi commercial debt, which could be available at a substantial discount in the secondary market, represents a worthwhile investment opportunity for Arab private investors, since the legal frame-work and other incentives for this group of investors have been available since the mid-1980s.

'X' as a political variable might include many important, desirable and urgently needed changes in the country's political structure. Debt could be swapped with democratic and internationally supervised elections, and many laws, decrees, and directives issued by the Revolutionary Command Council in contravention of the principles of justice could be abolished. There could be full compliance and institutional respect for human rights; legitimisation of political opposition; and legitimisation of various forms of truly independent NGOs and the civil society. In addition there could be an effective end to political monopoly by the Ba'th party and any one political party thereafter, and the army, police, and all internal security forces could be neutralised and constitutionally prevented from indulging in any political activities and parties. There would be guarantees for minority rights, and so on.

It would undoubtedly be in Iraq's long-term interests and future that debt swaps were made with as many of the above-mentioned political variables in mind as possible. However, the possibility is remote and it is unlikely that any of Iraq's major creditors (or for that matter those of any developing country) would be willing to pursue this course of swapping. First, unlike a debt–equity swap, there is no developed mechanism or indeed any real experience of this type of swapping. Second, apart from rhetoric and occasional enthusiasm, major official creditors pay little genuine attention to these political variables, and if they do they only use them as a component of their foreign policy. As for major commercial creditors, these matters have no place on their balance sheets. Third, other 'non-assets' swapping, such as the innovative debt-for-environment and debt-for-development swapping operations, have been very limited in magnitude, confined to NGOs initiatives and concentrated in a few countries.[67]

Another possible way to reduce debt is through conversion. This option could be pursued with the international companies that will eventually return to resume their work in the country. They need local currency to cover local expenses, and for these reasons part of their contractual payments are made in local currency. If they need more local currency than their contracts specify, they then have to borrow from the local banks. Instead of borrowing, therefore, they could buy what they needed through converting some of their or their country's old debt on Iraq. Unlike debt–equity swap, Iraq has had some experience with debt conversion. In 1986 Iraq and Italy agreed to convert US$10 million of Iraqi debt into local currency and the amount was used exclusively by Italian companies to cover their needs within Iraq.

Debt buy-back

Secondary market trading in commercial bank debts to developing countries has been growing rapidly since the mid-1980s. Commercial debt instruments are traded at their discounted face value, which depends on many factors, some related to the holder of the instruments and others to the debtor country itself. When the debtor has high risk and low creditworthiness, the instruments are usually offered at a significant discount.

A debtor country can reduce its total debt through buying back some of its commercial debt from this secondary market. Buy-back operations can be carried out independently if the debtor country is legally and contractually able to do so, or through a recognised mechanism such as Debt and Debt-Service Reduction (DDSR) programmes under the well-known Brady Initiative.

It is not possible for Iraq to benefit from these DDSR programmes unless it agrees at the outset to reschedule its official debt via the Paris Club, and unless its major commercial creditors agree to pursue the DDSR option. In other words, official debt rescheduling is a necessary but not a sufficient condition for DDSR.[68] Furthermore, the Brady Initiative, which was introduced in 1989, is not a permanent one and by the time Iraq returns to normality such a facility may not be available.

At the end of 1990 the stock of Iraq's commercial debt, both long and short term, amounted to US$10.7 billion. I am of the opinion that these debt instruments had been circulating in the secondary market at a substantial discount, and the Economist Intelligence Unit gave Iraq the absolute maximum level of country-risk rating of 100 during the last quarter of 1994.[69]

What Iraq *could* do is independently buy back those commercial debt instruments the loan agreements for which do not contain clauses preventing the issuers from buying them back. To my knowledge, none of the loan agreements contains such clauses. However, it is not always advisable for the borrowing state entity to buy back its debt.

A successful and meaningful buy-back requires a discreet, quick, substantive and well-organised operation, since the secondary market is very sensitive and the discount margin could easily be wiped out. Iraq has an additional advantage in this regard in that the borrowing state entities and their guarantor, usually the Rafidain Bank, know the whereabouts of the commercial papers, since the transfers of title for these instruments were subject to the approval of the borrowing entities and the guarantor. An operation of this type requires international financial expertise, which Iraq had in the past but which is undoubtedly lacking now.

Coercive debt repayment

By definition, this option implies a superimposed verdict with which Iraq could be forced to comply. This coercive action could be formalised either by the UN Security Council acting along the lines of UNSC resolution 687; or by the OECD countries taking a unified and united stand to impose a financial commercial embargo unless Iraq agreed to a dictated option by the OECD for debt repayment to the countries concerned.

A new UNSC resolution designed specifically to address Iraq's external debt is rather unlikely, and not only because all UN resolutions related to Iraq have been connected to the invasion of Kuwait. Any such new resolution will set a precedent by which the international organisation will have stepped into the forum of other venues specialised in debt matters, such as the Paris Club and the London Club. Furthermore, the demonstrated keenness of France, Russia and China for future business with Iraq does shed some doubt on the existence of a 'debt coalition' among the permanent members of the UNSC against Iraq. However, some British official sources have expressed views on the possibility of including Iraqi debt in the war-reparation clauses of the above-mentioned resolution.[70]

OECD countries might resort to exercising some pressure on Iraq through their Export Credit Group within the Berne Union. All OECD export credit and guarantees agencies were hit hard by the many recourse and compensation claims from companies and firms operating businesses in Iraq, and are therefore unlikely to be soft on Iraq. However, apart from exercising their rights not to make any new credit facilities or export guarantees, they are not empowered to impose coercive unilateral solutions. Even so, the potential for business opportunities in post-sanction Iraq might sweeten the position of such agencies against any coercive action.

The third coercive possibility is a unilateral decision by a creditor country. So far, only the USA and the UK can be identified as willing to take such an action, which would be for political considerations only, since Iraq owes both countries much less than other major creditors such as France, Russia, Italy, and so on.

Out of the three coercive possibilities, the UN forum might be used, not to impose debt repayment through war reparations but as a pretext for prolonging the sanctions.

Debt to Arab countries

To begin with it is important to differentiate between the following three categories: (1) funds from the GCC, mainly Kuwait, Saudi Arabia and the UAE; (2) debt to non-GCC Arab countries; and (3) debt to Arab and regional financial and multilateral institutions such as the Arab Monetary Fund (AMF), APICORP, Arab Fund for Economic and Social Development (AFESD), Islamic Development Bank (IDB), and so on. The bulk of Iraq's debt to Arab countries belongs to the first category, since the debt to the second category was usually paid (at least in part) at regular intervals, which means the stock value is not great. This also applies to the debt to Arab financial institutions.

Concerning the category of funds from the GCC, there seems to be no agreement on their exact amount. Philip Alger estimated the debt to the 'major Gulf states' alone at around US$40 billion.[71] In November 1988, a report based on information prepared by the American Embassy in Baghdad stated that Iraq had received 'between $30 and $50 billion from friendly Arab countries'.[72] According to Abdulatif Al-Hamad, 'the debt in-kind to Arab countries [is] estimated to be around $50 billion'.[73] Bradsher put the amount at US$40 billion.[74] Another estimate put the figure of US$50 to US$60 billion for Kuwait and Saudi Arabia alone.[75] Finally, Iraqi official sources put the figure of US$30 billion.[76] This is a very contentious issue. Iraq considers these funds as 'grants' while the countries concerned class them as debts.[77]

The following elements are, in very general terms, what constitute debt from the legal and operational perspective: (a) the legal instrument or documentation (i.e., commercial papers, agreements, protocols, etc.); (b) interest rates; (c) actual dates and amounts of disbursements; and (d) number and type of instalments, grace period and maximum repayments period.

The legal instruments or documentation specify clearly, among other things, that one party is the creditor while the other is the debtor, and that the amount involved is a debt or an obligation on the part of the debtor. However, it is not always necessary that debt be established on legal and documented evidences alone, since oral, non-documented and mutual consent may constitute forceful and solid foundations for proof of debt, especially in those societies where the moral code of conduct or customary law affirm such practices. The basic point here is whether under constitutional or customary law the amount involved should be considered and seen by both parties as loan debt.

Debt can carry a specified rate of interest or be interest free: the first is called *active debt* while the latter is *passive debt*. Again, the distinction between the two should be made by agreement between the debtor and the creditor.

Credit and/or loan should not be considered as debt unless and until they are disbursed or utilised. The actual disbursement of credit or utilisation of a loan establishes the beginning of debt occurrence and interest accumulation if the debt is of an active nature.

As to repayment structure, we note that debt has to be repaid in a specified period and in a certain manner: the number of periodic instalments, the amount of each instalment, the grace period, and the final maturity of the entire loan.

Of these four basic elements, the repayment structure of debt stands as the most important. If debt repayment is not established in all practical meanings, then the debt is non-repayable and this will be further enforced if the debt is of a passive nature. Sometimes the occurrence or the starting point of repayment is subject not to a specific time frame but to conditions or some other events. These may include phrases that are general and very widely used in Arab and Muslim societies, such as 'Allah karim' (God is generous), or 'insh'allah' (God willing), which imply an improvement in the debtor's ability to honour his obligation towards the creditor. When this improvement takes place, then debt repayment depends largely on the goodwill and the moral conduct of the debtor, the relationship between the two parties, and above all the prevailing economic conditions of the debtor.

If the above four elements can be proved on any amount, then it would be unreasonable or indeed impossible for Iraq to consider them as grants. Furthermore, the Iraqi claim is inconsistent: if these funds were grants then why did Saddam Hussain call them debts and request their cancellation?

One may also look at this issue from other angles. There were many instances when Arab countries, individually or collectively, extended financial assistance, on grounds of political and war conditions, to other Arab countries and none of this funding was seen, or considered, as debt. Furthermore, there was a widespread belief that these amounts were unlikely to be paid back. It has been reported, based on diplomatic sources, that 'Saudi Arabia no longer even bothered to keep the loans on its books'.[78] This means that one needs to differentiate between Saudi Arabia and Kuwait in this regard. Finally, some Arab governments have bilaterally cancelled large amounts of the debt and arrears of many developing countries, especially sub-Saharan Africa, which are facing dire economic difficulties. A fellow Arab country which has suffered unprecedented destruction is also entitled to similar acts of debt forgiveness and cancellation.

Regardless of their magnitude, however, these must be looked at as 'special case' obligations that are distinctly separate and treated differently from other debts. In this respect, one must bear in mind (i) the possibility of writing these amounts off, by renaming them for example, as grants, favours, assistance, solidarity funding, or war relief; and (ii) the fact that if they are to be paid sometime in the future, it is unlikely and absolutely unreasonable that repayment can be expected in the short to medium terms. The occurrence of these funds and their unknown maturities, if any, indicates that they have a very long-term horizon, if indeed they will ever be paid at all. In other words, the present value of their future repayment should be as low as possible.

Debt management

As already noted, most Iraqi expertise in debt management, though modest, had begun to disintegrate during the second half of 1987. Since then a qualitative decline in debt management has taken place, which has been further worsened by the prolonged sanctions and the decline in international contacts, and the consequent exclusion from the continued changes and developments in export credit systems, international financial markets and debt negotiations.

Iraq needs urgently to rebuild its debt-management system on sound and functional foundations, and the sooner the better. External technical assistance in this field is available from many national and international sources. Two examples of such facilities include, first, the

Debt Management and Financial Analysis System (DMFAS) software package, which is the main component of the technical cooperation provided by UNCTAD in the area of debt management; and, second, a programme initiated by the Canadian International Development Research Centre (IDRC). DMFAS is designed to strengthen the technical ability of the debtor country to record and monitor its external debt; to integrate debt management into the management of public sector finance; to related scheduled debt service payments to a comprehensive projections system for the external sector; and to improve the capacity to select appropriate external borrowing strategies.[79] The aim of the IDRC programme is to assist developing countries in better managing their external debt, and, in collaboration with the Commonwealth Secretariat, it makes available the Debt Recording and Management System (CS-DRMS). The programme is designed to improve legal and institutional frameworks governing debt management, and its implementation enables a developing country to access relevant information quickly and to formulate an effective debt-management strategy.[80]

5. Concluding remarks

Preliminary signals of debt stress in Iraq began to be observed during the second half of the 1980s, as evidenced by the debt-service-to-export ratio and an increasing weight of short-term debt. Even so, other debt indicators and ratios, net flows, and net transfers indicated that, bearing in mind the country's economic potential, this debt was a reasonably manageable short- to medium-term liquidity problem. However, the invasion of Kuwait, the subsequent destruction, and the ensuing severe and prolonged sanctions deepened the debt problem, increased its severity and converted it from liquidity to absolute unsustainability.

There will be no easy solution to Iraq's debt problem, and indeed the problem will be even further aggravated and will rapidly become increasingly unsustainable as long as the sanctions remain and the country is prevented from returning to normality and from resuming its oil exports. When that will be and under what conditions is an open question of the 'real politik' type, which is beyond the scope of this chapter. Iraq's options for addressing such problems are limited, and will remain so as long as the political stalemate between Iraq and the major international powers persist.

Debt rescheduling at either bilateral or, preferably, at multilateral levels is inevitable, and such action must take into consideration the financial requirements for reconstruction on the one hand, and the inflow of financial resources, export earnings, Official Development Assistance, and credit facilities on the other. If a multilateral approach can be pursued, this might eventually open the way for some degree of debt relief in the future.

APPENDIX 1

Estimation of disbursement, net flows
and net transfers, 1981–90

OECD data does not include either net flows, net transfers or disbursements, but it does include total debt stock, stock of long-term debt, amortisation of long-term debt, and interest payments on both short- and long-term debts.

However, by using OECD data and with the help of the following relations I will estimate the net flows, net transfers and disbursements for Iraq during 1981–90.

$$S_t = S_{t-1} + D_t - R_t$$
$$D_t = S_t - S_{t-1} + R_t$$
$$NF_t = S_t - S_{t-1} = D_t - R_t$$
$$NT_t = D_t - R_t - I_t = NF_t - I_t$$

where:

S is the stock of long-term debt, a given variable,

R is the repayment of long-term debt, a given variable,

I is the sum of interest payments for both short- and long-term debts, a given variable,

D is the disbursements of new money, variable to be estimated,

NF is the net flows, variable to be estimated,

NT is the net transfer, variable to be estimated, and

t is the time in years.

APPENDIX 2

1. The present value (end of 1997) of principal debt of US$42.097 (at the end of 1991) for each scenario is arrived at by multiplying the principal debt with factor (F) related to that scenario during the period (n) which is seven years and interest rate (r), according to the following relations:

    Scenario zero: $F = (1 + r)^n$ $r = 0$
    Scenario one: $F = 1 + (r*n)$ $r = 8$ per cent
    Scenario two: $F = (1 + r)^n$ $r = 8$ per cent

2. Annual Debt Service (ADS) is an equal amount payable annually during the entire chosen period ($n = 10, 15, 20$ and 25 years). It takes into consideration and includes amortisation of the principal debt (in our case the present value of debt according to the three scenarios) and interest payment ($r = 8$ per cent per annum), and is arrived at by multiplying the present value of debt by an annuity or Debt Recovery Factor (DRF), calculated according to the following formula:

    $$\text{DRF} = r (1 + r)^n / ((1 + r)^n - 1).$$

3. Total debt obligation for each repayment period is arrived at by multiplying ADS with the total number of years for that specific repayment period.
4. Sustainability requirement, for each scenario and under each repayment period, is defined here as the export level (E) which should be attained in order to satisfy ADS payments at each sustainability threshold (ST = 15 per cent, 20 per cent, and 25 per cent), and calculated by the following formula: $E = \text{ADS}/\text{ST}$.
5. (DS/E) ratios, for each scenario and under each repayment period, were calculated by dividing ADS over (a) the average of actual exports earnings (US$9.8 billion) during 1981–90; and (b) the maximum actual export earnings (US$12.3 billion) during 1981–90.

TABLE 2.2
Profile of the Iraqi debt ($ million)

	1980	1981	1982	1983	1984	1985	1986	1987	1988	1989	1990
Long-term debt											
I. OECD	2,041	2,651	1,984	2,906	3,952	5,594	6,834	8,769	8,082	9,410	7,421
ODA	181	166	146	158	149	191	255	408	388	324	320
Official Credit	1,720	2,420	1,838	1,975	2,781	5,185	5,716	8,143	5,546	6,182	4,909
Financial markets	120	50	–	773	1,021	218	863	218	2,148	2,904	2,192
II. Multilateral	83	15	79	271	312	313	374	368	356	362	368
III. Non-OECD	367	313	1,502	1,872	2,274	2,603	4,187	5,460	6,457	6,493	6,513
Subtotal: LT	2,491	3,043	3,565	5,050	6,538	8,510	11,394	14,597	14,895	16,265	14,301
Debt											
Concessional	557	487	146	163	2,433	2,807	4,463	5,891	6,864	6,842	6,871
Non-concessional	1,934	2,556	3,419	4,887	4,105	5,703	6,931	8,706	8,030	9,423	7,430
Short-term			2,406	2,169	2,512	4,329	5,606	5,810	5,214	6,512	8,544
Total debt	2,491	3,043	5,971	7,219	9,050	12,839	17,000	20,407	20,109	22,777	22,846
Debt service											
Long-term											
I. OECD	1,010	1,742	2,005	1,488	1,105	1,693	3,212	2,440	1,568	3,153	781
II. Multilateral	11	10	10	–	10	83	41	33	181	28	31
III. Non-OECD	95	87	10	10	9	90	55	25	26	26	68
Subtotal: LTD	1,116	1,839	2,025	1,498	1,124	1,867	3,309	2,498	1,774	3,207	880
Amortisation			1,923	1,399	965	1,499	2,524	1,940	1,235	2,502	601
Interest			102	99	159	367	785	559	539	705	279
Interest, STD			283	228	243	277	317	376	403	445	468
Total debt service	1,116	1,839	2,308	1,726	1,368	2,143	3,625	2,874	2,177	3,652	1,347

Source: OECD, *Financing and External Debt of Developing Countries*, Paris, various annual surveys.

NOTES

1 Hassanali Mehran (ed.), *External Debt Management*, Washington DC, IMF, 1985, pp. 10–20.
2 Ahmed M. Jiyad, *The Impact of Oil on the Iraqi Economy and Economic Relations*, research paper for the Fletcher School of Law and Diplomacy Tufts University, Massachusetts, USA, during author's Humphrey–Fletcher Fellowship, 1989.
3 The cumulative commitments of concessional financing to the end of 1991 were $845 million and the corresponding disbursements were $338 million. The OPEC Fund, *OPEC Aid and OPEC Aid Institutions: A Profile*, Vienna, 1992, p. 28.
4 Ahmed M. Jiyad, *Iraq's Foreign Investments*, Council of Ministers – External Economic Relations Committee (EERC), Baghdad, 1982 (in Arabic).
5 John Townsend, as cited by Abbas Alnasrawi, 'Economic consequences of the Iraq–Iran war', *Third World Quarterly*, vol. 8, no. 3, July 1986, pp. 869–95.
6 IMF, *Direction of Trade (DOT)*, 1981, pp. 211–12.
7 Data on military imports are from: US Arms Control and Disarmament Agency (ACDA), *World Military Expenditure and Arms Transfer 1986*, Washington DC, 1987, Table II, p. 105.
8 Oil had been used intensively with all of Iraq's trade partners, such as the USSR, Yugoslavia, Turkey, India, Brazil, France, Italy, Germany and others. The objective was to maximise the utilisation of export capacity after the expansion of the pipeline via Turkey and the completion of IPSA-1 and IPSA-2.
9 APICORP is one of the joint ventures established by OAPEC to foster cooperation among member countries in various aspects of the oil industry.
10 Frederick W. Axelgard, 'The United States–Iraqi Rapprochement', in Z. Michael Szaz (ed.), *Sources of Domestic and Foreign Policy in Iraq*, American Foreign Policy Institute, Washington DC, 1986, pp. 43–57.
11 Economist Intelligence Unit (EIU), *Iraq – Country Report No. 1*, London, 1989, p. 15.
12 Ahmed M. Jiyad, *The Impact of Oil on the Iraqi Economy*, op. cit., Table (3–7), p. 94.
13 *MEED*, 24 March 1989, p. 20.
14 This committee was created to replace the Follow-up Committee for Oil Affairs and Agreements Implementation (known as the 'Follow-up Committee'), an organ of the RCC, and was chaired by Saddam Hussein until mid-1979.
15 The G5 Plaza Accord was concluded between the governments of the USA, UK, Japan, West Germany and France. The aim was to reduce the value of the US dollar in order to improve the US trade balance.
16 Daily oil production figures are from the US Department of Energy, *Monthly Energy Review*, November 1988, Washington DC, Table 10.1b, p. 113. Oil export earnings were calculated by deducting non-oil exports from total exports. Non-oil exports data are from Ministry of Planning, Central Statistical Office, *Annual Abstract of Statistics*, 1986 and 1987, Baghdad, Iraq, Tables 8/1, p. 156. Figures on total exports are from IMF, *Directory of Trade*, op. cit.
17 The Berne Union (International Union of Credit and Investment Insurers) is an association founded in 1934 and is composed of a number of export credit insurance agencies from developed and developing countries. Aside from

coordinating export credit terms, the OECD Export Credit Group (ECG) within the Berne Union serves as a forum for exchange of information on debtor country situations and draws policies or takes positions towards that country. Representation in the Berne Union is at agency level while the OECD-ECG is on the official governmental level.

18 This opinion was reached and formulated not only while I was in Baghdad where, through my negotiations and engagements in debt issues between July 1987 and July 1988, I could observe the situation for myself, but also after meeting a number of Iraqi officials from ministries and other bodies who visited the USA during 1988–9. Furthermore, in a meeting of the Council of Ministers, Saddam Hussain accused the minister of ignorance and inefficiency. *Al-Iraq* Newspaper No. 4235, 12 December 1989. Incidentally, the same minister was at the 'Economic Bureau' of Saddam's Presidency Office before becoming a minister in the second half of 1987.

19 *MEED*, 23 June 1989, p. 15.

20 The composite risk rating is calculated on the basis of political, financial and economic risk. The highest overall rating – theoretically 100 – indicates the lowest risk, and the lowest score – theoretically 0 – indicates the highest risk. *International Country Risk Guide*, April, 1989, pp. S3–S5 (International Reports, a Division of IBC, USA).

21 Marian Houk, 'Iraq sees prospects brighten for financing reconstruction', *The Christian Science Monitor*, 17 April 1989.

22 *MEED*, 14 July 1989, p. 14.

23 The announcement was made in an interview on Baghdad television with Dr Mohammad Mehdi Salih, the Minister of Trade, on 4 March 1989, as reported in *MEED*, 17 March 1989, p. 20.

24 The Financial Protocols were annual financial facilities provided by the UK's Export Credit Guarantee Department (ECGD). Between 1983 and 1988 these protocols provided more than $1,700 million of credit to Iraq. See *MEED*, 24 March 1989, p. 20, and *MidEast Markets* (MEM), 29 May 1989, p. 16:11/11.

25 In addition to participating in most of these activities, I prepared two documents on the subject for circulation to Iraqi borrowing agencies to inform them on various aspects of these facilities. The two documents are: Ahmed M. Jiyad, 'Study of the OECD "Consensus" Agreement', External Economic Relation Committee, Council of Ministers, Baghdad, Iraq, July 1984 (in Arabic); and Ahmed M. Jiyad, 'Export Credit Financing', External Economic Relation Committee, Council of Ministers, Baghdad, Iraq, March 1985 (in Arabic).

26 Calculated from Statistical Office of the European Community (Eurostat), *External Trade, Monthly Statistics*, 2/1989, Table 10, pp. 134–73.

27 US Department of Commerce, Washington DC, as cited by Jonathan Crusoe, 'Iraq's flourishing US connection', in *MEED*, 25 August 1989, pp. 4–5.

28 Based on data from the Central Statistical Organisation (CSO), *Annual Abstract of Statistics*, Ministry of Planning, Baghdad, Iraq, 1986, Table 8/3, pp. 158–63; and 1989, Table 8/2, pp. 214–19.

29 There were many instances of European suppliers taking advantage of the terms agreed upon with Turkey and arranging their deliveries via Turkish intermediaries or, in the case of projects or service contracts, by entering into a joint venture with Turkish firms and companies.

30 When interest is paid annually, then nominal interest is equal to the effective interest and the borrower incurs no additional cost, but when the interest payment is made for less than a year's duration (such as semi-annually, quarterly, or monthly), then effective interest is higher than the nominal interest rate and the borrower incurs additional cost. The shorter the period for interest payments (e.g., monthly), the higher the effective rate and thus the additional cost.

31 A company which has a good profit margin on its project might shoulder a significant proportion of the spread in order to finalise the credit agreement as soon as possible. Some agreements even carry negative LIBOR, such as that concluded with one of the companies that implemented some of the projects with the Baghdad municipality, where the interest was set at a rate of 'LIBOR minus 2'.

32 AMF, *Annual Report, 1995*.

33 AFESD, *Annual Report, 1991*.

34 Variations and changes in debt statistics can be attributed to and explained by three major groups of factors or variables: *Direct factors* are those which have a real and immediate impact, and which include fresh credit, deferment of contractual dues in foreign currencies, amortisation, capitalisation of interest, and, finally, rescheduling. *Conversional factors* are related to debts denominated in currencies other than the US dollar, and involve the exchange-rate effect of outstanding debt stocks on the proportional composition of debt-currency structure on the one hand, and exchange rate position of each currency, in the debt-currency matrix, against the US dollar on the other, at the time of evaluating the outstanding stocks of debt. However, the exchange-rate effect materialises only at the date of actual payment of instalments, or when the related agreement contains some sort of indexation clause or formula which takes into account exchange-rate variations during a certain period of time before due payment. *Transactional factors* are those related to the status of debt, its 'owner', or its claim holder, such as debt-swaps and secondary markets operations.

35 Higher short-term debt obligations reduce a country's ability to service its long-term obligations, thus causing decline in the latter and increase in the former, and this leads to further deepening of the liquidity problem. A study suggests that countries with acute debt-servicing problems during 1981–2 had experienced an unproportional evolution of short-term debt in relation to available foreign reserves. See Donald Donovan, 'Nature and origin of debt-servicing difficulties: some empirical evidence', *Finance & Development*, (IMF/World Bank), Vol. 21, December 1984, pp. 22–5.

36 Using a discount rate of 10 per cent for ODA loans and the CIRR (Commercial Interest Reference Rate) for the Consensus loans, UNCTAD, *Trade Development Report, 1996*, p. 56.

37 Iraq concluded 10 project loans with AFESD between 1980 and 1990 at a total amount of KD59.725 million and disbursements of KD18.575 million (AFESD, *Annual Report 1991*, Kuwait). The outstanding balance of Iraq's debt to the IDB amounted to IsD113.4 million (IDB, *Annual Report 1412H / 1991–1992*, Jeddah, pp. 164, 174). Finally, the balance of outstanding Iraqi debt to AMF was AAD49.850 million (AMF, Annual Report 1995, Abu Dhabi).

38 A borrowing country in fact incurred additional costs other than interest rate. Some of these additional costs, such as management fees, are payable once and

are called up-front costs, while others are proportional, such as commitment fees levied on the remaining balance of the credit.

39 The figures for amortisation and interest could be different, since the data for 1980 and 1981 were aggregate of debt service of long-term debt only.

40 World Bank, *World Development Report 1989*, New York: Oxford University Press, 1989, p. 18.

41 World Bank, *World Debt Tables, 1994–95: External Finance for Developing Countries*, Washington DC, The World Bank, Vol. 1, p. 49.

42 Luisa E. Sabater, 'Multilateral Debt of Least Developed Countries', UNCTAD Discussion Paper No. 107, November 1995, p. 9.

43 The discount rates used by the World Bank to calculate the present value of future debt are many: the interest rates charged by OECD countries for officially-supported export credits; the most recent lending rates for the non-concessional multilateral loans; and average interest rates for export credit donated in currencies other than the French franc, German mark, Italian lira, Japanese yen, British pound, and US dollar. See World Bank, *World Debt Tables, 1994–95*, Washington DC, The World Bank, Vol. 1, pp. 49–50.

44 *The Economist*, 6 May 1995, London, pp. 83–4.

45 UNCTAD, *Trade Development Report, 1996*, p. 51.

46 It has been reported that, by the end of 1988, companies from various countries were working on projects with a total value of approximately $12 billion; *Middle East Executive Report*, May 1989, p. 16.

47 The value of military equipment alone, imported and paid for, was $102 billion. See Appendix I, 'Note from the Iraqi Minister of Foreign Affairs, Mr Tariq Aziz, to the Secretary General of the Arab League, 16 July 1990', in Pierre Salinger and Eric Laurent, *Secret Dossier: The Hidden Agenda Behind the Gulf War*, London: Penguin Books, 1991, p. 233.

48 This was the conclusion that emerged from a conference held in London in February 1989 on 'Ceasefire in the Gulf: Economic and Financial Prospects'. A summary of the papers appeared in the *Arab Banker*, 1989, pp. 8–13. See also *MEED*, 3 March 1989, p. 8.

49 I am grateful to *Middle East Economic Survey* for providing me with the text of the memo which was published in *MEES*, 13 May 1991, pp. D6–D9.

50 This extremely high figure reflects not only a bunching phenomenon but also indicates substantial arrears, in both value and occurrence, prior to 1990. On the basis of the previous trend, I would also expect that a significant portion of this figure is a short-term debt not paid during 1990 and 1991.

51 For discussion of sustainability indicators and other debt ratios, see Matthew Martin, 'A Multilateral Debt Facility: Global and National', in UNCTAD, *International Monetary and Financial Issues for the 1990s*, Vol. VIII, UNCTAD, 1997, pp. 147–76. See also Luisa E. Sabater, Multilateral Debt of Least Developed Countries, UNCTAD Discussion Paper No. 107, November 1995; UNCTAD, *Trade Development Report, 1996*, p. 51; and J.C. Berthelemy and A. Vourc'h, 'Debt Relief for Africa: From Criteria to Action', in EURODAD, *World Credit Tables: Creditor–Debtor Relations from Another Perspective*, 1994–95 edition, Brussels, EURODAD Publications, 1995, pp. 60–7.

52 The present value of debt, used by the IMF, the World Bank and UNCTAD, is the total of future debt service (including principal debt and accrued interest) adjusted

for concessionality of the debt stock. Since Iraqi debt has no concessionality and since no debt has been repaid since the end of 1990, the present value of Iraqi debt is compounded since that date.

53　See Arab Monetary Fund et al., *The Unified Arab Economic Report, 1996*, p. 142.

54　See EURODAD, *World Credit Tables*, op. cit: 1995, Table 3, p. 27.

55　Iraq has no arrangement with the World Bank to report its debt and to be included in the Bank's database.

56　See World Bank, *World Debt Tables, 1994–95, External Financing for Developing Countries*, Vol. 1, 1994, Table A1.2, p. 51.

57　Yemen, Jordan and other Arab countries also had arrangements for debt reduction, but these were made later in 1997 under different, also political factors. Egypt was also different because of the magnitude of debt forgiven, the proportion of the military debt, and its classification by the Paris Club as a middle-income country. This was not the case with Yemen, where in late 1997 the government reached agreement at the Paris Club for 80 per cent relief of the US$6.4 billion debt owed to Russia, with the remainder put for rescheduling on Naples terms (see Mutahar Al-Abbasi, 'The Education Sector in Yemen' in *Yemen into the Twenty-first Century: Continuity and Change*, eds. K. Mahdi and Anna Wuerth, forthcoming, Ithaca Press, Reading. [Editor's note]

58　For the provision and evaluation of these measures, see Percy Mistry and Stephany Griffith-Jones, *Conversion of Official Bilateral Debt*, UNCTAD/GID/DF/1, December 1992; Francesco Abbate, 'The Paris Club: a group of debt collectors or development-oriented institution?', *UNCTAD Bulletin*, No. 19, March–April 1993, pp. 4–6; and UNCTAD, *Trade Development Report 1996*, pp. 38–9. The 1991 Paris Club agreement for Egypt lowered the present value of debt/export ratio from an initial rate of 350 to 150 per cent. See Berthelemy and Vourc'h, op. cit., p. 62.

59　The inclusion of Russia in the G7 meetings in Denver, Colorado in June 1997, and the possibility of that country joining the Paris Club would further complicate Iraq's debt problem if it decides to seek debt rescheduling through the Club in the future. The magnitude of the Iraqi–Russian debt, its composition, currency structures and indexation, plus the provisions of the related agreements, make it a special and a complex relationship.

60　UNCTAD, *Trade Development Report, 1995*, Geneva, 1995, Table 6, p. 23, and UNCTAD, *Trade Development Report, 1996*, Geneva, 1996, Table 7, p. 29.

61　World Bank, *Global Economic Prospects and the Developing Countries, 1994*, Washington DC, World Bank, cited by EURODAD in *World Credit Tables*, op. cit., p. 44.

62　For brief information on the Paris Club see Peter M. Keller and Nissanke E. Weerasinghe, *Multilateral Official Debt Rescheduling – Recent Experience*, Washington DC, IMF, May 1998. For more detailed information on the Paris Club and the London Club see Alexis Rieffel, *The Role of the Paris Club in Managing Debt Problems*, Essays in International Finance, No. 161, December 1985, Department of Economics, Princeton University, Princeton, New Jersey, USA.

63　As expressed by a banker during the Confederation of British Industries' conference in London in February 1989, on 'Cease-fire in the Gulf: Financial and Economic Prospects', *MidEast Markets*, 6 March 1989, 16:5/5.

64 Figures were calculated from Peter M. Keller and Nissanke E. Weerasinghe, op. cit., Table 1.

65 On the latest developments in conditionality, see Devesh Kapur, 'The New Conditionalities of the International Financial Institutions', in UNCTAD, *International Monetary and Financial Issues for the 1990s*, Vol. VIII, pp. 127–38.

66 World Bank, *World Bank Tables, 1993–4: External Financing for Developing Countries*, Washington DC, Table A3.1.

67 World Bank, *World Development Report 1992, Development and the Environment*, New York: Oxford University Press, 1992, Box 8.8, p. 169; World Bank, *World Debt Tables 1994–95: External Financing for Developing Countries*, Vol. 1, Washington DC, The World Bank, 1994, Tables A6.2, A6.3, A6.4 and A6.5, pp. 165–9; and EURODAD, op. cit., p. 46.

68 Negotiating a DDSR agreement is a complex process that involves several stages: agreement in principle, agreement on terms sheet, securing waivers, creditor selection of options, securing collateral, signature, and exchange of instruments or closing. For brief information, see World Bank, *World Debt Tables 1994–95: External Financing for Developing Countries*, Vol. 1, Washington, DC, The World Bank, 1994, Box 2.2, p. 28 and Box A2.2, p. 69.

69 *The Economist*, 4 March 1995, p. 122.

70 *The Guardian Weekly*, vol. 152, no. 7, week ending 12 February 1995.

71 *MidEast Markets*, (UK), 6 March 1989, p. 16:5/5.

72 *Middle East Executive Reports*, November 1988, pp. 11–16.

73 Abdulatif Al-Hamad, 'The policy of official Arab aid agencies in Iraq's reconstruction", *Arab Banker* (London), February 1989, pp. 10–12.

74 Keith Bradsher, 'War damages and old debts could exhaust Iraq's assets', *The New York Times*, 1 March 1991, cited by Abbas Alnasrawi, *The Economy of Iraq: Oil, Wars, Destruction of Development and Prospects, 1950–2010*, London: Greenwood Press, 1994, p. 124.

75 Marion Farouk-Sluglett and Peter Sluglett, *Iraq Since 1958 from Revolution to Dictatorship*, London: I.B. Tauris, 1990, p. 273.

76 The figure was mentioned by Saddam Hussain during the Arab Summit held in Baghdad on 28 May 1990, and reported in Salinger and Laurent, op. cit., p. 30. Saddam, as reported, had referred to the amount as debts and requested their cancellation.

77 Apart from their amount, what is discussed here are funds provided for Iraq by the governments of these countries, and private funding or loans are not included.

78 Judith Miller and Laurie Mylroie, *Saddam Hussain and the Crisis in the Gulf*, New York, Times Books, 1990, p. 9.

79 UNCTAD, *UNCTAD Bulletin*, No. 20, May–June 1993, pp. 5–7.

80 Since the programme is computer-based, specific training on the use of the system is required. For further information see Shahid Akhtar and Antoine Raffoul, 'Third World debt and its management', *Canadian Journal of Development Studies*, Vol. XV, No. 2, 1994, pp. 265–75.

PART II

THE MAIN COMMODITY SECTORS: PROBLEMS AND POLICIES

3

The Oil Capacity of Post-war Iraq: Present Situation and Future Prospects

Fadhil Chalabi

1. Introduction and historical background

Iraq's oil is the least explored, developed and produced in the entire history of the world oil industry, despite popular belief that the country's oil potential is second only to that of the oil giant, Saudi Arabia. After the discovery of oil in Iraq in 1927, and until the early 1950s, the IPC consortium – which held concessions covering the entire territory of Iraq – did very little to find and develop oil, apart from the vast northern oilfield of Kirkuk and a scattering of other smaller fields. This was because of a conflict of interests within the consortium. Two powerful companies – British Petroleum and Esso – blocked the expansion of Iraq's oil in order to develop oil outside Iraq, in particular in Iran, Saudi Arabia and Kuwait, where their interests were greater. Even though the other partners – the French Compagnie Française de Pétrole (CFP), British–Dutch Shell and the American company Mobil – were interested in expanding Iraqi oilfields (since Iraq remained for them the main source of cheap oil), BP and Esso were able to introduce restrictive rules in order to protect their dominant position in the international oil industry.[1] For this reason, Iraq's production was kept at low levels and, until 1949, did not exceed 100,000 barrels per day (bpd), against 0.5 million barrels per day (mbpd) in Saudi Arabia and 250,000 bpd in Kuwait.[2]

Musaddiq's abortive Nationalisation Act in Iran and the consequent dramatic fall in that country's production pushed BP to reconsider its policy towards Iraq oil, with the result that the Iraqi government and the IPC consortium concluded an agreement in 1952. This, in effect, created a new situation whereby a very substantial increase in investment was made in Iraqi oil so as to expand its production and export capacity. Increasing its exploratory efforts in Iraq, the IPC group now developed the other giant oilfield, Rumaila, in the south. This led to an important upgrading of the country's reserves and a big increase in Iraq's production, which by 1961 had exceeded 1.0 mbpd, comparable to that of Iran at 1.2 mbpd and of Saudi Arabia at 1.4 mbpd.[3]

However, this expansionist phase of the Iraqi oil industry was short lived. Disputes between the government and the IPC consortium led eventually to a confrontation that, in turn, culminated in the promulgation of Public Law No. 80 of 1961. Any expansion in the Iraqi oil industry was effectively blocked by this law, which limited the activities of the IPC to the territories in which the companies were actually operating and that represented no more than 0.5 per cent of the total territory held by the concessions. The Iraqi government recovered the remaining territories (99.5 per cent) but for many reasons proved incapable of exploiting them during the decade of rapid expansion.

The law provoked various legal disputes with the companies, enabling them to prevent any independent company from developing oil in Iraq as long as the threat of legal action remained on the disputed lands that had been unilaterally relinquished. During the 1960s, many independent oil companies looked for opportunities in Iraq but were not prepared to confront the major oil companies, which were backed by their powerful governments. As a result, the Iraqi government was unable to implement the law in exploiting and developing oil from the land that it had recovered. At the same time, the IPC consortium had no incentive to develop and expand its activity in the very limited area allocated to it by the law. Thus, the huge oil potential of Iraq remained unexploited either by the government (Iraq National Oil Company) or by the IPC consortium.

Another option available to the government that might have led to a very vigorous expansion of Iraq's oil industry was not taken up for political reasons. An agreement could have been reached with the companies that were prepared to accommodate the government's demand to relinquish 90 per cent of the territories covered by the concession. Such an agreement could have helped create an Iraqi national oil company to exploit relinquished territories which, with the cooperation of many independent companies from Europe, Japan and the USA, could have exploited the relinquished territories without any legal obstructions from the major oil companies and their powerful governments. At the same time, the remaining territories could have been developed by the IPC group. With such an approach, Iraq could by now have reached a production capacity exceeding 6.0 mbpd.

Iraq missed another opportunity to develop its huge oil potential in 1965 when the government negotiated an initial agreement with the IPC consortium for ending the dispute. The draft agreement was frustrated by ill-judged and misplaced irrational 'oil nationalism', and by the insistence of politicians that oil exploitation should be exclusively in the hands of the state-owned national company, without any foreign participation. According to the draft, the oil companies would recognise Law 80 on two conditions. First, the government had to return 0.5 per cent of the land it had itself expropriated (which was allowed by Law 80) to the companies, including certain oilfields (mainly the northern part of the giant Rumaila oilfield) that they had discovered but not developed. Secondly, a joint venture called the 'Baghdad Petroleum Company' would be created, in which the Iraqi National Oil Company (INOC) would hold 33 per cent of the shares, with the rest being held by the companies within the IPC group (with the exception of Esso). The total area covered by this new joint company would represent about 7.4 per cent of the areas covered by the concessions. These conditions would have allowed Iraq's national oil company to exploit 91.6 per cent of its territories without any dispute with the consortium. INOC could have installed huge production capacity, utilising the tremendous oil potential of the regained land and the giant oilfields that have since been discovered, and this extra capacity could have led to great expansion from joint ventures with oil companies, as well as increased production from the areas that were left for the IPC Group.

Instead, Iraq's oil industry paid a very high price for the promulgation of Law 80 and the non-ratification of the draft agreement of 1965. The companies shifted their activities to new areas like Abu Dhabi and Libya, in addition to Saudi Arabia and Iran. At the same time, the law had the effect of isolating Iraq's oil and rendering the Iraqi government incapable of developing its huge oil reserves with the speed and efficiency needed to match developments in the international oil market. As a result, the country's oil industry suffered a long period of inactivity. No exploratory efforts were being made either by the government or by foreign investors,[4] and the country's reserves remained around the 30 billion barrels that had been estimated by the companies before the law came into force. This stagnation in the country's oil production was to the benefit of the other producers in the region, so that Iraq's share in total Gulf production was reduced from about 20 per cent in 1960 to around 10 per cent in 1973.[5]

It was only after the nationalisation measures undertaken (in totally different market conditions) by the Ba'th Government in 1972–74 that INOC began to pursue a very expansive exploration programme, helped by the (former) Soviet Union and a number of international oil companies from various countries, including Brazil, India and some European countries. The arrangements between INOC and these companies were based on service contracts without any equity concessions, and resulted in a great forward leap in Iraq's oil industry that took the country's estimated proven and recoverable reserves to more than 112 billion (bn) barrels (the present estimate). In the second half of the 1970s a large number of discoveries of major and giant oilfields were made, such as Majnoon (11.0 bn barrels), West of Baghdad (10.0 bn), West of Qurna (8.0 bn barrels) and others.[6]

In addition, Iraq made tremendous progress in its production and export capacities, which led to a dramatic increase in its production from an average of 1.3 mbpd before nationalisation to over 3.5 mbpd in 1980. At this point the eight-year war with Iran broke out. The Iraqi oil industry was one of the greatest war casualties, since the two deep-water terminals in the south, the mainstay of Iraq's oil-export facilities, were almost totally destroyed during the first weeks of the war.[7] In addition, Syria decided to shut down the pipeline that transported Iraqi oil to the Mediterranean. This resulted in a sharp drop in Iraq's production (to an average of less than 1.0 mbpd during 1981–4). Thereafter, Iraq succeeded in increasing its output, so that by 1990 its production and export capacity had reached 3.5 mbpd. More importantly, however, the war halted the government's plans for the development of the giant oilfields that had been discovered during the 1970s, as well as all exploratory efforts to find new fields.

2. Iraq's oil industry today
No sooner had the war ended and the country's oil activities been resumed than the disastrous invasion of Kuwait and the subsequent Gulf War took place, virtually destroying the oil industry. The heavy destruction of the oil installations inflicted by air strikes, combined with the UN oil

embargo placed on Iraq in August 1990 and the shutdown of its oil wells, severely downgraded the country's industry. The air strikes by the allied forces, involving 518 sorties that used 1,200 tons of explosives against 28 oil targets, caused terrible damage to the whole oil sector, in particular oil wells and their productive capacities.[8] All the surface facilities, including the storage tanks, de-gassing stations, water separators, main pumping stations, processing plants and cracking units in the refineries, were almost entirely destroyed. Among the installations that were seriously hit were the pumping station PS1, together with the pipelines that had pumped 1.6 mbpd of crude oil through the Iraqi/Saudi pipeline to the Red Sea. The strategic north/south and south/north pipeline within Iraq's territories was also heavily damaged, together with another pumping station in K3 in the north, which reduced the capacity of the Turkish pipeline to the East Mediterranean by almost half: from 1.65 mbpd to 0.8–0.9 mbpd. In the words of the UN Seebolt Consultant Report, Iraq's oil industry was reduced to a 'lamentable state'. Production capacity was drastically reduced.

Meanwhile, a new development was to have an important impact on Iraq's oil industry under the sanctions regime. The UN Security Council's Resolution 986, passed in 1995, provided that Iraq could sell oil to the value of US$2.0 billion every six months (or 'phase') to help it buy essential food and medicines. This sum was not, however, to be disposed of by the Iraqi government but was to be in held in an escrow account (under UN supervision) from which the cost of Iraq's imports would be paid. All purchase contracts were to be approved by the UN's Sanctions Committee, and 30 per cent of the amount was to be allocated for war reparations with a further sum to defray the UN's expenses in Iraq. The Iraqi government continued to reject the Resolution as an infringement of national sovereignty until, following much deliberation, an agreement was reached with the UN – the 'Memorandum of Understanding' – whereby the Iraqi government accepted the principles of Resolution 986. In 1998, the ceiling of permitted oil exports was raised to US$5.26 billion for each phase of six months.

With the passing of UNSC Resolution 1284 in December 1999 (to be discussed below), this ceiling was effectively lifted, which meant that Iraq's exports would henceforward be determined by its capacity to produce and export, rather than by a UN-imposed top limit on the value of oil sales.

In fact, as soon as Iraq started to export oil in 1996 in accordance with the UN's food-for-oil agreement, it was found that the country's production capacity, having been severely damaged, had reduced from 3.5 mbpd prior to the Gulf War to a mere 2.4 mbpd, according to normal sound petroleum practices. However, the government ignored these practices and decided to increase production well beyond real capacity levels and to the detriment of the productivity of the oil wells. By December 1999, Iraqi production had reached 2.8 mbpd, but the extent of the damage this has caused to the oilfields cannot immediately be assessed.

One important development in this respect was the agreement based on the terms of the 'food-for-oil' programme, that for each six-month 'phase' of US\$5.26 billion value of oil sales, US\$300m would be allocated towards the purchase of spare parts and equipment, to help Iraq rehabilitate its production capacity. This amount was later doubled. In practice, however, Iraq was not able to exhaust all the amounts allocated because of the delays in clearing contracts, caused by the UN Sanctions Committee under the pretext of examining equipment for any signs of intended 'dual use' (that is, for military purposes). Had the importation of the equipment allowed by the United Nations been secured and not held up by the UN Sanctions Committee, it is reckoned that under the UN sanctions system, Iraq could have boosted its production to 3.0 mbpd by the end of the year 2000 and even more by 2001. In fact, the approval process for contracts for the import of oil equipment and spare parts only began to be eased during 2000.

As far as export capacity is concerned, Iraq's oil transport capacity through to the East Mediterranean and the Gulf was drastically reduced by war. It should be recalled here that during the 1980–8 war with Iran, Syria closed the pipeline that had transported a volume of 1.4 mbpd of Iraqi oil through its territory to the East Mediterranean, prompting Iraq to increase the Turkish pipeline's capacity to 1.65 mbpd in 1987. As has already been noted, the Turkish pipeline's throughput capacity was reduced to 0.9 mbpd because of the destruction of the IT2 pumping station, in Iraqi territory, in 1991. With more spare parts and equipment, Iraq could restore the pipe's capacity to its original level by early 2001.

In the meantime, the Iraqi government has been able to rehabilitate and greatly expand the loading capacity of the Al-Bakr deepwater oil terminal in the south from 0.6 mbpd (the estimated capacity immediately after the Gulf War) to 1.2 mbpd. With these two outlets (that is, the reduced capacity of the Turkish pipeline and the renovated Al-Bakr deepwater terminal) Iraq was able by December 1999 to reach an export level of 2.2 mbpd – to the surprise of oil industry observers. Iraqi engineers managed to overcome many difficulties caused by the air strikes through the expedient of cannibalising equipment from other industrial units.

Again, it can be expected that if the smooth implementation of the UN Resolution will allow Iraq to import more equipment and spare parts, the Iraqi government may be able to increase the Al-Bakr loading capacity still further, besides completing the rehabilitation of the Turkish pipeline's capacity. The Iraqi oil ministry has even announced plans for the partial rehabilitation of other deepwater terminals in south Al-Amaya, which were thought to have been totally written off during the Iran–Iraq war. At the same time, Iraq and Syria have been discussing the partial rehabilitation of the Syrian pipelines of Banyas and Tripoli, which transport oil to the East Mediterranean. [Editor's note: since this chapter was written, Iraq is believed to have resumed limited oil exports through the Syrian pipelines, enabling Syria to buy Iraqi supplies on heavily discounted terms and to export equivalent quantities of its own oil. This arrangement, which began in late 2000, is outside the food-for-oil agreement, and details are therefore not made public.]

3. Future targets of Iraqi oil

Iraq's oil potential is enormous, but its future will depend not so much on physical and technical feasibility as on regional and international geopolitics, especially with regard to the UN sanctions regime. Although the oil embargo was imposed to drive Iraq out of Kuwait, it has been maintained and some aspects will continue in force even after economic sanctions have been lifted. The future of Iraq's oil will also depend on the international market and the world supply/demand balances.

In a conference held in March 1995,[9] the Iraqi government envisaged an expansion programme for oil after the lifting of the sanctions, with a target of 6.0 mbpd capacity, which could be achieved in five years, and with a total investment outlay of US$30 billion to be provided by foreign investors. According to the government's plan, this expansion programme could be achieved in three stages, with Iraqi total production reaching 4.2 mbpd in the first stage, 5.0 mbpd in the second and 6.0 mbpd in the third. In fact, if the agreements and contracts for oilfield development negotiated by the Iraqi government with various European and Asian companies were to be implemented, around 3.0 mbpd of new capacity (above the pre-sanction capacity) could be added within five to six years from the commencement of work.

A recent detailed study[10] by the Centre for Global Energy Studies (CGES), in collaboration with Petrolog & Associates, showed that the current known recoverable reserves in Iraq may only be the lesser part of Iraq's *real* reserves and that its *discoverable* reserves (i.e., probable but not yet known) could amount to about two hundred billion barrels. This puts Iraq and its oil at the same level as Saudi Arabia. Based on the findings of the study, the Iraq government's target of reaching 6.0 mbpd is not only feasible, but could very well be below what Iraq is able technically and physically to achieve. The study also placed Iraq's feasible production capacity in the range of eight mbpd and 12 mbpd, which could be achieved in eight and 12 years respectively, following the lifting of sanctions. The lower production scenario is based on the development of existing and known (but undeveloped) reserves, whereas the higher level shows additional capacities from new reserves yet to be discovered. These estimates by the CGES of Iraq's production potential represent what could be physically and technically feasible, under normal conditions and assuming that the expansion programme would not face any constraints of a political or financial nature.

In fact, such constraints will be crucial in determining Iraq's oil industry in the future. If investment in the oil industry was prevented by unfavourable political, financial or market conditions, Iraq's production capacity after the lifting of sanctions would at best reach the pre-war capacity of 3.4 mbpd, a level that could not be sustained for more than a few years before beginning to decline. This means that depending on the political climate, Iraq's oil production capacity could range from a low of 3.3 mbpd (which could not be sustained for long) in heavily constrained situations, to a high of 7.0 to 12.0 mbpd in cases where Iraq was absolutely free of any internal constraints (such as political instability) or external constraints (caused by the UN sanctions regime or unfavourable market conditions).

4. Oil and Iraq's post-war economy

Given the destruction caused to its economy by two devastating wars,

Iraq has every reason to expand its oil industry as much as is physically feasible within the shortest lead time.

Even before the disastrous invasion of Kuwait, the war with Iran had already depleted Iraq's resources because of excessive expenditure on military operations and the acquisition of imported military hardware. The Iraqi government itself stated that the value of imported weaponry actually used and destroyed during the war with Iran amounted to over US$100 billion; a figure which far exceeds the total Iraq oil revenue accrued to the country during the eight-year conflict. According to many estimates, the war with Iran led to a sharp decline in Iraq's GDP, which, coupled with the increase in population, meant a dramatic fall in the living standards of the individual Iraqi citizen. For example, it has been calculated that in terms of 1980 dollars (the year of the outbreak of the Iran–Iraq war), Iraq's post-war per-capita GDP had declined in 1988 to half of what it had been in 1980.[11]

What is perhaps more important is that the war led to heavy external debts, the servicing of which is likely to wipe out a great proportion of the country's oil revenues and leave it with only meagre amounts for development. Based on the government's Official Memorandum submitted in 1991 to the UN Compensation Commission in Geneva, Iraq's external debts stood at US$47 billion by the end of 1991, plus interest at a yearly rate of 8 per cent.[12] If this interest is added, Iraq's cumulative serviceable debts have probably now reached at least US$87 billion, and the longer the economic sanctions remain in effect, the greater the debt burden will become.[13]

Against this background of a severely ailing economy following Iraq's war with Iran, the invasion of Kuwait and the Gulf War then inflicted such damage upon the economy that it is hard to see how it can recover. As well as the heavy destruction of the infrastructure by the air strikes, the estimated cost of which ranges between US$100–$200 billion, the backbone of the economy was quite broken by the oil embargo. From August 1990 (when the embargo was imposed) until the end of 1997 (when Iraq began selling oil for humanitarian needs under the UN's 'food-for-oil' programme), the country's cumulative oil revenue losses exceeded US$130 billion. For a single-commodity economy like that of Iraq – whose dependence on oil revenue to meet all its import requirements and governmental expenditures is almost total – the loss of this oil income was enough to bring the country very close to collapse.

According to many estimates, Iraq's per capita income in real terms (1980 dollars) stood after the war at less than its 1950 level. Estimates of Iraq's real per capita income at 1980 prices are put at US$485 in 1993, compared with a 1979 level of over US$4,219.[14]

In its April 1991 Memorandum to the UN, the Iraqi government estimated that to restore part of the country's economy it would require, in foreign currency terms, about US$67 billion (including the cost of surging external debt) for the period 1991–5; that is, an average expenditure of more than US$13–14 billion per annum.[15] However, the Memorandum seems grossly to have underestimated the reconstruction costs of the destroyed infrastructure, including the oil sector. Most of the vital sectors of the economy, including power-generating plants, oil refineries, oil-pumping stations, industrial units, railway lines, airports, petrochemical plants, steel and cement plants, and so on, were destroyed during the 43 days of air strikes. According to many estimates, the replacement value exceeds $200 billion.[16]

It is true that Iraq will not have to spend as much for reconstruction as indicated in these estimates. In spite of severe working conditions and the lack of spare parts, substantial repairs have already been made by Iraqi engineers and technicians, whose achievements have exceeded all expectations. Many of the facilities have been partially rehabilitated in a much shorter period than estimated and at practically no cost in terms of foreign currency. Even so, huge levels of funding will still be required, since a lot of the impressive and speedy repairs were achieved by dismantling numerous industrial units and cannibalising their spare parts. Full rehabilitation costs of the oil sector, which was one of the major targets of the air strike, could amount to as much as US$5 billion. Therefore, in order to restore Iraq's essential infrastructure, no less than US$15–20 billion will be required during the first five years following the lifting of sanctions.

As if this tremendous haemorrhage of resources was not enough, the Iraqi economy has been further shackled by the UN-imposed war reparations, set at a maximum rate of 30 per cent of its oil revenues; if fully implemented, this could seriously hinder any capital investments for development. If these reparations are actually imposed, any hopes for the recovery of Iraq's economy will, even on the best of assumptions, be extremely meagre, if possible at all.[17]

5. The need for a new oil policy

All these estimates of the drain on the Iraqi economy are, of course, theoretical. They will depend to a very great extent on the political climate that will be created following the lifting of the sanctions. However, no matter how all these burdens are to be alleviated, Iraq will have an ever-increasing need for more money from its oil, simply to meet the requirements for a minimum restoration of its ruined economy. In the circumstances, the need to expand its oil industry is crucial, and once sanctions are lifted, the oil sector will be placed at the top of the Iraqi government's priority list.

It was mainly for this reason that the Iraqi Government changed its policy and has now opened its oil industry to foreign investors. This is in sharp contrast to the previously prevailing 'oil nationalism' (in the sense that all oil operations had to be performed by the state without any foreign participation), which followed the earlier promulgation of Public Law 80 in 1961 and the nationalisation measures taken in the 1970s. Since 1992, Iraq has been conducting negotiations with many foreign companies from France, Russia, China, Italy and elsewhere to develop the country's giant oilfields, which were discovered in the 1970s but which were not developed because of the wars.

All these agreements are generally based on the terms of production-sharing agreements (PSAs), the concept of which is that the foreign investor will undertake to spend all the capital needed to develop the oilfields and to reach a certain level of production. The foreign investor will then have the right to recover his expenditure in kind (oil) up to a certain percentage of the targeted production (called *cost oil*). The remaining production (called *profit oil*) will be shared between the Iraqi government and the investor in the form of certain negotiated percentages, determined according to the nature of the oilfield and the amounts of investment involved. The percentage share of profit per oil barrel by the companies might range between 8 and 12 per cent).

Apart from the two agreements made with Russia's Luk Oil and with China's CNOC, which were signed and ratified by Iraq's National Assembly, no other agreements have yet been signed by the Iraqi government. It seems that the government intends to use oil as a political instrument with which to push the countries to which the companies belong into lobbying for Iraq, with a view to lifting sanctions.

[151]

However, recent press reports indicate that with regard to foreign investments, Iraq may shift its oil policy from production-sharing agreements to what are called 'buy-back' agreements (as are found in Iran). According to this latter formula, foreign investors are remunerated with crude oil for the cost of their investments as well as for their profit margins. This is in sharp variance with the PSA, which is intended to last for a long period (normally 20–25 years): the foreign investor in the 'buy-back' arrangement is compensated with crude oil that is to be lifted over a limited period.

6. Iraq's oil industry and political constraints

The UN's sanctions policy on Iraq, which has been in force since August 1990, is the most important of the political constraints. As long as the sanctions remain, nothing can be done to implement investment projects for expanding Iraqi's oil industry. According to UNSC Resolution 687, the removal of sanctions is conditional upon the elimination of all weapons of mass destruction developed by Iraq, and upon the assurance that Iraq will not be able to rebuild the destroyed weaponry in a way that would threaten the region in future. Although the Iraq government formally agreed to this Resolution, its performance in cooperating with UNSCOM (the UN Special Commission for Inspection) has been extremely poor. A common practice has been to conceal or provide false information, or to acknowledge the existence of previously denied information only under pressure or exposure, and so on.

This has been extremely damaging to the government's credibility, and more than eight years after UNSCOM started its work, the committee still believes Iraq to be hiding chemical, biological and even forbidden ballistic missiles. This has helped reinforce America's hard-line position against the lifting of sanctions on Iraq in accordance with its so-called 'dual containment' policy regarding both Iraq and Iran, and has at the same time weakened the position of France, Russia and other countries within the UN Security Council who, driven by self-interest, are trying to help Iraq by speeding up the removal of sanctions. It is worth mentioning, however, that some of Iraq's complaints concerning UNSCOM were justified. As was later revealed, the Commission was used as a vehicle for spying and intelligence work serving foreign strategic interests. Furthermore, Iraq claimed that UNSCOM and the Security

Council were not appreciative of the fact that the bulk of prohibited weaponry had been destroyed.[18]

The formal US position is that sanctions should be maintained until Iraq fulfils all the conditions of the UNSC resolutions, but in practice it calls for maintaining sanctions as long as Saddam Hussain's regime is in power; the only exception has been its acceptance of limited oil sales for humanitarian purposes in accordance with Resolution 986. Previous statements by the US Secretary of State have strongly confirmed this position, which is clearly in contravention of the UN Resolutions. According to Paragraph 22 of UNSC Resolution 687, the UN will decide to lift the oil embargo once it has given Iraq a clean bill of health with regard to the elimination of mass-destruction weapons, irrespective of who is in power.

Meanwhile, there is nothing to suggest that any important political change will take place in Iraq in the near future, either in connection with the regime itself or even in the regime's undertaking of reforms that might improve its ruthless and irrational image. At the time of writing, President Saddam Hussain is still in power, a decade after his humiliating defeat in the Gulf War and despite the appalling destruction inflicted on the Iraqi economy and its infrastructure, and the increasing impoverishment of the Iraqi population. He will probably remain in power for some time to come, unless (as will be discussed later) there are major geopolitical changes in the area and/or a change in US policy. Apart from its tight and efficient presidential security system, the Iraqi regime is weak and incapable of maintaining control of the whole country, let alone of the rule of law. The north of Iraq is out of its control, and the USA and Britain have imposed a 'no-fly' zone in the south.

Yet, against all the odds, the regime has been able to survive, not only because of its unrestrained ruthlessness but also because of the many regional and international political developments from which it has benefited, including the conflicting interests and lack of agreement among Iraq's neighbouring states. For these states, the continuation of the status quo with a weak Iraq represents a lesser evil. Turkey, for example, is taking an increasingly positive attitude vis-à-vis Iraq, not only because of the very sensitive Kurdish problem but also because of its desire to restore its lost economic advantages (from oil and non-oil trade and from pipeline transit fees). Turkey's cumulative losses caused by the sanctions are claimed to be in the order of US$27 billion. On the other hand, Iran – a historical rival of Iraq in the Gulf – sees a political bonanza in a weak and broken Iraq, especially in increasing its own presence in the Gulf. For its part, Saudi Arabia perceives a financial bonanza in a continuing embargo on Iraqi oil, since it effectively replaces Iraq's oil in the world market. Since the imposition of sanctions and until the end of 1996 (the year Iraq began exporting oil under the UN 'food-for-oil' programme), it had made cumulative financial gains exceeding US$100 billion.[19]

Furthermore, the Iraqi regime is benefiting from a conflict of interests among the major world powers. The international coalition that was formed during the Gulf War has been faltering. Russia, France, China and many other European countries are now increasingly keeping their distance from US policy on Iraq, and have started to adopt their own agendas. Additionally, the absence of a valid alternative to the present regime has been causing apprehension regionally and among the international powers, since it is felt that any sudden change in Iraq may trigger a process of chaos, as well as political and even formal territorial disintegration. The so-called Iraqi Opposition groups (based outside Iraq) have proved totally ineffective.

On the other hand, the continuation of the present stalemate in Iraq is actually serving American strategic interests, since by becoming in effect the sole real guarantor of this region's security, the USA gains enormous political and economic advantages. It has, in fact, succeeded in isolating Iraq and, to a lesser extent, Iran, seeing in both countries a potential threat to the region's stability. The US military and political presence in the Gulf has been greatly enhanced, while, for the most part, the additional oil money accruing to the Gulf as a result of the embargo on Iraq's oil is recycled back to the USA in the form of military and non-military exports to the region.

Saudi Arabia's oil revenue has practically doubled as a result of the embargo, from an average of US$21 billion during the pre-war period to over US$40 billion after the Gulf War. Furthermore, the blocking of Iraq's enormous expansion potential has served to maintain the high level of oil prices, without which the oil industry, especially in the USA and the UK, cannot live. Thus, given the state of the oil market, especially in the post-Gulf War period, the full rehabilitation of Iraq could have led either to a drastic change in the oil power balance within OPEC – for which Saudi Arabia would have had to pay the highest price – or to a world oil-price collapse. Neither result would have been favourably received by the United States. In fact, the OPEC price mechanism, which otherwise would have fallen apart, has been allowed to function only because of the absence of Iraqi oil expansion.

From the above analysis it is obvious that in maintaining or in breaking the Iraq deadlock, the two major players are the United States and Iraq's own regime. Given the very small likelihood of any substantial alteration in their respective strategies, there can be no expectation that the Iraqi oil embargo will soon be fully lifted. This will only happen if a real change in the region's geopolitics or some development in the world oil market makes Iraq's full return necessary for achieving the demand/supply balance. As will be discussed below, given the market fundamentals for the coming few years, Iraq's oil does not appear to be needed.

Furthermore, it is unlikely that major changes in the area's geo-political power balance will occur in the near future. Although the Iraqi stalemate has the effect of disrupting the Iran/Iraq balance in the former's favour and of creating some fears in the Gulf regarding Iran's 'expansionist' policy, there is nothing to indicate that this development would be enough to cause a change in the US position. It is not expected that the USA will show much concern about the inherent dangers in a continuing stalemate in Iraq, even though this could affect regional geopolitics, especially those resulting from a de facto territorial disintegration, as in the northern part of Iraq where the Kurds have two authorities totally independent from the central government. Nor does the USA seem aware of the danger inherent in the absence of any real, central, governmental authority in the south.

It was hoped, once Iraq obtained a clean bill of health from UNSCOM, that the other powers (namely Russia, France and China) would push for the implementation of Paragraph 22 of Resolution 687, which provides for the removal of sanctions. Here again, and for the sake of Iraq, it is doubtful whether these countries would enter into any real confrontation with the USA. It may, therefore, be concluded that for the foreseeable future the present stalemate will continue and sanctions will not be removed in a way that will allow Iraq to implement its oil expansion programme.

UNSCOM ceased to exist after many confrontations between the UN Security Council and the Iraqi Government, and particularly when the chairmanship of the team was taken over by Ambassador Richard Butler. As a result of the latter's report on Iraq's non-compliance with the UN arrangements, Operation 'Desert Fox' took place,[20] whereupon there was an outcry within the UN about the legitimacy of the US/UK action against Iraq. The UN Security Council then established three panels to re-examine the case of Iraq and to find ways and means of alleviating the misery of the Iraqi people. Following the panels' investigations, the UK took up a new initiative in an effort (a) to allow Iraq to open up its own oil industry to the international oil companies that had previously negotiated with the Iraqi government to develop a number of oilfields; and (b) to help the Iraqi government to rehabilitate its industry and restore its pre-sanctions status.

The UK initiative, supported by the USA although France, Russia and China abstained, took the form of a new Security Council Resolution no. 1284. This resolution aimed at suspending sanctions for a certain period, renewable as a quid pro quo for the cooperation of Iraq in eliminating its weapons of mass destruction. The resolution is very complicated and its many loopholes could well impede its efficient implementation. Nevertheless, it has been a very important development as far as Iraq's oil is concerned, and for the time being there is no alternative. Although the Iraqi government did not formally reject the new resolution, it made obvious its unwillingness to accept the re-entry of the new UN inspecting team (UNMOVIC) into Iraq. For reasons related to the survival of the political regime in Baghdad, the government appears adamant in withholding information on chemical and biological weaponry, but by doing so it is closing an open door for the development of Iraq's huge oil potential, which is so badly and urgently needed for the reconstruction of the country's devastated economy.

The political constraint on Iraq's expansion programme could well continue, even after the lifting of sanctions. In accordance with Resolution 1051, the UN system provides for the prerogatives of the Sanctions Committee to continue even after the sanctions, especially with regard to approving imports of equipment and materials that could be considered as having dual use – both military and non-military. In fact the UN Committee's monitoring could obstruct the expansion of Iraqi oil; for example, in a situation where certain major powers, such as the USA and the UK, might see such an expansion as a threat to their own interests or to the stability of oil prices. However, all will depend on the political climate, the nature of the new regime, and its regional and international politics. Under favourable political conditions, Iraq may be given a freer hand in implementing its oil-expansion programme in cooperation with the foreign companies. By contrast, any new regime in Iraq that fails to gain full international support will probably find that the future of the country's oil is likely to be much less bright than its potential would indicate.

7. Iraq's oil and the world oil market – the medium term

Much of Iraq's future oil production will also depend on the state of the world oil market and the extent to which Iraqi oil will be needed to achieve a supply/demand balance. Given the present and likely market fundamentals and the planned new capacities within and outside OPEC, Iraq oil will not be *physically* needed for many years to come. OPEC's existing spare capacity of about 4.5 mbpd (two-thirds of which is held by Saudi Arabia and the rest by Kuwait and the United Arab Emirates) is expected to be complemented with an additional OPEC and non-OPEC planned capacity. The oil industry in many OPEC countries (including Iran, Libya, Algeria, Nigeria and others) has already been opened up to foreign investors, with a view to expanding production capacity. Assuming that OPEC will continue its policy of producing no more oil than is required by the market to complement supplies from outside OPEC (swing or residual production), Iraq's oil will be needed by the year 2005 only if the incremental 'call' on OPEC oil exceeds 7–8 mbpd. This is unlikely to happen, given a weakening world demand for oil and a continuing growth in oil production outside OPEC, especially in the former Soviet Union and the Gulf of Guinea.

Over the period 1988–98, the average annual increment in the 'call' for OPEC oil was less than one mbpd, or a 3.7 per cent increase per annum. Even if we assume that the same rate of increase in OPEC and non-OPEC production continues (which is doubtful, as we shall see), the incremental call on OPEC oil would be around 6.0 mbpd by the year 2005. This is less than the likely future OPEC capacity increase, and is even less if new non-OPEC capacity is added. Such an increment would be too small to create a real need for Iraq's oil beyond the country's current capacity, if we assume that UN sanctions will continue.

Clearly this is not sufficient to determine the need for Iraq's oil in the near future. Nevertheless, in spite of the increase in OPEC and non-OPEC capacity (which will exceed the increase in the call on OPEC oil), the present narrow margin of spare capacity of 4 per cent of total world consumption of oil will continue to exist for some years to come. In the absence of additional Iraqi oil, it will be concentrated largely in one major producing country, namely Saudi Arabia. This margin is not large enough to ensure market stability against the many hazards of disruptions in world oil supplies caused by technical problems, weather conditions, explosions, political upheavals, and so on, and it also means an unnerving global overdependence for supplies on one single country. The Saudi Arabian share in total world oil trade increased from 14.7 per cent in 1989 to over 20 per cent in 1998, and will continue to increase if Iraq's additional oil is kept out of the world market.

8. Iraq's oil and long-term market requirements

There are many uncertainties about long-term trends in the supply of and demand for oil, and the need for Iraqi oil. The Iraqi production required from OPEC will, of course, depend on how much of world demand can be met by increases from OPEC and from production from outside OPEC, in the event that the organisation continues to be the world residual supplier. More important is the *extent* of world need for Gulf oil, and how this need will be distributed among the major producers, especially Saudi Arabia and Iraq. Capacity increases in non-Gulf OPEC countries will have much less impact.

Many widely divergent forecasts of world supply and demand, as well as estimates of the need for Gulf oil, have been made for the years 2010 and 2020. According to forecasts by the International Energy Agency (IEA), the need for Middle East oil could reach as much as 41 mbpd by the year 2010, which is more than double the OPEC Gulf states' current production of around 18 mbpd. Conversely, as Table 3.1 indicates, a comparative forecast made by the US Department of Energy (USDOE) suggests far smaller need, and leaves the impression that the IEA estimate is visibly exaggerated.

The OPEC Secretariat, on the other hand, has reached various estimates for OPEC as a whole, depending on the level of oil prices (high OPEC prices presaging lower demand and vice versa). Table 3.2

[158]

shows the OPEC Secretariat's reference-case scenario, whereby in the year 2010 the entire call on the whole of OPEC would amount to less than 40 mbpd.

TABLE 3.1
Oil demand and Middle East OPEC supply, 1996–2020 (mbpd)

		2010	2020
IEA	World demand	94.8	111.5
	ME OPEC supplies	*40.9*	*45.2*
USDOE	World demand	93.5	112.8
	ME OPEC supplies	*28.3*	*41.6*

Source: IEA, *World Energy Outlook 1998* and USDOE, *Outlook 2000*.

TABLE 3.2
OPEC forecast of world oil demand and OPEC supply
for 2010 and 2020 reference case (mbpd)

	2010	2020
World oil demand	87.9	99.0
OPEC supply	*39.6*	*51.2*
Non-OPEC supply	*48.3*	*47.8*

Source: OPEC.

If these forecasts are to be believed, it is difficult see how adding new capacities on such a scale would be feasible without Iraq. Although Saudi Arabia's potential is still large, it could ultimately be limited. Saudi Arabia's oil reserves of about 260 billion barrels would be enough for ninety years of production at current rates, but this very comfortable R/P ratio will become tight if the additional oil required to meet increased world demand comes mainly from this country. By contrast, and as discussed earlier, Iraq's enormous oil potential has, for political reasons, been very much underdeveloped, and technically speaking, the production forecast made in the CGES report noted above places Iraq's capacity at twice that which existed prior to the Gulf War; that is, 6 mbpd in a relatively short time. Other non-Gulf OPEC producing countries have limitations on capacity expansion.

However, energy forecasts made in the past have mostly proved to be wrong. They are based on assumptions related to the prevailing conditions under which forecasts are made, and these are bound to change over time. In fact, forecasts are subject to too great a margin of error to be used as a reliable tool for policymaking. One aspect of their unreliability is the estimates of future oil supplies from outside OPEC. Most oil forecasts assume that because of their meagre resource base, these supplies will level off and eventually decline, so that their recent increases cannot be sustained. However, the huge technological progress achieved over the last ten years has challenged geologists regarding the real size of the oil resource base.

The history of oil production, especially in the USA and the North Sea, has shown how technology can sustain production from limited oil resources. Along with technology, the level of non-OPEC supplies will also depend on future price levels; the higher the price, the better the opportunities will be for non-OPEC oil to continue its increase. Future technological progress could have the effect of continuously upgrading reserves and oil recovery factors. Abundant new oil reserves have been discovered recently in the Caspian Sea (especially Kazakhstan), the Gulf of Guinea, the Middle East and elsewhere. New technology and high OPEC prices have both been instrumental in promoting these discoveries.

On the other hand, future world demand for oil will depend on many unpredictable factors such as global economic growth, environmental restrictions, fiscal policies of the consuming countries, balance of payments constraints of the oil-importing countries, and so on. The IEA's assumption that world demand for oil will increase at an average rate of 2 per cent until the year 2010 cannot be sustained. Much of the need for this oil will also depend on OPEC policies, and on whether the organisation will continue to be the world swing oil producer or whether it will adopt a more market-oriented strategy.

Current economic, technological and environmental trends suggest that world consumption of oil will slow down. Future world economic growth, the main driver of energy consumption, may tend to be at lower rates, and hence there will be lower oil consumption. More significantly, economic growth, especially in the industrialised world, is being driven mainly by the information industry and services. Neither of these sectors are energy-intensive, and they represent a fundamental change in the growth components, thus reversing the previous trend of high economic growth, sustained by energy-intensive industries, by which the economy was driven in the 1960s and 1970s.

Another downward pressure on world demand for oil stems from the huge advances in technology, especially in transportation, which allow for much greater efficiency of fuel utilisation. New devices, such as the hybrid engine in motor vehicles, can more than double the mileage per gallon of petrol. In the USA, where petrol (gasoline) accounts for some 50 per cent of total consumption, such devices tend to slow down utilisation of this product in the largest oil-consuming country in the world. The information revolution also reduces traffic volume. Communication through the Internet allows for vast amounts of work and transactions without displacement. Even more significant is the shift from petrol to electrically charged fuel cells for powering vehicles. Since the energy produced by such cells avoids the burning of fossil fuels, they are thus totally free of air-polluting effects.

Environmental concerns, which are gaining important political support, constitute another reason for the declining trend in world oil consumption. Although barely feasible in the medium term, the Kyoto Protocol's targets for reducing airborne emissions of carbon dioxide have generated political pressures against the consumption of coal and oil and, to a much lesser extent, natural gas. (Carbon dioxide emissions are highest from coal and lowest from natural gas.) The European Union's carbon-tax debate has already been reflected in higher taxes on the consumption of oil as a means of reducing such emissions. Increasingly, too, high taxation on oil products, which raises the price paid by the end consumer to exorbitant levels, is adding further pressure on oil demand in the industrialised countries. This is especially obvious in the case of petrol sales in Europe, where the tax component reaches more than 80 per cent of the price paid by the consumer. Constant review of the tax component of the price paid by the end consumer for a composite barrel reveals that in many European countries taxes on oil products have been rising continuously. The tax component in Western Europe increased from around 30 per cent in 1980 to 63 per cent in 1998.

Finally, high prices for crude oil in the world market discourage consumption. With OPEC maintaining artificially high price levels through production cuts, economic, political and technological trends are expected to dampen the growth of demand sufficiently for forecasts such as those of the IEA to be regarded as unfounded. Furthermore, these trends encourage the growth of supplies from outside the Gulf, including those emanating from non-Gulf OPEC countries, and especially the hugely reduced costs of finding, developing and processing oil. Over the last decade or so, the costs of exploration drilling and upstream development have become much cheaper (by 40 to 50 per cent). More importantly, technological progress, such as three-dimensional and four-dimensional seismic surveys, has made identification of oil reservoirs almost certain, thus eliminating the risk of finding dry holes or water and no oil.

At the same time, the high-price policies pursued by OPEC have been instrumental in increasing supplies from new high-cost areas that had been considered uneconomic until the price shocks of the 1970s and 1980s. Increasing volumes of new oil – that is, from outside OPEC, the USA and the Former Soviet Union (FSU) – have since flooded on to the market, at the expense of the Gulf share in world production and trade. For example, North Sea oil, which is expensive to produce, increased to more than 7.0 mbpd from negligible amounts 20 years ago. The share of new oil in world production has grown from less than 6 mbpd in 1978 to more than 27 mbpd by the end of the 1990s.

From the above analysis of trends in world oil supply/demand, it looks as if the world will live on an abundance of oil: supplies are growing much faster than demand. In other words, time is not in favour of Iraq oil, and the sooner the government creates the conditions for foreign investments, the better it will be for the devastated Iraqi economy. Otherwise, there is a strong likelihood that, in the long term, Iraqi oil wealth will be without markets and forever buried underground.

9. Iraq's oil and political uncertainties

With no likely change of position on the part of the two major players, namely the USA and Iraq, the stalemate seems set to continue at least for the foreseeable future. However, political changes inside Iraq should not be totally excluded. Incidents such as the assassination attempt on the President's son in 1996 indicate that the system is not invincible, in spite of the abhorrent control that it exercises.

Various political scenarios and their impact on Iraq's oil are possible, some of which are considered below.

The present regime and status quo continue

This presupposes much the same rule, without any major change. The USA will continue to oppose President Hussain, and this will mean maintaining sanctions, resulting in a persistently deteriorating economy. However, as with other 'pariah' states, there could be a gradual, though small, erosion of the sanctions – similar to the situation with Cuba and North Korea. This would help the country survive although at a very low standard of living and with hardly any economic development to speak of. Under this alternative the oil scenario looks bleak. In order to increase the value of oil exports allowed under UNSC Resolution 986, the Saddam regime would try to stretch the humanitarian oil sales, at the same time continuing its efforts to win enough international support for the sanctions to be lifted totally. The dilemma for Baghdad is that, regardless of the steps it takes, it must go back to the Security Council to seek approval for what it wishes to do next. Here it will be confronted by the USA and the United Kingdom, whose hard-line policy aims to keep the country ostracised and weak, mainly through the operation of a comprehensive sanctions policy. Given this scenario, there would be no hope for the implementation of any contract or agreement between Iraq and foreign companies.

The regime continues in power but accepts UN Resolution 1284 and cooperates fully and genuinely with UNMOVIC

This might lead to the suspension of sanctions and eventually to the opening up of Iraq's oil industry to foreign investors. This scenario suggests that the future of Iraq's oil industry could hardly be brighter, since many international oil companies are interested in Iraq's oil due to its very low cost. Even more significantly, the likely involvement of US and UK oil companies in developing Iraq's oil industry would create vested interests in both countries that would push for adoption of a more lenient attitude towards Iraq. However, this scenario becomes less feasible when one considers that the USA, despite its support for the UK initiative and for the UNSC Resolution, may be less keen on the expansion of Iraq's oil industry if it means the oil balance will be jeopardised. This is because such an expansion could exacerbate OPEC's (and especially Saudi Arabia's) problem in maintaining a stable oil price at levels that the USA finds acceptable. In addition, it is also unlikely that the Iraqi government would accept and cooperate with UNMOVIC, especially given its very poor performance so far with UNSCOM.

The regime is overthrown by army officers similar in background and upbringing to the present elite

Many observers consider this to be a plausible scenario, the success of which would depend on a quick move that would bring about an abrupt and total end to the regime and its leader. In the event of success the main concern of a new government would be to survive, and in the early days of its rule it would be largely preoccupied with providing security and with launching massive economic reconstruction. As for the oil industry, this scenario could offer it opportunities, but only after considerable delay and until the political situation became clear. Much would depend, in the first place, on how soon this new political group could ensure its control over Iraq, but, over and above the domestic problems, the new regime would still have to contend with the UN sanctions regime. In these new circumstances, sanctions might either be lifted or else eased gradually because of the pressure generated by fierce international competition and support for their firms among the big powers, whose intention would be to play a part in the lucrative Iraqi oil industry. One question that could arise here concerns the future of the accords that have been negotiated (but not signed) with major European oil firms, and whether these would be honoured or not by the new regime.

A military coup d'état takes place without total control and authority over the country being secured

The adverse political and social conditions that have been created in Iraq under the Ba'th regime, and especially since the two Gulf Wars, could spark off acts of vengeance, sectarian conflicts, and even civil wars. If the regime is not totally eliminated, its nature and its leadership is such that, even if the price of its political survival might be a devastating civil war, it will not give up. If such changes are not counteracted, it could take a long time for the country to be put together again, especially as many regional powers might take advantage of the situation and try to influence the course of events. Given that such conditions could continue for years, there would be no hope for any development of the oil industry, and it would continue to remain in limbo.

[164]

*A scenario of rehabilitating Baghdad's present regime cannot be
totally excluded, although present conditions assign a degree of
remoteness to the possibility*

Middle East politics cannot be predicted, and a different climate that
would accommodate such a rehabilitation might be created by various
unforeseen changes in regional geopolitics. Given the scenario outlined
here, favourable conditions that would provide incentives for foreign
investors in Iraqi oil could be secured for the expansion of Iraq's oil
industry. In addition, and assuming the total lifting of UN sanctions
and the regime's ability to regain international and regional support, Iraq
would be controlled by a strong government with sufficient resources
for it to establish centralised powers. It would thus be enabled to
implement plans for expanding the oil industry, helped by foreign
investors. The major powers, apart from the USA and Britain, seem
to be perfectly prepared to live with the present regime, once it has
finally fulfilled the conditions of UNSC Resolution 687 concerning the
complete elimination of weapons of mass destruction. This position is
unequivocally adopted by France, Russia and China, and is supported
by such regional powers as Turkey, Egypt and even Iran.

It is also clear that for these countries other UN resolutions, such as
that concerning protection of human rights in Iraq, are not relevant and
are regarded as internal matters in which the international community
has no right to interfere. Here again, the USA – whose change of attitude
and policies is crucial – is the key to the whole equation. Because
Saddam Hussain's image has been so demonised across the political
spectrum in the United States, where US foreign policy is concerned,
it will be probably be difficult to shift American attitudes towards
Saddam's regime.

The US position can perhaps be altered, but only if there is
sufficient inducement to justify such a change and if the impact of
internal US politics is moderated. No change in American policy can be
effected without Iraq offering incentives and concessions by granting to
the US oil incentives as well as political advantages. However, in the
event of a deal being struck between the Iraqi regime and the United
States, nobody can tell what would happen to the agreements and
understandings that have so far been reached with the Europeans
and with other oil companies. This is because, apart from the firm
agreements concluded with both Russia and China and ratified by the

Iraqi 'parliament', none of the other draft agreements have been ratified by the government of Iraq.

NOTES

1 Edith T. Penrose, *The Large International Firm in Developing Countries: the International Petroleum Industry*, London: Allen & Unwin, 1968, pp. 157–9. In this book Penrose gives a clear and detailed description of how the stronger partners (BP and Esso) were able to prevent the smaller partners from lifting more oil in Iraq. In 1948 the Heads of Agreement document governing the planning of increased output introduced two instruments to block any expansion in Iraqi oil that could favour the other partners, especially CFP. Most important was the 5/7ths rule, which gave these two companies the power to determine how much Iraqi oil could be produced. The other instrument was the five-year planning of Iraq's production, so that no more oil than planned could be produced.
2 OPEC, *Annual Statistical Bulletins.*
3 Ibid.
4 However, during this long period of stagnation in Iraq's oil industry until the nationalisation of the mid-1970s, there were two developments: the first was the conclusion in early 1968 of an agreement between the Iraqi National Oil Company (INOC) and the French oil company ELF-ERAP, according to which an area north-east of Basra (near the border with Iran) was explored. This led in 1971 to the discovery of a relatively important oilfield (Abu Ghirab), production from which started at modest rates before the government took it over. The second, more important, development was the agreement concluded with the FSU to develop the North Rumailah, of which only the south part had so far been developed and produced by the IPC Group. Although the Group had discovered the northern part of the giant Rumailah oilfield, it had remained undeveloped and was taken over by the government in accordance with Law 80 of 1961. North Rumailah production began in spring 1972, a few months before nationalisation of the IPC.
5 During the period 1961–72, Saudi Arabia's production increased from 1.48 mbpd to 6.0 mbpd; Iran's from 1.2 to 5.0 mbpd; Kuwait's from 1.7 to 3.3 mbpd and the UAE's from zero to 1.2 mbpd. Iraq's production remained at 1.46 mbpd. (OPEC, *Annual Statistical Bulletins.*)
6 The irony is that all these giant oilfields were within the areas relinquished to the government and not in the areas designated for the Baghdad Petroleum Company.
7 The deep-water terminal Khor-Al-Amaya (with its 1.65 mbpd loading capacity) was considered to be virtually written off, whilst the capacity of the heavily damaged Al-Bakr terminal (of a similar loading capacity) was drastically reduced.
8 Muhammad 'Ali Zayni, *Al-iqtisad al-'iraqi fi dhil nizam Saddam Husayn: tatawwur am taqahqur* [The Iraqi Economy Under the Regime of Saddam Hussain], London: Mu'assasat al-Rafid lil-Nashr wa-al-Tawzi', 1995, p. 301.

9   See *Middle East Economic Survey* (MEES), 20 March 1995, which published a
    report on this conference as well as the text of the unofficial Iraqi memorandum.
10 Centre for Global Energy Studies, *Oil Production Capacity in the Gulf*, Vol. 4,
    Iraq: Book 4, Parts 2 and 3, 1997–8.
11 Abbas Alnasrawi, *The Economy of Iraq: Oil, Wars, Destruction of Development
    and Prospects, 1950–2010*, Westport, Conn.: Greenwood Press, 1994, p. 152.
12 Zayni, op. cit.
13 Adding cumulative interest to the original amount of external debts might be
    legally contested by the Iraqi government, giving rise to a case of *force majeure*;
    that is to say, the imposition of sanctions makes it impossible for the government
    to pay. However, discussion of this issue could be controversial and depends on
    political conditions prevailing in Iraq after sanctions, as well as to the extent to
    which Iraq can be aided in recovering its economic health.
14 Cf. Alnasrawi, op. cit.
15 *Middle East Economic Survey* (MEES), 13 May 1991.
16 See, for example, those made by the Brookings Institution and the Arab Economic
    Report (issued in 1992 jointly by the Arab Monetary Fund and the Arab Fund
    for Economic and Social Development).
17 The extent to which the UN resolution on reparations will be implemented
    depends on political conditions following the lifting of sanctions. History shows
    that war reparations tend to be counterproductive and it would not be in the
    interests of a future Kuwait to insist on them, since a severely impoverished Iraq
    could perpetuate a source of future conflict. It is worth noting here that the
    war reparations imposed on Germany by the Treaty of Versailles at the end of
    the First World War (and insisted upon by the French under Clemenceau) were
    contrary to the recommendations of the then British advisor at Versailles, John
    Maynard Keynes. Keynes argued that such harsh terms imposed on Germany
    would cripple its economy and create the conditions for renewed conflict. Keynes
    could not have been more prophetic if we consider the ensuing collapse of the
    German economy and the sense of humiliation that gave rise to Hitler and the
    Nazi movement.
18 In fact the former Chairman of UNSCOM, Ambassador Rolf Akeus, was more
    objective and neutral than many members of the teams. Prior to the defection of
    Hussein Kamil (the President's son-in-law), he was on the point of concluding
    that UNSCOM's work was nearing completion. However, the huge amount of
    information on the hidden and prohibited arms revealed by Kamil radically
    changed the situation, and relations between UNSCOM and the Iraqi government
    became acrimonious.
19 Because of the embargo on Iraqi oil and the existence of excess capacity in Saudi
    Arabia, the latter increased its production from 5.4 mbpd (according to the
    then OPEC production quota agreement) to 8.5 mbpd to compensate for
    the disappearance of Iraqi oil, and then settled for 8 mbpd. Saudi Arabia had
    captured about three-quarters of Iraq's oil-market share. The above-mentioned
    financial gains for Saudi Arabia resulting from the oil embargo on Iraq, are
    calculated on the assumption that Iraq's oil would continue to flow into the
    market without the UN oil embargo at the pre-invasion capacity of exports of
    about 2.9 mbpd, multiplied by the monthly average price for Iraq's oil during
    the period from August 1990 to the end of 1996. Iraq's share in the OPEC quota

system prior to the invasion was 14.5 per cent of total OPEC production. In fact, if this percentage were maintained, Iraq's losses would be even greater, taking into account the annual increase in the total OPEC production.

20 Tension between the Iraqi government and UNSCOM, chaired by Ambassador Richard Butler, entered a new critical phase in December 1997 with his report to the UN Security Council that Iraq was not fulfilling the terms of the agreement reached with the UN Secretary General, Kofi Annan. Following Butler's report, the USA and UK launched a very heavy air strike called Desert Fox, in which great damages were sustained and many civilian lives were lost. The operation was unilateral (that is, not within the mandate of the Security Council), and this provoked protests among other members of the Security Council. As a result, UNSCOM itself became defunct.

4

Agricultural Policy Issues and Challenges in Iraq: Short- and Medium-term Options[1]

Mahmood Ahmad

1. Introduction

Historically, it is the agriculture sector which, after oil, has played a dominant role in Iraq, and which – since it provides food, employment and income for the largest section of the population – is considered the mainstay of the Iraqi economy. However, even before the embargo the overall performance of the agriculture sector was weak, despite the sizeable investment in the irrigation infrastructure, the wide and ever-increasing subsidy on fertilisers and chemicals, and the levels of credit provided to farmers throughout the 1980s (not to mention the cost of natural resource degradation incurred by such policies). During this period, increases in the yields of major crops (especially food crops) were generally below the population growth rate. With the agricultural reform programme operating at less than the desired levels, combined with deteriorating terms of trade and a declining natural resources capital, growth rates in the agricultural sector were insufficient to meet the growing demand.

Iraq's agricultural economy has been hard hit by the sanctions imposed in the 1990s by the United Nations. Productivity has diminished, mainly because of the shortages of necessary inputs and the lack of support services such as credit, extension and research. The performance of the sector is also constrained by inappropriate macro and sectoral policies that severely limit the sector's incentive to produce and market strategic agricultural crops. The government has instituted a price support programme for strategic crops, has tried to remove technical constraints on plant protection, seed availability and machinery repair, and has attempted to encourage private investment in agriculture. Even so, the sector is on the verge of collapse unless short- and medium-term assistance is forthcoming.

[169]

Iraq has the potential to sustain well-diversified agricultural production, since it is one of the few countries in the Arab region that is endowed with land, water, and the workforce necessary for developing a prosperous agricultural sector. Financial resources were also plentiful in earlier decades. For reasons of technical and institutional inhibition, non-conducive macro and sectoral policies, natural resources degradation, costly wars, and the sanctions of the 1990s, this potential has never been fully tapped. However, and in contrast with the relative neglect of agriculture during the era of the oil-dominated economy, the sector is now under pressure to provide badly needed food, income and employment opportunities for the economy. To spur production of key agricultural commodities, the government is relying on a myriad of agricultural policies, though under sanctions these have had a mixed record in meeting the challenges of developing agriculture.

The present contribution takes stock of the agricultural situation in the country and reviews the key agricultural policies pursued by the government to enhance agricultural production in this difficult situation. It reviews, and considers the effects of, the macro and other enabling environments facing the agriculture sector and evaluates how an across-the-board consumer subsidy contributes to monetary expansion and hyperinflation on the one hand and negatively affects the production of key commodities on the other. It also enlarges on the strategic policy options available to the government both in the short term (under sanctions) and the medium term (partial or full removal of sanctions).

2. The agricultural system

Farming and production

The agricultural sector is characterised by smallholdings that range in size from three to ten *donums* of land (a *donum* is one-quarter of a hectare) in irrigated areas, and from 40 to 120 *donums* in the rainfed, northern areas.

Most of the cultivated area in the northern part of the country is rainfed, with the rain falling in various micro-climatic zones. Rainfed agricultural production is practised in the governorates of Ninevah, Irbil, Duhuk and Sulaimaniyya, and the region is distinguished by three rainfall regimes: high, medium and low. Vegetables and fruits predominate in the high-rainfall zone in the north (700–1,000mm), wheat occupies most of the area in the medium-rainfall zone (400–700mm), and in the low-rainfall zone (under 400mm), barley is the main crop grown. In areas with less than 200mm of rainfall, pastures prevail. In the central and southern regions, irrigated agriculture is mainly practised in the areas that normally receive relatively little rainfall. In the south, the land is low-lying and often below sea level. It therefore creates flooded marshlands associated with a high water table that gradually drops northwards. The soils in this region remain highly saline in places where large drainage and reclamation projects were completed in the past to deal with brackish water.

The northern region accommodates varied agricultural activities including production of field crops (wheat, barley and maize), vegetables (*Solanaceae* and *Curcurbits*) and fruit (mainly grapes). Heavy investments have also been made in the poultry, dairy, and fisheries industries. Animal production, for which feedstuffs are mostly imported, is confined mainly to large commercial units. Otherwise, livestock in Iraq is raised under predominantly traditional systems, which include small farmers who meet their household requirements of milk and meat by raising small herds of cattle, sheep and goats in villages and small towns. Under this system, animals are mainly fed on crop residues and are not given any concentrates. Nomads and transhumance tribes in the northern Jazira steppes also raise sheep and goats in relatively larger flocks, and therefore depend mainly on rangelands. However, in dry years they move their animals closer to crop-growing areas, where they are grazed on crop residues and agricultural by-products. The severe shortage of feed has encouraged many farmers to move towards integrated patterns of farming in which the production of forage crops, such as *barsim* (alfalfa), is assuming increasing importance. Furthermore, the sharp decline in cereal yields under the monocropping system has prompted the introduction of planting leguminous crops in rotation to restore soil fertility.

There are no distinguishable farming systems pursued at the farm level in the central and southern regions. Agricultural practice is mainly based on mono-cropping without the adoption of any crop rotation. This tendency is being encouraged by government directives that require the planting of 80 per cent of reclaimed land and 50 per cent of unreclaimed farmland with cereals (wheat and barley). Rice is normally planted in smallholdings, while double cropping of wheat/rice or barley/rice is occasionally practised, as for example in Najaf and Qadissiyya.

Mixed farming is practised by a few farmers, as in the Wasit region, and involves growing corn and barley as feed for livestock and sheep, alfalfa for fodder, and vegetables and fruits as cash crops. However, farm holdings in this region are often small and do not allow for the variety of these crops to be accommodated on any one property.

Horticultural crops are largely grown in the central region, especially in Kerbala, with date palms predominating, while vegetable production under drip-irrigated plastic tunnels is practised in the western desert region. Further east in Babil, the typical cropping sequence involves growing cereals (wheat and barley) in winter, followed by corn in summer, or growing winter vegetables followed by summer vegetables. In the central region, the Salah El Din governorate is an important cereal- and vegetable-producing area, with significant animal production (75,000 sheep and goats, 50,000 cows and 3,200 buffaloes). Here too the severe shortage of feed has encouraged a move towards integrated farming patterns in which forage production is becoming increasingly important. As in other areas, the decline in cereal yields as a result of mono-cropping has encouraged the alternation of leguminous crops (vegetables) with cereals to combat declining fertility.

Performance of the agricultural sector
Before sanctions the share of agriculture in GDP had been decreasing continuously, mainly due to the growth of other sectors, but 1992 figures indicated an increase to 35 per cent in agriculture's share in GDP. According to the 1987 census, about 30 per cent of the population live in rural areas and about 12.5 per cent of the labour force were engaged in agriculture, forestry and fisheries.

Cereal production in the past registered an unstable pattern, due to unpredictable rainfall, non-availability of essential inputs and non-conducive macro and sectoral policies. Table 4.1 indicates trends in area extent, production, yields and procurement prices for strategic crops in Iraq. The best grain harvest for decades was in 1990, with 1.2 million tons of wheat and 1.9 million tons of barley,[2] and in this instance increases in procurement prices might have contributed to acreage increase. However, during the period 1967–90 the most frequent yield levels for wheat and barley ranged between 400 and 800 kg/ha, low yields that contributed directly to the country's food and feed deficit. The main reasons for continuing low yields during that period include

TABLE 4.1
Average yield, area, production and purchase prices for cereals in Iraq, 1980–95

Years	Wheat				Barley				Paddy				Maize			
	Yield	Area	Production	Price	Yield	Area	Production	Price	Yield	Area	Production	Price	Yield	Area	Production	Price
1980	173	5,647	976	60	192	3,554	682	57	697	239	167	112	454	131	59	68
1981	186	4,847	902	72	225	4,108	925	65	708	229	162	129	377	104	39	68
1982	204	4,728	965	86	197	4,583	902	72	667	245	163	216	355	80	28	80
1983	164	5,126	841	92	152	5,503	836	81	486	227	111	158	259	108	28	90
1984	89	5,271	471	100	84	5,744	482	86	601	181	109	200	223	138	31	210
1985	224	6,266	1,406	115	230	5,795	1,331	105	607	245	149	250	300	116	41	235
1986	206	5,041	1,036	145	171	6,108	1,046	120	970	211	141	300	431	123	53	275
1987	148	4,881	722	159	128	5,823	743	120	668	294	196	300	415	148	61	450
1988	212	4,383	929	170	245	5,874	1,437	120	631	223	141	400	315	245	78	486
1989	142	3,450	491	270	105	6,276	663	180	738	314	231	500	477	218	103	550
1990	250	4,783	1,196	600	232	7,980	1,854	250	673	339	229	1,000	563	286	171	700
1991	147	10,069	1,476	800	80	9,649	769	500	491	186	189	1,700	657	297	296	700
1992*	173	5,800	1,006	4,000	269	5,600	1,509	2,000	722	360	260	5,000	603	520	313	2,500
1993	168	5,400	916	5,000	171	5,200	899	2,500	617	348	215	7,000	558	520	290	3,700
1994	168	5,600	1,080	35,000	179	5,600	1,002	2,000	611	360	220	75,000	548	520	285	17,000
1995	200	5,800	1,160	105,000	160	4,800	768	60,000	605	650	393	400,000	555	430	238	75,000

Notes:
Yield = kg/donum
Area = 1,000 donums
Production = 1,000 tons
Price = ID/ton
*From 1992 data for North (Irbil, Suleimaniyya and Duhuk not included).
Source: Preliminary figures, Ministry of Planning, Central Statistical Organization.

lack of proper cultural practices, farm inputs, shortage of water, increasing salinity and waterlogging, emergence of plant and livestock diseases, infestation of insect pests and weeds, crippled farm machinery, the gradual destruction of natural resources, loss of poultry and livestock, and non-availability of seasonal expatriate farm labour.

The areas under wheat and barley in the northern region registered similar trends before the sanctions, touching their lowest point in 1988–9 and starting to show an upward turn again as a result of liberalising the market in the latter part of the decade. Under sanctions, the northern wheat-growing area jumped to a record level of 3.2 million *donums* in 1991, but returned to higher than historical levels in the next four years, with generally upward trends. The area under barley also increased in 1991, returning to historical trends but at a lower level.

The actual production for wheat and barley also registered similar tendencies. However, it is important to note that before the sanctions the productivity of these two crops was unstable with a downward trend, but that following sanctions productivity started to improve at stable and upward trends indicating growth of these two crops on a sustainable basis.

A media report on performance of the agriculture sector published in 1997 depicted a tremendous growth in selected crops during 1996. The production for corn was projected to increase by 460 per cent, cotton by 600 per cent, grains by 15 per cent, and paddy by 16 per cent, in spite of the fact that the area under rice was reduced by 33 per cent during 1996. It seems that farmers were responding to relative prices and shifting comparative advantages as a result of the appreciation in value of the Iraqi dinar during 1996. During that year, the dinar rose from its lowest value of ID3,000/dollar to touch ID1000/ dollar. This level prevailed during most of the year, thereby providing farmers with a favourable price incentive for key commodities (wheat, barley, rice, maize and cotton) in 1996 as compared with 1995. In 1996, for instance, prices of corn in the open market remained around ID100,000 to ID110,000 per ton, compared to the incentive price from the government of ID150,000 per ton. Similarly, the farmers were able to market cotton and rice to the government at ID400,000 and ID500,000 per ton during 1996. With these prices, farmers were marketing more of their products and hence the overall supply situation seems to have improved. However, in the absence of reliable statistics it is very difficult to establish the true picture.

FIGURE 4.1
Wheat and barley area in the northern region of Iraq

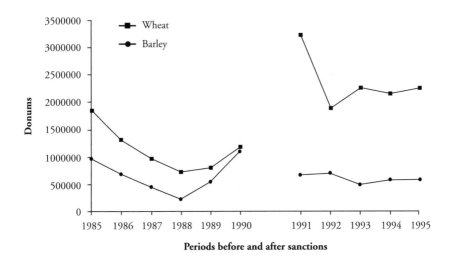

FIGURE 4.2
Wheat and barley production in the northern region of Iraq

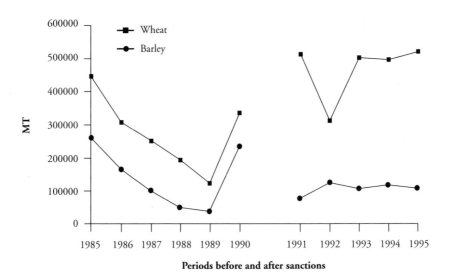

FIGURE 4.3
Wheat and barley productivity in the northern region of Iraq

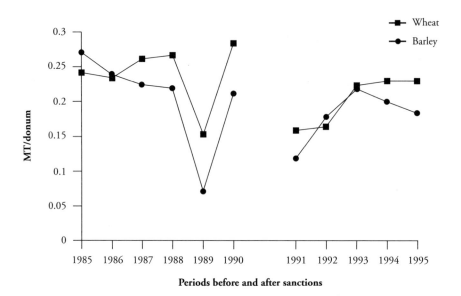

Periods before and after sanctions

FIGURE 4.4
Productivity: wheat and barley in Iraq

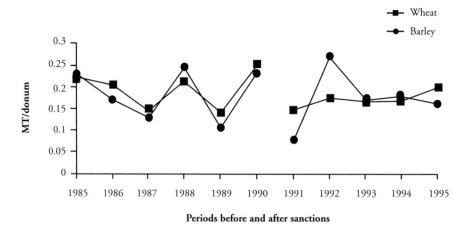

Periods before and after sanctions

The productivity improvement in the north can be attributed to labour-intensive farming, low levels of technology, and better farm-management practices. In addition, the policy environment is more conducive to sub-sector growth. Furthermore, the improved performance of the agriculture sector during the sanctions period may be due to the demand placed on this sector as the main source of employment and income provision, which thus makes it a vital sector. At the same time, because of the absence of a public rationing system in the north and the limited coverage of targeted international food aid, the entire population depends on local supplies to meet the demand. To conclude, the farmers in the north were operating under fewer of the policy distortions that had previously tended to allocate resources inefficiently, and that had often provided disincentives to produce more and invest in agriculture.

Shortage of wheat in the central and southern part of Iraq represents potential market access, but this will often depend on the prevailing political situation.

Future economic prospects

The future growth of the economy, viewed from the perspective of 1996 when this piece was written, depends on how soon Iraq will be able to sell its oil to pre-sanction levels. It is commonly believed that the economy will turn around very fast once the embargo is lifted. This might be true, but all will depend on how Iraq deals with the grim financial and economic realities confronting the country.

Iraq faces three main factors that deplete it economically and which, when combined, will continue to haunt the economy and slow down the country's chances of early recovery. The first pertains to the huge cost of rehabilitating the infrastructure that has been destroyed during the last few years, an amount estimated at US$50 to US$100 billion. The second relates to the enormous foreign debt that has been accumulated, roughly estimated at US$100 billion, while the third factor constitutes the liabilities for war reparation, estimated at US$78 billion until the year 2005, or an annual payment of US$8 billion. At the time of writing (1996), Iraq's capacity to repay will depend on how soon it returns to the oil market, on the volume of oil exported, and on the oil prices prevailing in the global market.

It will take time to replace the country's industrial and social infrastructure, but rehabilitation and development efforts can be more effective if Iraq starts to develop prudent economic policies, and if it promotes a strong campaign of self-reliance that places greater emphasis on the optimum utilisation of natural, human and capital resources. Iraq has, after all, performed remarkably in rebuilding some of its infrastructure using solely its local human and financial resources. A large part of the economy already operates under a fairly competitive environment, and this needs to be supported with enabling macro and sectoral policies that should not wait until sanctions have been lifted.

Prospects for economic recovery following the resumption of limited oil exports under Security Council Resolution 986 remain bleak. Sales revenues are insignificant compared to the needs of the starving economy, and short- and medium-term economic reforms are needed to boost the economy towards greater self-reliance.

The agricultural sector will continue to be a main sector in the economy in terms of meeting food-security needs, and in generating income and employment opportunities. It can be reasonably assumed that increased resources will become available for the sector after the present sanctions have been lifted, with an enabling environment to develop the sector on a sustainable basis.

3. Major policy issues and constraints

Since 1990, and in order to face the sanctions imposed on the country, the government's policies concerning agricultural production have changed direction. With the aim of increasing and diversifying agricultural production, both quantitatively and qualitatively, and with the sole objective of achieving self-sufficiency and reducing imports, new policies have been introduced and a series of measures have been effected at the different sub-sectoral levels.

In the short term, the government's objectives include increasing the production of strategic crops like wheat, barley, paddy rice and corn, in order to contribute to the country's food-security situation and reduce food import costs. The government is also placing greater emphasis on increasing fruit and vegetable production to meet the population's nutritional requirements. Policy options being considered are to remove some of the technical and policy constraints, to include increasing the price of strategic crops, enhancing plant protection measures and assuring seed availability, providing for machinery repair, and encouraging private investment in agriculture. The success of agriculture depends on supportive macroeconomic and microeconomic policy environments. Removing technical constraints is important, but the benefits are often lessened if the policy environment is not supportive.

Macro issues

The oil sector had been the driving force for the economy. While its share fluctuated according to movements in the international oil markets, the share generally fell in the 1980s, and, even before the imposition of sanctions, Iraq's oil-based economy was suffering from structural problems.

The average annual growth of GDP, estimated at 4.4 per cent during the period 1965–75, had increased to about 10.5 per cent in 1980 before the onset of the Iran–Iraq war. The effect of the war was severe and, combined with the fall in oil prices, GDP took a constant downward trend during the second half of the 1980s. The economy was now marked by high inflation, a distorted exchange rate and a dominant public sector. Tables 4.2 and 4.3 present the trends of key economic indicators before and after sanctions in real terms.

TABLE 4.2
Selected macro indicators before sanctions

Item	1985	1986	1987	1988	1989	1990
GDP (1992 prices)*	19,367	18,148	14,917	8,921	7,250	7,280
Per capita GDP*	1,246	1,145	908	525	413	403
Private consumption expenditures*	10,123	10,118	7,670	4,592	3,873	3,675
Government consumption expenditure*	5,540	6,329	4,728	3,021	2,066	1,919
National saving*	1,201	-811	541	164	384	786
GDP (current)**	14,994	14,652	17,600	17,034	20,407	22,848
Agricultural GDP**	2,160	2,173	2,518	2,791	3,346	4,613
Agricultural GDP (%)	14.4	14.8	14.3	16.4	16.4	20.2

Notes: *Expenditure on GDP are in millions of constant 1992 US dollars. The exchange rate applied is an annually weighted average non-official rate used by ESCWA. The Iraqi dinar depreciated rapidly in unofficial markets during the 1980s, and declined much faster during the first half of the 1990s following the introduction of sanctions. **These figures are in millions of Iraqi dinars at current prices.
Source: *National Accounts Studies of the ESCWA region. Bulletin No. 15*, UNESCWA, 1995.

Whereas the agricultural sector registered a low and erratic growth, the industrial sector showed a steady increase over time since it was given high priority, especially in relation to war-supporting industries.

However, the abundance of oil and other resources was not reflected in the overall performance of the economy, and this was probably due to the very high cost of the war, and also to the serious decline in the supply of labour. Iraqi youths were drafted into the army and were initially replaced by foreign labourers, but the latter were subsequently induced to leave the country because of the war conditions, and the deteriorating foreign exchange position. Ongoing development activities were thereby seriously affected. In brief, shortage of both labour and financial resources marked the war economy of Iraq.

TABLE 4.3
Selected macro indicators after the sanctions

Item	1991	1992	1993	1994	1995
GDP (1992 prices)*	3,042	2,917	1,191	917	880
Per capita GDP*	164	153	61	46	43
Private consumption expenditure*	1,466	2,081	853	–	–
Government consumption expenditure*	1,073	452	185	–	–
National saving*	130	12	–	–	–
GDP (current)**	21,313	59,348	122,997	30,0456	–
Agricultural GDP**	6,047	20,844	–	–	–
Agricultural GDP (%)	28.4	35.1	–	–	–

Notes: *Expenditure on GDP are in millions of constant 1992 US dollars. The exchange rate applied is an annually weighted average non-official rate used by ESCWA. The Iraqi dinar depreciated rapidly in unofficial markets during the 1980s, and declined much faster during the first half of the 1990s following the introduction of sanctions. **These figures are in millions of Iraqi dinars at current prices.
Source: UNESCWA, *National Accounts Studies of the ESCWA region. Bulletin No. 15*, 1995.

The poor performance and widespread distortion of the economy forced the government to start a reform programme with policies that were geared towards transferring to a market economy and increasing the role of the private sector. A number of state enterprises were either abolished or sold to private investors. The imposition of sanctions seriously affected these economic reforms, and resulted in a shrinking of real GDP, hyperinflation, and the complete disintegration of already weak financial markets.

During the embargo era, the inherent macro imbalances were magnified to very high proportions. Exchange-rate distortions negatively affect the economy at large and in particular the food and agriculture sector, which depends heavily on traded goods. Excess money supply fuels spiralling inflation, which is reflected in a worsening of the unstable exchange-rate markets. The economy is still operating under multiple exchange rates, and these have an adverse effect on relative prices and cause allocative inefficiency. The sudden swings in the currency market seriously affect consumer budgets and producers' profits, resulting in a negative impact on potential private investment.

One of the key policy reforms is to eliminate distortion in the exchange market and to adopt a better fiscal regime and strict monetary discipline. A stable exchange rate is a necessary condition for building confidence in the economy and, therefore, for any possible future growth. However, depreciation of the dinar cannot be controlled without correcting other macro imbalances. The government needs to rely on a prudent fiscal instrument in raising revenues and cutting back on expenditure rather than on adopting a policy of monetary expansion through printing money. There are hardly any statistics on other macro variables that might indicate the direction of the economy and areas where correction is needed.

As the economy started to be unmanageable, the government realised that it needed to take harsh and unpopular measures, such as reducing subsidies, raising taxes, introducing greater monetary discipline, freeing up the exchange market, contracting the public sector and encouraging the private sector. Monetary policy that previously relied heavily on printing money to balance the budget is now pursuing a relatively conservative policy to combat hyperinflation. Increasing taxes and raising charges for public utilities that were once almost freely provided at cost, were among the first measures undertaken to revamp the budget, and other measures have included giving citizens the right to own, trade and open bank accounts in a foreign currency. Iraqi citizens are currently permitted to receive remittances from abroad and cash them in dollars or their equivalent to open-market value in local currency.

These policy reforms are beginning to indicate amelioration in the economy, as reflected in the recent relative improvement in the value of the dinar. After reaching a record low value of ID3,000 to the dollar in early 1996, the dinar improved to ID400 to the dollar in June 1996, following agreement on the 'food-for-oil' deal. However, by mid-1997 it had decreased to 1,100ID per dollar.[3] Prices of consumer goods also fell in 1996, and were expected by traders to continue to respond to changes in the value of the currency.

Trade and marketing regimes

The overall terms of trade for Iraq started deteriorating after the Iran–Iraq war. After the imposition of the UN sanctions, agricultural exports and imports fell significantly. Agricultural exports dropped from US$124 million in 1987 to a mere US$4.4 million in 1993.[4] The export of dates, which had previously been at the top of Iraq's list of agricultural exports, dropped from 157,000 tons in 1987 to 20,000 tons in 1993, reflecting the fact that they became a staple food following the imposing of sanctions.

During the same period, agricultural imports fell from US$2.6 billion to US$1.1 billion. Iraq had been a traditional importer of cereals. In 1989, imports of wheat and flour reached 3.4 million tons, but declined to 797,000 metric tons in 1993 due to the inability to finance imports.

In 1991 the trade flow became limited largely to imports from Jordan and other neighbouring countries: these, comprising mainly fruit and vegetables, were predominantly directed to the open market. Imports of cereals had been taking place on a limited scale since 1992, and some food aid had also been reaching Iraq after being approved by the UN.

Pricing, marketing and trade policies for cereals in general, and for wheat in particular, assumed higher strategic importance after sanctions. The government has therefore monopolised trade in all grains and oilseed crops, limiting its intervention in other crops (mostly vegetables and fruit) to price control. Prices are set by a central pricing committee that announces a weekly list of wholesale and retail prices after taking into consideration the prevailing conditions of supply and demand. The Ministry of Trade buys all cereals from farmers through procurement centres that have been established throughout the country, and runs a fairly successful distribution programme for the basic food needs of the Iraqi population. The long-term cost of the programme may, however, be high in terms of depressing producer prices of key agricultural commodities and of generating galloping inflation, and in the writer's view the programme has been more than necessarily overstretched.

MACRO POLICY CONSTRAINTS

These include the following: (a) the embargo-induced restriction on oil exports; (b) distortion in the exchange rate; (c) an unsustainable budget deficit; (d) a low public investment and implementation rate; (e) too large a role still played by the government; (f) the huge financial burden of public-sector institutions, which are often overstaffed, inefficient, low in morale and often in financial deficit; (g) weak legal and institutional structures for encouraging private ownership and investment.

POLICY ACTIONS AND RECOMMENDATIONS

The following are suggested:

(a) to achieve macroeconomic stabilisation, the government must decrease its fiscal deficit to sustainable levels and eliminate the multiple-exchange-rate regime;

(b) monetary and fiscal disciplines are needed to curb inflation and restore macro stability;

(c) structural adjustment needs to be initiated, with the main focus on price and trade policy reforms, privatisation of public enterprises, and regulatory reforms for attracting private investment;

(d) the government should speed up its current efforts to privatise input supply and other services;

(e) a preliminary study is recommended to determine those parastatals which are the greatest government budget liabilities, and to prioritise them for quick privatisation;

(f) the government must ensure that investment funds are increasingly directed toward private-sector agriculture in a competitive market environment;

(g) a legal, regulatory and organisational structure should be established by the government to support a competitive market structure; and

(h) government support and services need to be provided to small- and medium-size private enterprises.

Sectoral policy issues

Crop production and livestock policies

During the 1970s and 1980s, the major objectives of the country's agricultural policies were manifested in the dominant role of the state-owned agricultural institutions – mainly state farms, collective farms and production co-operatives. The idea was to limit the role of the private sector (landlords and individuals) in rural economic and political life. However, the poor performance of the agricultural sector forced the government to review its land-reform programmes and agricultural policies. With the aim of increasing efficiency and production, private-sector-led growth that minimised the role of the government in private farming was encouraged, and public–private agricultural projects were established. Later on drastic changes took place. In 1983 the government issued Law No. 35, which raised the ceiling for private landholding and permitted the lease of state land to entrepreneurs on a long-term basis.

The thrust of production policies in Iraq during the 1970s and 1980s aimed to achieve efficient utilisation of available resources, diversification of production, improvements in productivity, increased incomes for farmers, and the meeting of domestic food requirements. Before the war with Iran, the government was providing heavily subsidised farm equipment to farmers through farmers' cooperatives. Seasonal immigrant farm labour was easily available, and many landowners chose to live in cities, leaving the cultivation of land to tenants. Good storage and marketing facilities were available, irrigation pumps were widely installed, and vaccinations for poultry and livestock were available at subsidised prices. An extension service was also available through more than 3,000 field-extension agents. In short, farmers enjoyed strong government support. Unfortunately, however, productivity gains did not match the resources provided to the sector.

Further policy reforms commenced in 1987 with the objective of revitalising the private sector and encouraging investment in agriculture. The government sold most of the state farms and agricultural projects to the private sector or to public–private companies, with a view to improving the sector's efficiency and performance. In the late 1980s, productivity gains started to become apparent but were arrested by the crisis triggered by the embargo, and conditions forced the government to renounce its reform programme and return to public controls.

The government's policies under sanctions are two-pronged: to provide farmers with incentive prices in key or strategic commodities (such as wheat and barley), and to enforce a strict acreage and procurement policy. The government sets the procurement prices for these crops, purchases the farmers' produce, and sells it directly to consumers at much lower prices. Trading of the rest of the crops (mainly vegetables and fruit) is market-based, but their prices are set by a central pricing committee which announces a weekly list of wholesale and retail prices, taking into consideration the prevailing supply and demand conditions. The marketing of grains and oil crops is a government monopoly, while the marketing of other agricultural products such as fruit and vegetables is free from government control.

Producer prices for wheat, barley, paddy and maize are supported by the government and, except for rice, are procured through a compulsory delivery system. Wheat is the most important crop in Iraq. Its pricing policy has undergone several upward revisions since the enforcement of sanctions. The purchase price of wheat, which was ID270 per tonne in 1989, increased to ID600 in 1990, then to ID800 in 1991, to ID2,000 in 1992, to ID5,000 in 1993, and to ID35,000 in 1994. In 1995, the price again increased, to ID105,000, and this was the prevailing price for the 1996–7 growing season. The wheat price was fixed at this high rate to compensate for the rising cost of inputs and to encourage farmers to increase wheat production.

Farmers in irrigated areas are obliged to plant wheat and barley on 50 per cent of their holdings. However, this arrangement does not apply for holdings of less than ten *donums*. The percentage of forced wheat and barley production increases to 80 per cent of the holdings in rainfall areas. All farmers have to deliver their entire production of barley and wheat to the government, calculated to a minimum of 200 kg per *donum*. Private trade in wheat and barley is prohibited. The same arrangement was applied to corn production, with effect from 1991 (see note 44 in Chapter 1).

Livestock production is an important sub-sector of agriculture, contributing significantly to GDP and to the country's food security, although the productivity of this sector is low. One of the major factors limiting growth and development is the scarcity of animal feed, which results in malnutrition and susceptibility to animal diseases. Superimposed on prevailing climatological and ecological factors are the effects of embargo and sanctions. Because of limited financial resources the improvement programmes have been stopped. Moreover, due to government policies of augmenting grain production for human consumption, large areas of rangeland have gone into cereal production. In order to further enhance wheat and barley production, the government has officially allowed these crops to be grown in the rangelands.[5] For this reason, livestock/crop integration is decreasing and is mainly dependent on stubble grazing.

Agricultural inputs are provided by the Ministry of Agriculture and Irrigation and its public agencies. Fertilisers, chemicals, insecticides and farm machinery are provided through the Agricultural Supply Company, while improved seeds are provided through the Mesopotamia Company for the Improvement and Production of Seeds. These inputs are provided to farmers at subsidised prices, with the aim of promoting agricultural production and providing cheap food for consumers, although the government has recently reduced its subsidy for some of these inputs. They are therefore provided in very limited quantities and can only be used for strategic crops. Because prices on the open market for some of the subsidised inputs are so high, farmers are often tempted to sell them on the open market.

The situation is more difficult for small farmers. Hardly any subsistence farmers have the ability to purchase fertiliser from the market, and no fertiliser has been distributed by the humanitarian programme during recent seasons. Production of fertilisers has also been seriously affected by the war and by sanctions. Annual production of nitrogen dropped from about 1,000,000 to about 450,000 metric tons before and after 1991. The figures for phosphorous and nitrate are 425,000 and 100,000 metric tons, respectively.[6]

Policy constraints in plant protection are clearly related to the imposition of sanctions. The implementation of aerial control operations is now more difficult due to the extension of the southern 'no-fly' zone to the 33rd parallel, and in the autonomous region of northern Iraq the problems are compounded by the imposition of additional sanctions by the Iraqi government and by internal conflict. Another complication includes the policy of free or highly subsidised input supply, as practised prior to the imposition of sanctions. Despite the shortages this has largely been continued post-sanctions, in part with FAO support. If it becomes possible to phase out input subsidies, then considerations of security of land tenure, credit availability, and marketing and pricing policies will all become increasingly important for the farmers if they are to adopt more sustainable pest-management strategies. Further, the long-standing poor performance of the extension service, and the lack of real coordination between research and extension, will also remain a major constraint to the development of more rational pest-management strategies.

The policy environments facing the agriculture sector in the northern region of Iraq outside government control are different from those in the central and southern regions, and the north-western areas. In the north the main cereal crop grown is wheat, while the region is rich in a variety of fruit orchards. Unlike the situation in the central and southern governorates, there is no government intervention concerning crop production and, similarly, there is no provision of subsidised farm inputs by the government.[7] The farmers are not obliged to hand over the production of strategic crops to the authorities and the trade is generally free. This also stems from the fact that there is no food-rationing system of the sort that often lures governments to adopt distortionary policies such as acreage and price control and compulsory procurement.

Production is responding to market signals and there is active trade, while food is available in the markets, though at exorbitant prices. The food shortages in the southern and central provinces and the price differentials between the two regions (and often in export markets in Iran and Turkey), continue to provide market access to farmers producing a marketable surplus. This often results in a steep price increase in the local markets, and consumers are the net losers. However, at times it is only the political reality on the ground that determines the degree of market access, and farmers are left with decreasing prices, often to the benefit of consumers. The trade between the northern provinces and the central government has also been active, due to mutual economic gains and comparative advantage. Cheap oil (at 0.002 cents per litre) is exported to the north in order to import northern wheat at half the international prices. The authorities in the north gain, since they impose certain local taxes on agricultural produce going out of the region and are also able to import badly needed oil at less than competitive prices.

However, the region suffers from increasing deforestation, mainly because of woodcutting for provision of energy, which results in land degradation. Frequent strife among rival factions contributes to resource depletion. During the war, large numbers of mines were planted for which no maps are available, and this increases the general feeling of insecurity.

The policy impact

The impact on strategic crops of current government price and acreage policies, based on data collected for average farm producers as well as the ones with improved technology in Iraq, is presented in Tables 4.4 and 4.5. The analysis should be viewed as only preliminary. In the case of wheat, analysis showed that farmers are receiving 34 per cent of the world price, and even when these prices are adjusted for subsidy on traded inputs, the result is that farmers are receiving only 43 per cent of international prices. It needs to be pointed out that this analysis holds only if we consider world prices as a yardstick of efficiency prices, or if farmers are paying world prices for traded inputs. However, in view of the shortages of inputs, farmers are buying a sizeable part of the traded inputs from the open market. Moreover, in many cases they sell the subsidised inputs in the open market. Nevertheless, the availability of inputs has been one of the main constraints to developing production, and hence the response to output prices might to a great extent be subdued. This means that the impact of increased prices would be more on the area planted than on production, since productivity is depressed by the unavailability of traded inputs.

TABLE 4.4

Preliminary estimates of incentive structure facing the agriculture and domestic resources cost (average for Iraq with optimistic yield)

	Wheat	Barley
NPC	0.34	0.27
NPI	0.22	0.21
EPC	0.43	0.37
Domestic resource cost	0.67	1.57

Source: Field research by the author and local experts.

TABLE 4.5

Preliminary estimates of incentive structure facing the agriculture and domestic resources cost (average for Iraq with average yield)

	Wheat	Barley
NPC	0.34	0.29
NPI	0.22	0.21
EPC	0.43	0.50
Domestic resource cost	0.99	3.12

Source: Field research by the author and local experts.

The government is also promoting cultivation of barley in both irrigated and non-irrigated areas, through a price support policy that increased the price of barley from ID120 per ton in 1988 to ID60,000 per ton at the time of writing, almost maintaining the price in terms of free market US dollars. The impact of price increases for barley on area and production has been better, in view of the increasing importance of barley for livestock and poultry production caused by the refusal to allow imports of concentrates under sanctions.

The impact of the current government policies on barley can also be seen by examining the private and social profitability and key policy indicators. The nominal protection of 0.29 indicates that the farmers are only receiving 29 per cent of the world price, which is even lower than that for wheat. Adjusting for subsidy on traded inputs, the effective protection indicates that farmers are receiving only 33 per cent of international prices. With the present pricing structure, farmers are facing negative profitability. A negative social profit may be taken to indicate that production of barley does not enjoy a comparative advantage, and accordingly government support policies for barley need to be reviewed.

On balance, government-instituted price and acreage control policies are definitely bringing more areas under cultivation for strategic crops, but productivity gains have not been forthcoming under these policies. It is very difficult to envisage productivity growth when technical packages, such as fertilisers and good-quality seeds, are in short supply or are simply not available to the majority of the farmers. Poor performance of the agriculture sector, even under relatively favourable policy environments, can be partly attributed to lack of input supply, such as certified seeds, high-crop varieties, herbicides, pesticides, fertilisers, crop protection, agricultural machinery and spare parts. Fertilisers are currently applied to cereal crops only, and represent one-fifth of the recommended rates. Other factors affecting cereal production include poor land preparation and lack of pesticides and herbicides. Existing machinery is old and in poor shape. It is estimated that around 5 to 10 per cent of harvest losses are due to worn-out harvesting machinery. Further, the policy of promoting cereals in marginal lands and rangelands might even decrease the long-term sustainability of the sector to produce more.

A comparison reflects on the relatively free market environment under which the northern farmers operate. The output price is very close to international prices. With the exception of seed, the input prices are quite competitive since most of the inputs are imported (smuggled) from neighbouring countries. This means farmers are allocating resources to the crops with a favourable comparative advantage and are using the resources more efficiently. The productivity of wheat has, if anything, improved. The region is able to export wheat in order to import badly needed oil. Marketing of wheat seems to be a real problem for farmers wishing to expand the area under cereals.

POLICY CONSTRAINTS IN CROP PRODUCTION AND LIVESTOCK

Among the factors affecting crop and livestock production are: (a) lack of appropriate pricing, marketing and trade policies; (b) a productivity gap; (c) the non-availability of good-quality, fungicide-treated, weed-free seed; (d) increased planting in marginal lands; (e) the absence of appropriate crop rotations; (f) lack of input availability; (g) a highly subsidised input supply that limits rational and sustainable pest management; (h) small and fragmented land-holdings; (i) low productivity of the livestock sector; (j) traditional systems of raising livestock; (k) shortages of feedstuffs leading to malnutrition; and (l) sanctions-affected local production of vaccines and other veterinary supplies.

POLICY ACTIONS

Suggested policy actions and recommendations for crop and livestock production include:

(a) improvement of productivity through the provision of enabling policy environments, and improved technology transfer through research and extension;

(b) reliance on crops that will have a comparative advantage once sanctions are removed;

(c) strengthening the seed-production programme;

(d) removal of shortages of the most important inputs (such as improved seeds, fertilisers, pesticides), and mechanisation;

(e) gradual removal of the subsidy on pesticides;

(f) emphasis on better farm-management practices that can lead to a sustainable farming system;

(g) improvement of marketing infrastructures and financial institutions, in order to provide credit;

(h) formulation of a survival strategy concentrating on disease control and eradication;

(i) provision of a sufficient feed base, particularly during the dry season;

(j) improvement in the performance of indigenous sheep and goats through genetic and animal husbandry methods; and

(k) improvement in the carrying capacity of rangelands.

Natural resource management policies
Because the country is endeavouring to increase food production under sanctions conditions in order to meet the increasing demands of a rapidly growing population, natural resources in Iraq are under severe stress. The supply-oriented strategy of self-sufficiency for wheat and barley has led to area expansion, bringing more marginal land under cultivation and reducing the extent of the area left fallow. In other words, land and water resources are being over-exploited, and this will result in an extensive decline in future productivity. The efficient management of natural resources means developing agriculture on a sustainable basis. Therefore it is impossible to separate environmental policy from economic policy and development strategy. An integrated strategy towards economic development will be required, so as to address the problem through a holistic approach to future expansion of the sector.

A previous lack of demand-management practices has also contributed to the present low efficiency in water use. At the same time, the improvements in water availability that resulted from the somewhat uneven technology introduced in the past have diverted attention from demand management, and have also reduced the emphasis on low-cost alternatives such as improvements in efficiency, conservation and reduction of waste through maintenance. Policies for demand management should focus on the efficient use of water, supplementary irrigation, appropriate maintenance of the irrigation infrastructure, better land use and tenure policies, and the application of appropriate water charges.

After energy, irrigation water represents the main subsidy on the input side. These subsidies constitute a financial drain on the government, especially under present conditions. However, in 1995, the government issued Law No. 12, which placed water charges of ID750 and ID500 per *donum* per year respectively for reclaimed and non-reclaimed irrigated land. Farmers also are obliged to maintain their on-farm drainage networks, or else maintenance will be carried out by the government at their expense. As a result, the government was able to maintain 11,000 km of the drainage system in 1996. Given that water generates considerable rent, the current irrigation tariff per *donum* is still below the operation and maintenance costs, nor does it lead to the rationing of agricultural water use. On the other hand, water charges are area-based rather than being crop- or volume-based, and therefore any increase in water use has zero marginal cost. A future water-pricing policy could be based on detailed water-tariff studies, taking efficiency, equity and environmental issues into consideration.

Moreover, current pricing policy for wheat and barley has a direct bearing on the sustainable use of rangelands in Iraq. With the collapse in the exchange rate of the Iraqi dinar under sanctions, the support price for wheat and barley is encouraging farmers to bring more area under these crops, and the policy is resulting in continuous cropping rather than cereal/fallow rotations. The situation is further exacerbated by allowing farmers to plant wheat and barley on rangelands. Such government policy has to be reviewed to give due consideration to long-term productivity and conservation of the natural resources base.

POLICY CONSTRAINT IN NATURAL RESOURCES MANAGEMENT

Current constraints in the management of natural resources include the following: (a) agriculture is being consumed by waterlogging and salinity; (b) clogging with weeds of the canal system is affecting agricultural productivity; (c) there is higher water consumption at lower water-use efficiency; (d) the low cost of irrigation water constitutes a disincentive to adopting modern technology, combined with inappropriate institutional arrangements and absence of a regulatory framework; (e) public awareness is lacking; (f) depletion and quality-degradation of freshwater increase cost and depresses profits; (g) there is the problem of deforestation in the north; (h) cereal expansion policy is causing rangeland and ecosystem degradation.

POLICY ACTIONS/RECOMMENDATIONS IN NATURAL RESOURCES MANAGEMENT

The following actions/recommendation in dealing with natural resources management are suggested:

(a) giving emergency assistance in the maintenance of the irrigation infrastructure;
(b) carrying out pre-investment studies in the water sector;
(c) offering economic incentives to adopt water-saving technology, such as sprinkler and drip irrigation, and improving water-harvesting structures in rainfed areas;
(d) setting up short-term irrigation pricing charging, cost recovery and, in the long term, full-cost pricing;
(e) encouraging farmers to participate in the development and management of irrigation water;

(f)   campaigning for public awareness of the rational use of water and in its inter- and intra-sectoral use;

(g)   developing a resources conservation strategy with greater focus in/on the north;

(h)   internalising externalities to enable progress towards full-cost pricing;

(i)   establishing an appropriate policy for cereal/livestock/rangeland management integration;

(j)   adopting an integrated approach to natural resources management.

Food security and poverty alleviation policies

The concept of food security relies on three components: food availability, food stability and food accessibility. The production of food, as analysed above, has been erratic and the food gap tends to be of considerable magnitude, especially for wheat and rice (Table 4.6). With the declining food-security situation, Iraq will continue to look to international markets to meet its domestic needs, and to do this, the country needs to have access to its resources (mainly oil) in order to enhance its purchasing power and enable it to buy food from oil revenues. It is therefore crucial to lift the embargo imposed on Iraq to enable it to utilise its resources and achieve food security for the whole population.

TABLE 4.6
Per capita production and consumption of cereals in Iraq, 1970–95

Crops	Production (kg/capita/year)	Consumption (kg/capita/year)
Wheat	77	145
Barley	56	56
Rice	13	29

Source: 'Agriculture at a Glance', Working Paper prepared for FAO Programming Mission, October, 1996.

The adoption of sustainable production systems requires rational use of available land and water resources. Preservation of the country's natural resources capital will determine the long-term sustainability of the sector. A first step towards this end is to avoid agricultural practices that degrade natural resources, such as monocropping and the non-rational use of land and water, and to halt the deforestation that is occurring in the north.

At present, access to good-quality food is very difficult for an average Iraqi. To meet food-security needs under sanctions, the government provides minimal food to all Iraqi families through a food-rationing programme, which involves procuring strategic food crops from the farmers and selling them to consumers at much lower prices. This entails a huge subsidy to the consumer and an enormous fiscal cost to the government. The food distribution policy that has been in place since sanctions provides a food basket across the board to consumers that meets only one-third of their food-energy needs. Initially, a basket of food consisting of wheat flour (7 kg), rice (1.50 kg), vegetable oil (0.750 kg), sugar (0.5 kg), tea (0.1 kg), and some other non-food items was provided to a family of five at a price of ID48.[8] Government policy subsequently increased the price for the same basket of goods to ID500 – this move indicated a substantial reduction in subsidy but was still not sufficiently large to make any substantial reduction in the financial cost to the government. Under the more recent 'food-for-oil' deal, the food rations have been increased.

The quality of this food is often below standard, but still offers considerable relief to the large part of the population that lives under conditions of poverty. The distribution system is efficient, equitable and functions very well. However, government intervention at both the producer and the consumer level is also one of the factors adding to the country's spiralling inflation and the programme's high cost, pre-empting productive investment in agriculture and other related sectors, and contributing to the declining productivity of the sector. A programme for alleviating poverty that meets the basic needs of the impoverished sections of the population can be more cost effective and sustainable if channelled through the system as a targeted subsidy to the poor.

FOOD-SECURITY POLICY CONSTRAINTS
Constraints on the food-security policy include: (a) the embargo, which limits imports; (b) unstable production; (c) a policy bias towards cereal production; (d) the consumer subsidy on wheat which depresses producer prices; (e) lack of access to food, due to high prices; (f) sanctions-induced poverty, which is estimated to be widespread and leads to destitution and to malnutrition in children; and (g) a food-rationing programme that is costly and difficult to sustain in the long term.

POLICY ACTIONS

Possible actions in relation to the food security policy include the following:

(a) a targeted subsidy that enhances food security only for the needy;
(b) bringing producer prices in line with international prices;
(c) redirection of programmes by providing safety nets only for the poor;
(d) focusing in the medium term on policies which will increase economic growth;
(e) introducing growth/productivity improvement in the agriculture sector that can reduce rural poverty;
(f) investment in human-resource development, with a greater focus on women.

4. Strategic policy options

Agricultural policies in Iraq need broadly-based reform in order to realise efficiency, equity and sustainability in the agriculture sector within the compass of an integrated programme. In the short term, the sector's policy and institutional reforms should include a more open economy, more efficient use of land and water resources, improving of productivity through a viable research and extension programme, and rational use of measures for plant production and protection. Crops for which production should be increased include strategic crops like wheat, barley, paddy rice, and corn, along with fruits and vegetables.

The efficient use of land and water is the key to the performance of the agriculture sector, and is also closely related to the issue of greater food security and environmentally sustainable growth. Excessive grazing and timber-cutting in the northern provinces, primarily due to fuel shortages, are denuding them of their protective vegetative cover, causing desertification and land degradation, while free or highly subsidised inputs continue to constrain the possibility of implementing rational and sustainable pest-management strategies. These factors have contributed to the low resource productivity and the deteriorating food-security situation. The subsidy on input (water and pesticides) does not provide farmers with the correct economic signals or give any idea of the true scarcity value of meagre resources, and often results in over-production and over-consumption of certain crops because of the lower cost of production. The social cost (which includes the production subsidy), the depleting cost and the environmental costs are much higher, and are not paid for by the producers and consumers of the commodity. Removal of these subsidies and the adoption of improved technology, along with institutional reform in the land and water sector, will enhance productivity, contribute to resource conservation and reverse its degradation. Additionally sufficient funds will be generated to rehabilitate the faltering infrastructure, which is the key to the sustainability of the sector growth and food security.

In the medium term (partial or full lifting of the embargo), a prudent economic policy will encourage the allocation of resources among those crops that enjoy a comparative advantage, are water efficient and can compete in the global market. It would be desirable to achieve self-reliance rather than self-sufficiency. Such a policy would ensure long-term food security, through promoting the export of high-value crops and utilising their proceeds to finance imports needed to bridge the food-security gap, if any. Further, under changing global environments (GATT and the regional groupings), Iraq needs a sound domestic policy environment which promotes trade and channels investment flows, while protecting its natural resource capital. The water deficient regions in the Near East also provide an excellent opportunity for the sector to trade in 'virtual water'.

Policy adjustments might be needed to address problems of food security and poverty alleviation in the wake of rising food prices under sanctions. Unlike many other countries in the region, Iraq has the comparative advantage of being able to grow irrigated wheat. Domestic Resource Cost (DRC) analysis suggests that rainfed and irrigated wheat has low DRCs (less than one), and policies should be encouraged that will strengthen its cultivation in order to meet Iraq's needs and most probably the needs of the export markets.

In the short term, a competitive price regime and improvements in yield will enhance the production base, eliminate smuggling and directly alleviate poverty for a large section of the rural population. In the long term it would be desirable to close the food gap with a rational policy that makes the country economically self-sufficient in food production. There is considerable potential to enhance rainfed wheat production through the introduction of water harvesting and supplementary irrigation, along with other improved farm-management practices. The irrigated areas can concentrate on producing high-value crops, which have a comparative advantage, and which can also compete in regional markets. In order to compete in local, regional and global markets, production policies should be flexible enough to meet changing market demand. This will also require a concerted effort to create a viable and modern market-information system, capable of disseminating price forecasts and other statistical information concerning the outlook for the commodity market.

5. Strategic policy recommendations

Emerging challenges and constraints facing agriculture in Iraq call for a rethinking of the agricultural development strategy. At present, there is an urgent need for rehabilitating the sector and for putting greater emphasis on the optimum utilisation of its natural, human and capital resources. Key themes underlying the short-term (emergency) and medium-term policies include the following:[9]

(a)   Implementation of a *macroeconomic stabilisation programme* with a view to restoring confidence in the economy, the most important move being removal of multiple exchange rates and a reduction of expansionary monetary policy.

(b)   The carrying out of *sectoral policy and institutional reforms*, with the aim of gradually removing consumer subsidies and instituting a targeted subsidy for the poor; reducing the cost of support programmes and input subsidies; bridging the productivity gap; providing incentives to grow crops with comparative advantage; further liberalising trade and revamping agricultural credit institutions; providing effective research and extension services; and strengthening the technical and institutional capacities of the Ministry of Agriculture and Industry in the area of planning and policy analysis in order to meet the challenges of the new role of government in the economy.

(c)   Enhancement of *productivity and balanced growth*, which is paramount to national, regional and household food security. In this connection, due consideration should be given to improving cereal production through programmes for increasing the current low yield levels of wheat, barley, rice and maize. Actions focusing on the development of wheat-based cropping systems for the cereal-producing areas of the central and northern regions, and rice-based cropping systems for irrigated areas in the southern and central provinces should be given appropriate priority. In this respect, high priority should be given to the development of high-yielding seed varieties of good quality, application of appropriate crop rotations for each zone in order to reduce weed and pest infestation, and improvement of soil productivity. Livestock development policies that take into consideration disease control and eradication, along with rangeland improvement, should be emphasised.

(d) Promotion of the *rational use of natural resources* to sustain long-term productivity of the agricultural sector. Improving rangelands and halting the deforestation of large areas in the north should attract a high priority. Rehabilitation of the irrigation infrastructure to arrest waterlogging and salinity and to maintain irrigation canals is urgently needed. Demand-management policies in water resources development, including a rational pricing policy, are also needed. A natural resources conservation policy, based on detailed sub-sector reviews, is highly recommended as a way to develop the agriculture sector on a sustainable basis.

NOTES

1 The views expressed in this paper are those of the author and do not necessarily reflect the views of the Food and Agriculture Organization of the United Nations.
2 *Country Report: Middle East and North Africa*, London: The Economic Intelligence Unit, 1996.
3 The dinar resumed a relatively slow rate of depreciation and was negatively affected by the recurrence of tension and military conflict in the late autumn of 1997. By March 1999, it stood around ID2,000 per dollar and it has fluctuated close to that rate into the second half of 2001.
4 Figures taken from the FAO statistical data base on the Internet.
5 K.H. Shideed, 'Crop/Livestock Integration Technology Adoption, Agricultural Policies and Property Rights under Rainfed Conditions', 'Mashreq/Maghreb Policy Workshop', Tunis, 8–10 October, 1995.
6 Figures were taken from data provided by the Iraq Ministry of Agriculture and the FAO office in Baghdad.

7 Food and Agriculture Organization, 'Future Role of FAO in Post Embargo Iraqi Agriculture', May 1996.
8 'FAO–Iraq Strategy for Agricultural Development', *Horizon 2010, World Food Summit,* follow-up, 1997.
9 Ibid.

5

Iraq's Manufacturing Industry: Status and Prospects for Rehabilitation and Reform

Tariq Al-Khudayri

1. Introduction

During the past 17 years or so, Iraq's economy and infrastructure as well as its social fabric have been severely damaged. A decade of armed conflict with neighbouring countries has resulted in heavy debts and a severe economic blockade, along with internal political tensions provoked by an unprecedentedly bureaucratic totalitarian regime. A whole generation has been traumatised, the untroubled coexistence of the country's multisectarian or confessional and ethnic communities has been disrupted, and there has been a large-scale emigration of the educated and technically qualified sector of the population. As well as a severe decline in the standard of living, health and education systems have suffered badly, as has the country's important physical infrastructure, including industrial production. The manufacturing sector in particular has sustained heavy damage, not only because of the bombing of factories and off-site facilities during the wars but also because of the disruption in implementing industrial development programmes. These latter are essential for integrating the country's production facilities and strengthening its technological capacity.

The following paragraphs are designed, as far as available information permits, to give a brief account of the status of manufacturing industry in Iraq and some thoughts with respect to its future and the prospects for its rehabilitation and development.

2. The status of manufacturing industry

Review of Iraq's industrial policy development and framework

Prior to the 1950s, small- and medium-scale industries in Iraq were developed solely on the initiative of the private sector. A formally planned industrialisation development programme was launched only after 1952, when the state introduced its own investment schemes to complement, and support, the activities of the private sector. The government's intention was to invest in, and develop, an industrial base that the private sector could not achieve by itself, because of its financial and technical limitations. The aims of this initiative were: (a) to contribute to economic growth and employment generation; (b) to make maximum use of local resources by investing in resource-based industries; and (c) to encourage viable import substitution industries.

Following the 1958 Revolution, government policy became firmly based on the principle of central planning. However, the public sector lacked a dominant role until 1964, when 27 major manufacturing establishments were nationalised. The motives behind nationalisation were largely political rather than socio-economic, and the principle of a centrally planned economy and maximum state intervention in manufacturing became more deep-rooted after 1968. Since then, the public sector has been responsible for the greater share of industrial investment, employment and output. It should be noted, however, that public ownership imposed severe constraints on industrial enterprises, resulting in a low or even negative return on investment. While poor management of industrial establishments can be attributed partly to political interference, which in turn may stimulate bureaucratic control, the latter is, in general, the result of a complicated public administration system typical of those prevalent in developing countries.

Increased oil prices in the 1970s were a major factor behind the high investment in industry in general, especially in the public sector where development of heavy and large-scale industries was achieved at a total cost of over US$14 billion during 1975–80.[1] The outbreak of the first Gulf War in 1980 brought about severe damage to most of the production facilities of these industries, as well as to others. It also disrupted oil exports, reduced revenues and, in turn, led also to a near halt in public investment in new projects, apart from those related to essential strategic industries, such as oil refining, fertilisers and power plants. Outside loans were sought for these projects, thereby adding to the dramatic rise in foreign debts that had started to build up during the 1980s. Furthermore, the war needs of the military services led to a shortage in skilled manpower.

Apart from the consequences of war, the poor performance of the public-sector enterprises resulted in a heavy financial burden on the state budget. This, combined with the Iran–Iraq war's draining of the country's dwindling oil revenues and an international shift that favoured a market economy (supported by Western countries that might extend political and even military backing for its war efforts), prompted the Iraq government to indulge in a privatisation programme during 1987–8. More than 50 public manufacturing enterprises in various sub-sectors were sold to the private sector through an auction system that ensured higher revenue, although it was not without faults. The government also issued directives that stipulated the decentralisation of public-sector industries and that provided for more financial and managerial autonomy and responsibility, with the aim of improving their efficiency and viability.

In its efforts to create a favourable climate for higher investment in manufacturing, the government also published Law No. 46 of 1988 (which incorporated the legal and regulatory framework for Arab investment), as well as other laws and directives for the promotion of industrial development (such as the Revolutionary Command Council's Directive No. 774).[2] All these were designed to encourage private investment and create a healthy environment for entrepreneurs. Aspects addressed by these directives included, among others:

(a) permitting the import of essential inputs with no limitation or control by the Central Bank if no foreign exchange transactions were involved;

(b) allowing the private sector to compete in all branches of manufacturing, including those sub-sectors that had been confined to the public sector since the 1960s and 1970s; and

(c) raising the ceiling of the capital for private joint-stock companies to ID7.0 million, and for companies with limited liability to ID1.5 million (having been held at ID70,000 since 1964).

Further legislation published in 1989 exempted all existing and future industrial enterprises from all taxes for a period of ten years.

In spite of all these laws and directives, the public sector continued to have a direct impact on industrial activities. This was probably because there was insufficient time and an unsuitable climate for any effective application of the privatisation measures and incentives offered to national and Arab entrepreneurs before the invasion of Kuwait in August 1990 and the start of the second Gulf War crisis. One can reasonably argue that the private sector had had only limited experience in all aspects of industrial development. In addition, the period for normalisation and stability that followed the launch of the reform measures after the first Gulf War was too short for any significant improvement, either in the efficiency of privatised companies or even in public-sector enterprises.

The legal and regulatory structure

Generally speaking, and leaving aside the large-scale and heavy industries developed by the public sector, the structure of the manufacturing sector is characterised by small- and medium-scale industries (SMIs). Small and medium-size enterprises employing fewer than 30 workers have been active in specific sub-sectors of small-scale industries (food and beverages, clothing and leather footwear, building materials, wood and plastics products, metal fabrication, and so on). Most of these enterprises, especially those employing fewer than ten workers, are family- or individually-owned businesses, and their production facilities typically have a very low level of mechanised processes and technology (and hence a high level of manual operations). Around 45 per cent of them are involved mainly in industrial services (maintenance and repair, and so forth), while the rest, especially among those employing between ten and 29 workers, are usually joint-stock companies or take the form of partnerships and family concerns. Their production facilities are adequately mechanised (even though they usually incorporate appropriate technology[3] or consist mainly of assembly lines and simple operations such as mixing and packing).

Large enterprises are defined as those employing more than 29 workers and with a fixed capital value in non-power generation equipment exceeding ID100,000 in value (as at the time of the prevalence of the official exchange rate of US$3.2 per dinar). They are usually either joint-stock companies or are owned by the public sector, and in general their production facilities are highly mechanised and incorporate modern or advanced technology.

Using the above definitions, and without taking into consideration the technology content when differentiating between different categories, an industrial census for all registered manufacturing enterprises presents the following picture:

TABLE 5.1
Structure of manufacturing industry in 1988

Enterprises according to size and ownership	Number	Employment		Salaries		Output	
		No.	%	1000ID	%	1000ID	%
Large enterprises							
public sector	142	126,324	82	240,166	90	2,185,065	82
private and mixed sector	498	27,777	18	26,059	10	479,353	18
Sub-total	640	154,101	100	266,225	100	2,664,418	100
Small enterprises employing less							
than 10	39,460	91,295	95	90,455	94	1,176,672	94
employing 10–29	280	4,459	5	6,082	6	74,919	6
Sub-total	39,740	95,754	100	96,537	100	1,251,591	100
Grand total	40,380	249,855		362,762		3,916,009	

Source: Based on official data in the present author's 'Industrial Development in the Republic of Iraq' (in Arabic), presented at the conference on Industrial Development in the Arab Countries, organised by the Arab Organisation for Industrial Development, Tunis, 1989. It is unlikely that the data includes any information on the military manufacturing facility.

In 1988, according to Table 5.1, large enterprises employed about 62 per cent of the total workforce in manufacturing industries, paying 75 per cent of its total wages and salaries, while their output constituted 68 per cent of the overall manufacturing output. At that time, the employment, wages and output of the public-sector enterprises represented 51 per cent, 66 per cent, and 56 per cent respectively. Table 5.2 represents a simple comparative assessment of the performance of the different categories of enterprises, based on Table 5.1 above:

Usually – or at least since the late 1980s – it is only the large industrial enterprises that are considered for the real assessment of the manufacturing sector. According to a government-conducted industrial survey in 1991, large industrial enterprises employed about 128,000 workers and their output was estimated at ID83 million. At that time, the public sector owned 107 out of a total of 574 large enterprises and employed 78 per cent of their total workforce, while the private sector owned 427 enterprises and employed 12 per cent of the total workforce.[4] The rest belonged to the mixed sector (which represents joint-stock companies owned jointly by the public and the private sectors), and to the cooperatives sector.

TABLE 5.2
Comparative assessment of enterprise performance

	Productivity	Average wage	Wage share of output
	ID per employee	ID	%
Large establishments	17,290	1,728	10.00
Public sector	17,297	1,901	11.00
Private and mixed sectors	17,257	938	5.40
Small establishments	13,071	1,008	7.71
employing less than 10	12,889	991	7.69
employing 10–29	16,801	1,364	8.12

Source: Figures calculated from Table 5.1.

We would argue that, for technical and practical purposes, only those enterprises that meet both the minimum fixed-capital value and the employment criteria should be considered for the assessment of the manufacturing sector. However, many researchers consider that all those enterprises whose employment exceeds ten workers, irrespective of the capital invested in equipment, should be considered as part of the manufacturing sector. The difference of approach on this matter has led to discrepancies in the statistical data pertaining to the structure and performance of the manufacturing sector. Thus, in the absence of any technical assessment of the production facility of small-scale industries, the criteria for categorising the relevant enterprises may not be completely adequate.

3. Performance of the manufacturing sector

While the manufacturing sector has grown or developed more or less steadily over the past 25 years, its contribution to Iraq's gross domestic product (GDP) has, for various reasons, fluctuated considerably since the early 1970s. The contribution of the mining sector (mainly oil) to GDP has moved between two extremes, due not only to high rises and falls in production, but also to the world-market situation, while an equally important factor has been the dependence of oil-exporting and industrial activities generally on domestic and regional peace and stability. The wars instigated by the Iraqi regime and the latter's politics have led to destruction and idleness or, at the very least, have resulted in a limited capacity utilisation of a large number of manufacturing plants.

The performance of the manufacturing sector as a whole, however, has been inconsistent, due partly to some of the factors indicated above, but also for other reasons. These include the overwhelming bureaucratic control and hegemony of the public sector, the manufacturing ability of which suffered from inappropriate policies regarding employment, price control and over-protection procedures, all of which brought about low productivity in certain sub-sectors.

During the 1980s, the performance of the manufacturing sector was poor in comparison with the preceding decade. In the second half of the 1970s, according to available statistics, manufacturing value added (MVA) grew at an average annual compounded rate of fifteen per cent (at 1975 prices). This was even higher than the high growth rate of GDP of the period. MVA reached its highest level in 1980, the year in which the first Gulf War started, going into negative growth thereafter until 1984, falling annually by an average rate of 2.2 per cent (also at 1975 prices) before starting to climb again after 1984. In the meantime, overall employment in the manufacturing sector for all categories of enterprises declined from a peak of 254,000 in 1980 to 218,000 in 1984. During this period the military-service draft was in force, and the war, which had in its first few weeks caused such severe damage to the manufacturing facilities, was still in progress.

Toward the end of the 1980s, the annual growth rate went up, registering 4.2 per cent in 1988 and 5.1 per cent in 1989. It then slid back to 4.6 per cent in 1990, the year that marked the beginning of the crises that led to the second Gulf War. The MVA growth rate during those three years was higher than the growth rate of GDP, but lower than the average rate achieved by other West Asian and North African countries except Egypt.[5]

The whole economic environment started to improve after the eight-year war with Iran ended in August 1988; this, coupled with the international recovery of oil prices in early 1989, led to anticipation of a reconstruction boom period. The government drew up an ambitious development plan to improve prospects for industrial growth during the 1990s, and for this purpose, among others, a wide-ranging cooperation programme was initiated with various advanced manufacturing and specialised high-technology companies from several developed countries. Liberalisation of trade was expected to be important for securing finance and mainly advanced technology, although oil revenues would play the major role. Needless to say, the predominance of oil in the country's economy has the effect of constraining the industrialisation process through both demand and supply-side effects. This is because oil-export revenues, by providing a means for production through the importing of intermediate inputs, capital goods, spare parts and technologies that ultimately encourage higher local consumption, effectively promote further reliance on imports of finished products and inputs for highly protected industries.

After 1990, a sharp decline in MVA and manufacturing output took place, not only because of the military operations of early 1991 that followed the crisis provoked by the invasion of Kuwait, but also because of the economic blockade imposed by UN Security Council Resolutions Nos. 661 and 687. Employment and productivity also declined, but there were other reasons too.

Table 5.3 illustrates the serious decline in the manufacturing sector in real terms to reach the very low level of 1994. Even so, its contribution to GDP was not greatly affected and was in fact increasing after 1990. This could be attributed to the fall of the contribution of the mining industries sector, especially with the dramatic decline in oil revenues due to the economic blockade.

Because of the continuing sanctions dictated by the UN's numerous resolutions, the manufacturing sector's performance is expected to go on deteriorating, in spite of the implementation of Resolution 986 which was designed to provide for the import of food and medicine and hence does not benefit the sector. Moreover, there will be more constraints imposed on the recovery and development of this sector in the future, as long as UN Resolution 1051 continues to be rigidly enforced.[6] This resolution imposes direct surveillance on the manufacturing facilities, and will probably lead to interference in relevant activities, including the procurement of necessary intermediates and inputs, whether imported or locally produced. As such, delays in production and alterations in production lines and technology can be expected.

In general, the performance of all manufacturing branches over the past two decades or so has been poor in comparison with their performance

TABLE 5.3
Development indicators of manufacturing industry, 1980–94

	1980	1985	1990	1994
GDP (millions of 1990 dollars)	98,479	73,401	64,898	19,529
GDP per capita (1990 dollars)	7,571	4,791	3,590	980
MVA (millions of 1990 dollars)* (all establishments)	6,983	7,445	5,735	1,753
MVA (millions of dollars – current prices)** (large establishments)	2,070	3,676	3,623	606
Gross output (millions of dollars)** (large establishments)	5,155	7,162	7,560	1,387
Employment (thousands of employees)** (large establishments)	177	174	134	117
Productivity (dollars output per employee) (large establishments)	29,100	41,100	56,400	11,860
Productivity (dollars MVA per employee)	11,686	21,089	27,250	5,153
Annual average wage in dollars	3,700	5,242	3,552	1,933
Wages and supplement % of gross output	13	13	6	16
Intermediate inputs % of gross output	60	49	52	56

Notes: *These MVA figures were based on national income accounts that usually cover all industrial activities, including small manufacturing enterprises. **MVA, Output and Employment figures were extracted from the industrial census covering large enterprises as defined earlier, and where the cost of non-industrial activities is excluded.
Source: *UNIDO Industrial Development Global Report*, Oxford: Oxford University Press for UNIDO, 1996; official rates of exchange apply to all years.

during the 1970s. However, certain sub-sectors were less influenced than others by the events of the 1990s. Compared with other branches, chemical, non-metal mineral, and metal-engineering industries had developed during the 1970s and 1980s at a higher rate than the rest, and had achieved higher performance levels.

During the 1980s, with some manufacturing facilities suffering from war operations and while, for strategic reasons, the government's priorities lay in emphasising the development of certain sub-sectors, a different pattern of performance emerged in various branches. After 1990, the UN sanctions began to take effect, and manufacturing operations became confined to those industries that could rely mainly on local inputs or that had adequate stocks of already imported intermediates. Even so, a new pattern of performance still prevailed (see Table 5.4).

Chemical industries hold the highest share in MVA in comparison with that of all other branches, and they have done so for at least three decades. This is due mainly to the contribution of oil refining, and to a lesser degree to fertilisers and some industrial chemicals (at least during the 1980s). By 1994, non-metal mineral (mainly cement and other construction materials) and food-industry branches, although trailing far behind the chemical branch, were making some contributions to MVA. The main reason why these three branches have continued to operate effectively is the availability of their raw materials locally and the possibility of having access to some imported inputs. However, it should be noted that productivity has been falling in all sub-sectors since 1990, as a result of the lack of essential inputs, spare parts and components needed to supplement the manufacturing facilities. This has led to a very low level of capacity utilisation.

TABLE 5.4

Development of MVA* in the main manufacturing branches, 1980–94

Manufacturing branch	1980		1985		1990		1994	
	Value	%	Value	%	Value	%	Value	%
Food industries	382	18.4	661	18.0	570	15.7	85	14.0
Food products	183		396		306		59	
Beverages	91		125		139		19	
Tobacco	108		140		125		7	
Textiles industries	330	15.9	383	10.4	480	13.3	49	8.1
Spinning and weaving	246		248		362		20	
Wearing apparels	42		53		47		7	
Footwear and leather products	42		82		71		22	
Wood and furniture	11	0.5	14	0.4	15	0.4	1	0.2
Paper and printing products	77	3.7	85	2.3	128	3.5	28	4.6
Chemical industries	703	33.9	1,490	40.6	1,460	40.3	245	40.4
Petroleum refining and related products	432		908		892		146	
Industrial chemicals	67		151		167		79	
Other chemicals	187		389		362		6	
Plastics and rubber products	17		43		39		14	
Non-metal mineral industries	212	10.9	601	16.5	589	16.3	104	17.2
Glass, pottery and earthenware	22		36		32		4	
Other mineral products	190		565		557		100	
Metal and engineering industries	357	16.9	441	11.4	379	10.0	94	11.6
Iron and steel	7		20		17		24	
Metal products	53		47		56		28	
Non-electric machinery	160		149		111		13	
Electric machinery	122		185		139		26	
Transport equipment	15		40		56		3	

Notes: *Value of MVA in million dollars at current prices, for large establishments that employ more than ten and the capital investment of which in other than power-generation machinery exceeds ID100,000.
Source: *UNIDO Industrial Development Global Report*, Oxford: Oxford University Press for UNIDO, 1996.

Overall, employment in manufacturing also fell over the same 25-year period, and even during the 1980s. After a significant increase in total employment from 186,000 in 1975 to 254,000 in 1980, it went down by 14 per cent by the mid-1980s. The 1970s' increase was highest in the non-metal mineral industry and chemical industry. The reduction during the 1980s could be ascribed to the enlistment of workers and technical staff for military service, and to war casualties, but further reductions in the workforce after 1990 can also be attributed to emigration, especially in the case of skilled labour and technical staff, and this trend has continued. Statistical information based on official census data for large establishments (except for military and related engineering industries) indicates that total employment was 126,000 in 1992 (against 177,000 in 1980). The textiles branch had 39,000 employees, while the chemical and non-metal mineral industries were close behind, with 23,000 and 24,000 personnel respectively, followed by the metal industry and food industry branches.

The share of wages and salaries in MVA averaged 20 per cent, with the highest share in textiles, clothing and footwear, while the lowest was in the petroleum-refining, industrial chemicals and construction materials sub-sectors. Of course, under such abnormal conditions, a high wage content of MVA cannot be considered as an outright indication of low productivity and/or degree of mechanisation. Idle capacities are to be expected, due to the shortage of previously imported inputs and intermediates.

The 1970s showed strong annual growth in manufacturing exports, though modest in absolute terms, which reached its peak of over ID62 million in 1979. It then dropped in the mid-1980s to a insignificant level (of less than ID one million), due once again to the first Gulf War and its effects on production and access to markets within or outside the region. Exports in the 1970s consisted mainly of such commodities as cement, sulphur and other industrial chemicals, fertilisers, and some oil-refining products. With the new production facilities that were commissioned in 1979 and 1980, it was expected that exports would increase significantly and would also include certain basic petrochemicals, iron and steel, phosphate and mixed fertilisers, farm machinery and implements, and so on. Such an outcome, however, did not materialise, even when the war operations ceased in 1988. Manufacturing exports were thus mostly confined during the 1980s to small quantities of fertilisers, petroleum products and cement that found their way overland, as well as to some natural gas that was exported to Kuwait under a long-term contract in the late 1980s. However, no manufactured export took place after 1990, with the exception of small smuggled quantities mainly of petroleum derivatives and fertilisers, and the limited quantities of petroleum-refining products allowed for sale to Jordan under the UN sanctions ruling.

Since 1990, manufacturing output as well as MVA have, according to a government report, been sliding, with production facilities having become idle in almost 80 per cent of the enterprises by 1992. Further deterioration of the production capacity can, of course, be expected as long as UN sanctions continue. It is estimated that Iraq's manufacturing industry relies on imports for almost 80 per cent of its inputs; that is, intermediates, spare parts, and even raw materials. During the 1980s, the annual imports of manufacturing inputs, including capital goods for non-military industries, fluctuated between ID1.0 billion and ID3.8 billion, with an annual average of ID2.5 billion.

This analysis of the manufacturing sector does not cover military industries, since official information and/or data on the status and performance of these industries is not available. However, general information regarding war machinery and the few studies on the subject of armaments published outside Iraq, all indicate that the country's military industry is technologically well advanced, and that it may well have comprised a large manufacturing and research and development facility that cuts across a number of sub-sectors. In support of the latter point, one may note that in recent years official statistics have not, as they used to do up to the early 1980s, included performance indicators for specific sub-sectors, mainly in engineering industries. This gives the impression that the military industry may have acquired full access to certain manufacturing units and production lines to supplement its own activities, and that for security reasons their content was not revealed. However, it is not clear how fully those manufacturing units were integrated with those of the military industry, or whether only some of their components and activities were linked and/or coordinated with the functions of the latter. Nor it is possible to determine the degree of reliance of the military industry on imports. In any event, the military manufacturing facilities and R&D capacity, which must have developed significantly during the 1980s and up to the start of operations in the second Gulf War, could certainly be rehabilitated to augment future manufacturing activities.

4. Prospects for improving and developing the sector

Iraq's policymakers will face the heavy task of rebuilding the nation after more than two decades of civil strife, two devastating wars and a highly destructive international economic blockade. These have led to unrest, displacement, and the emigration of a large part of the population, and have, in addition, traumatised a generation, which now faces acute health problems and declining educational standards and moral values in an atmosphere of despair augmented by a declining physical and social infrastructure. Not only will the policymakers have to attend to the formulation of long-term strategies and to the provision of a framework for socio-economic and industrial development programmes; they will have also to address the implication of rapid changes at the international level, where the transformation of the world economy is marked by rapid advances in high technology, in globalisation, and in the increasing interdependence of trade flows and investment.

The task of the policymakers will start with identifying priority areas as well as the means and mechanisms for dealing with all the aspects involved, including the constraints that will face even a modest development plan. This will apply to the manufacturing industry as well as to other economic and social sectors, and because of the interrelationships and links between and involving all these sectors, the parameters of an overall development plan need to be drawn before the problems of the manufacturing sector are addressed. For the latter, one might foresee an indicative plan of action that would address its problems via a two-stage implementation process encompassing distinct activities and incorporating different, yet related, functions. These are examined briefly in the following paragraphs, taking into account the overall development goals of this sector and its interrelationship with the other sectors.

Rehabilitating the manufacturing sector

At this stage, the main objective of policy is to assist in stimulating post-war and post-sanctions social and economic rehabilitation and reconstruction. This may be viewed as a multi-phase and multi-functional activity. Its priorities would include the assessment of areas of immediate need, such as food processing, pharmaceuticals, clothing, construction materials, and so on, and the restoration of industrial production.

The principal activities and relevant functions proposed for this stage involve:

(a) reconstructing and recommissioning existing viable production facilities and essential infrastructure;
(b) updating major production lines and eliminating bottlenecks to ensure an effective synchronisation of their operation and utilisation of by-products;
(c) implementing short, intensive vocational and technical training programmes;
(d) alleviating environmental damage; and
(e) redefining and implementing an appropriate legal and regulatory framework.

All these activities should be preceded by a detailed survey, and accompanied by the following: (a) a well-conceived subcontracting arrangement for the efficient utilisation of locally produced intermediates and by-products; (b) a quality-control system that, among other things, safeguards consumer interests; and (c) an innovative approach to streamlining employment and price-control measures in the light of the prevailing social and economic conditions. The survey should also point out the merits of, and logistics for, rehabilitating those manufacturing facilities, after a thorough appraisal of their technical components and a realistic assessment of their viability. The logic for this approach is evident in the following assertions.

First, the two wars have had a devastating impact on industrial enterprises and facilities. Some plants, or parts of their production and utilities equipment, were completely destroyed although a few were efficiently rehabilitated and maintained. Others, even though unharmed or somewhat less severely damaged, remained idle for a long period of time with no proper maintenance. They may therefore be amortised and scrapped in parts, or else will require costly repair and readjustment for adequate operations.

Second, many of the public sector's extensive heavy industrial plants that were constructed in the late 1970s at a very high cost (and in some cases at an even higher cost than expected, according to internationally accepted norms), were directly targeted and bombarded during the initial weeks of the first Gulf War. Many were still undergoing their test runs or were in their commissioning stages. Most of those plants, which are situated in Khor al-Zubair, and which include petrochemical, iron and steel, and fertiliser plants, never went into commercial production again, although they escaped total destruction and were repaired and mothballed. Others, which had been on-stream for a long time, were totally destroyed, including the Abu Floos fertiliser plant at Abu El Khasib, the Hartha pulp and paper plant, the aluminium fabrication plant at Nasiriyya. Among plants that were severely damaged were the cane sugar factory at Al-Amarah and the Shuwaiba oil refinery at Basra.

Third, the major damage during the second Gulf War was to power plants, and to oil and gas production and refining facilities, and some of these, along with other important facilities, were given priority in the repair and maintenance schemes that have been implemented during the past few years. The drive to reconstruct and operate the essential production facilities and equipment that had been damaged resulted in a dedicated initiative on the part of the national technical staff and skilled labourers to restore operating systems and to produce spare parts and components to combat the disruptive effect of the sanctions. However, these initiatives did not create new business within the enterprises, the necessary intermediate manufacturing inputs being, in most cases, unavailable locally. This situation led to a slow-down in manufacturing operations in most sub-sectors until eventually a near halt to industrial output took effect.

Fourth, some production facilities continued to operate on an intermittent basis, using their old stock of previously imported as well as recently smuggled inputs to stay in business. This situation applies to public as well as private- and mixed-sector facilities, even when local funds are available. This is due to the heavy dependence on imports of essential inputs that rely, in turn, on the availability of foreign exchange, both of which are affected by the UN sanctions. If the sanctions are lifted, the problem of foreign exchange would still prevail because of the heavy reliance on oil revenues for the reconstruction and rehabilitation of other social and economic sectors. These would have to compete with industry for the balance of that revenue after national foreign debts and war reparations have been deducted.

Rehabilitating the old production facilities might be regarded as an important step in support of the country's socio-economic development task. In the first instance, however, financial support for the reconstruction and recommissioning of the industrial sector's existing manufacturing facilities may not be abundant. This is despite its likely positive impact on the recovery and growth of the economy through creating employment and providing essential products to the local market, and thereby participating in the process of social development. To salvage old production facilities, the Iraqi government and/or financial institutions and entrepreneurs may initially be directed towards selected industries and plants, according to well-designed criteria that take into consideration:

(a) the type of industry, and its importance in providing essential finished products to satisfy the urgent needs of the population, as well as intermediate inputs to other productive sub-sectors;
(b) the condition and level of technology of the production facility in question, and the magnitude of the cost of revamping it;
(c) available labour requirements and skill levels;
(d) the degree of reliance on imports for essential inputs, and the output manufacturing costs in comparison with the import prices of finished products;
(e) the financial requirements and financing schemes attainable; and.
(f) the future viability of the existing manufacturing facility as an export-oriented industry, and/or its importance in developing other industries that lend themselves to further integration on the national level.

The major outcome of rehabilitating the manufacturing facility would be the improvement of production, the development of human resources, the provision of a suitable environment for higher investment, and a greater involvement of the private sector in industry. In order to achieve its goals, the rehabilitation process will have to address a number of issues. Recommissioning and revamping the physical production infrastructure are not enough on their own. Updating the technology and techniques constitutes a major task in areas where, because of extreme depreciation and technology obsolescence, replacement of equipment is deemed necessary, or where a low level of capacity utilisation, attributable to bottlenecks and inadequate synchronisation of production lines, requires additional units. Otherwise, low efficiency will still prevail, resulting in excessive operational costs, especially when the production facilities have been idle for a long period of time.

Revival of manufacturing activities depends not only on the rehabilitation of the existing physical facilities, but also on the efficiency with which available resources are utilised, and how well the reform schemes are conceived in related areas. The following represents a brief account of specific issues to be addressed by a comprehensive reform programme that will need to be implemented in conjunction with, and in support of, the rehabilitation process.

First, in order for the public-sector enterprises to be transformed smoothly into more commercial entities, an adequate economic reform programme should aim at a higher degree of autonomy and management responsibility. Second, the stagnant situation prevailing in the country in general, and the situation caused by the idle capacity of the manufacturing sector in particular, coupled with the draining away of skilled labourers and high-management echelons, make it necessary that a speedy but effective programme for the development of human resources is launched.

Next, the comprehensive statistical data and information on which policy formulation and longer-term projection of economic growth should be based will probably be available at the Central Statistics Organisation, although updating may be required. However, for the rehabilitation process in respect of the manufacturing sector, detailed industrial surveys will be needed in order to assess the conditions and requirements of the existing production facilities, at least of the large enterprises and their essential infrastructure. An important function of such surveys would be to determine the technical manpower situation, in view of the large-scale emigration of the educated and entrepreneurial classes and of the displacement of personnel and workers that results from internal politics and old age.

Fourth, it needs to be remembered that a whole new generation has grown up during the war years with no experience of technical work because of their lengthy service in the army and the lack of proper on-the-job training. Vocational and technical training programmes, as well as the standards of university graduates, have also suffered. Accordingly, industry may have to rely on a lower-quality labour force while waiting for 'home-grown' skilled people to emerge from a rehabilitated training system. In the meantime, a practical approach should be followed that will foster links between industry and academic institutions as a way of upgrading technical and technological capabilities and capacity in the near future.

Another problem that may have to be addressed concerns the depressed living standards of the middle class to below subsistence levels, a high rate of unemployment, and low income in the face of extremely high rates of inflation. This may well require a national approach to the socio-economic crisis by way of a comprehensive reconstruction and rehabilitation programme. However, because of the likely burdens on the government treasury – given that the international debt service and war reparation payments will probably siphon off more than 50 per cent of the oil revenues – the long-term reconstruction and rehabilitation programme calls for extra efforts in incorporating adequate schemes to mobilise and secure financing for its components. The manufacturing sector might have a better chance of securing financing from international capital markets if its rehabilitation programme is set on a track that will appeal to foreign investment and financial assistance.

Another essential factor will be peace and stability, supported by a proper independent judicial system. This will help to reinstate confidence and provide a healthy environment that will be inviting both to local entrepreneurs and also to foreign financial loans, assistance and investment. In this way it will benefit both the rehabilitation programme and industrial development. However, peace and stability will be achieved only by rebuilding the national identity and forging a consensus among the multi-confessional and ethnic factions of the society. This will be the major and most crucial task for any government that comes to power. A representative government of the people will have to be responsible for providing social protection and support to all citizens but most particularly to the low-income classes, so as to avoid widening income disparity and social injustice and in order to insure peace and stability. An important mechanism to be constructed for this objective would be a well-designed social security system.

Furthermore, in order to secure a well-balanced rehabilitation programme, the government may have to redefine its role of providing a healthy environment for industrial development by taking into consideration the shortcomings of its role as investor, operator and marketing agent for the manufacturing sector. In the course of strengthening its part in providing a healthy environment and industrial support services, some of its functions may have to be transferred, carefully and in stages, to the private and mixed sectors. Its role in managing relevant or specific functions and in supervising or monitoring the whole sector will have to be maintained, but in close consultation or even cooperation with the private sector and the appropriate non-governmental organisations (NGOs). Such functions and activities could include the rehabilitation of the education system to include vocational training facilities for developing human resources and a skilled workforce for industry, as well as for building high-technology services. Other functions important to the manufacturing sector and which merit consideration for restructuring include legal and regulatory frameworks, industrial-support information services, and control and certification of product specification.

Based on what has been discussed above, it would be encouraging to see changes in favour of a larger role for the private sector, starting with the privatisation of certain public-sector enterprises. It is important to note, however, that effective privatisation depends on relaxing financial controls, restructuring finance and credit, and creating a real stock market, not a superficial one. It should be remembered that in the late 1980s, when the government was encouraging and promoting the private sector, the latter's access to domestic or public credit and international finance was limited. Moreover, even before the crisis of the 1990s, Arab investment had been slow in coming. The main reasons, apart from political uncertainty, could be traced to legal and regulatory shortcomings, the absence of an authentic financial market and commercial arbitrage, and a stifling bureaucracy. It will be necessary to address these issues seriously, along with the relevant commercial laws and even the labour laws, which need revisions to suit the prevailing unemployment and living conditions while being flexible enough to ensure higher productivity in the long term.

Last but not least, a wider structural reform may have to be pursued to facilitate the transition toward a more market-oriented economy. A timetable with specifically tailored structural changes should be carefully drawn up to ensure gradual market deregulation and economic liberalisation, while taking into account the immediate social needs and dilemmas through which the country is passing. Thus, while specific industries and some manufacturing enterprises may continue to be controlled or owned by the public sector so as to minimise the negative impact of a sudden shift from social considerations to cost considerations in management and employment policies, a gradual shift from central planning and protectionism may be essential for enhancing the efficiency and competitiveness of enterprises in all sectors.

In order to be effective, the rehabilitation process should be directed towards removing political interference, reducing subsidies, and eliminating the dependence of public enterprises on the central state budget. Readjusting ways of thinking about such ideas among industrial management personnel in both the public and the private sectors will require awareness campaigns and probably an educational programme to enable such personnel to carry out their responsibilities with a clear understanding of the objectives of the reform plans.

Major constraints

The rehabilitation and restructuring programme briefly outlined below is proposed for consideration by the interested parties concerned, and should be subjected to an in-depth review and elaboration in the light of prevailing conditions. However, it should be pointed out that any plan for implementing such a programme or any of its components cannot be drawn up until it has become clear that political stability and restructuring are being seriously pursued, and that the sanctions imposed by UNSC Resolution 687 have been lifted. In this context, it is important to note that there are a number of constraints facing the effective implementation of any process of rehabilitation and reconstruction or recommissioning. These include:

(a) UN Resolution no. 687, and its negative effect on imports of essential manufacturing inputs, on manufacturing export, financial resources, and so on;
(b) lack of medium- and long-term financing facilities;
(c) a decline in the level of skills and educational standards;
(d) absence of confidence and a healthy environment for private investment;
(e) limited access to essential information and data;
(f) a high degree of bureaucracy in public administration; and,
(g) UN Resolution no. 1051, which imposes direct supervision and continuous surveillance and scrutiny on all manufacturing facilities and industrial support services, thus subjecting developmental programmes to delays and alteration of production lines.

TABLE 5.5
Comparison of industrial development in certain countries

Years	1980			1994		
Indicators	Output in million US$	MVA in million US$	Productivity in US$	Output in million US$	MVA in million US$	Productivity in US$
Countries						
Iraq	5,155	2,070	11,686	1,387	606	515
Some oil exporting countries						
Algeria	9,023	3,644	11,684	9,147	4,084	91,819
Iran	15,870	8,186	17,411	13,302	5,839	9,120
Libya	1,177	358	19,577	2,530	784	26,395
Saudi Arabia	9,586	5,283		25,813	6,780	
Venezuela	30,213	14,461	32,530	25,881	10,643	23,863
Turkey, various Arab countries and Israel						
Egypt	6,986	1,769	2,023	19,566	6,405	5,549
Jordan	917	406	11,819	2,966	862	10,603
Tunisia	3,579	939	7,525	12,287	3,818	14,723
Turkey	29,413	10,837	13,617	64,237	27,459	25,991
Israel	14,332	6,490	24,733	33,047	12,030	34,518
Some industrializing countries						
Argentina	55,936	24,511	18,208	217,063	88,366	89,820
China	232,460	88,577	3,632	451,906	139,031	2,245
Indonesia	13,226	4,371	3,499	79,062	28,605	6,954
Malaysia	13,181	3,623	8,060	68,789	18,560	15,317
Philippines	17,369	4,861	4,552	34,753	12,694	12,334
Some OECD countries						
Canada	167,211	59,803	32,187	286,933	100,322	58,465
Greece	20,906	6,129	16,204	29,836	10,412	33,368
France	453,636	161,552	30,101	685,694	268,611	65,118
Portugal	17,932	5,602	8,087	54,783	13,384	14,917

Source: *UNIDO Industrial Development Global Report*, Oxford: Oxford University Press for UNIDO, 1996. Values at current prices and official exchange rate.

5. The restructuring of manufacturing industry

Table 5.5 illustrates how badly the performance of Iraq's manufacturing sector has deteriorated since 1980 in comparison with that of some other developing countries and even of those that are regarded as the least developed among them. The main reason for this phenomenon is, of course, attributable to the two Gulf Wars, the events and consequences of which have blocked further development of the whole sector and hampered the efficiency of its existing physical production and infra-structure facility. Accordingly, the structure of manufacturing industry has not changed greatly since the early 1980s. It still, for example, consists mainly of small- and medium-scale industries, lacks highly diversified or well-integrated production lines, has minimum inter-phasing links to other economic sectors, and is still dominated by the public sector, whose bureaucratic and highly centralised management has been counterproductive.

Once the rehabilitation programme outlined earlier in this essay gets underway, a comprehensive plan for restructuring the manufacturing sector should be launched. The objectives of this restructuring process would be to broaden the industrial base by pursuing a balanced development programme aimed at: (a) diversifying manufacturing production while ensuring a high degree of integration, and employing the latest advanced technology while ensuring the maximum utilisation of locally produced intermediates and by-products; (b) developing export-oriented industries the main inputs of which are available locally; (c) adopting a rational approach in de-emphasising non-viable import-substitution industries and in phasing out non-viable, isolated assembly-production lines unless they lend themselves to integration into other economically productive sectors; and, (d) implementing a well-prepared procedure for privatising most (but appropriately selected) SMI enterprises in stages, while improving public sector efficiency.

All the above are functional activities that should be preceded by redefining the development objectives of the manufacturing industry and its linkages with other productive sectors, and take into consideration international trades and financing conditions as well as investment and consumption trends. A realistic plan of action would have to be designed at the same time, with a view to identifying priority areas for implementation. In this context, a series of other important issues can be addressed, including, among others:

(a) the restructuring of industrial enterprises;
(b) the efficient functioning of less-bureaucratic industrial-support institutions;
(c) effective policies in respect of environmental issues;
(d) a sound environment for continuous consultation on all matters between the policymakers or the public sector and the private sector via NGOs;
(e) an adequate financing system and a stock market;

(f)   full access to internationally linked information systems; an internationally recognised standardisation and product-certification system; and

(g)   an effective link between industry and academic institutions.

A stable political environment and a proper judicial system will be essential prerequisites for the success of the industrial restructuring plan. Moreover, in designing different components of the plan, the policymakers have to be abreast of state-of-the-art developments in respect of relevant industries, and aware of changes in international trade and industrial structures as well as of the relevant functions and policies of the World Trade Organisation.

Elaboration on the above calls for in-depth analysis which is beyond the scope of this essay. However, a few remarks regarding certain relevant issues are justified.

The events of the past two decades and the prevailing conditions, which have shown how heavily the economy and society of Iraq rely on imports for essential inputs and for survival, indicate the necessity for making structural changes in the economy in favour of agriculture and manufacturing. For manufacturing, the leading sectors should be based on oil and on agriculture or agriculture-related industries, whose downstream activities might create interconnections or a network of integrated production lines to enhance food products, textiles, and consumer goods industries. The establishment and/or expansion of heavy industries, and the efficient utilisation or processing of other local natural resources – including sulphur and phosphate deposits as well as other non-metal mineral resources providing the raw material for other consumer products, construction, and industrial chemical industries – are essential for economic diversification and the development of exports.

Manufacturing industry in Iraq has not been significantly diversified or well integrated, and its production has been directed mainly to the local market. From the 1950s, plans had been drawn up for developing oil-based downstream industries, such as petrochemicals and fertilisers, and energy-intensive industries including iron and steel, aluminium, cement, glass, and so on. Due to the advantages of locally available, low-priced feed-stock, these industries were considered the growth leaders in manufacturing, and are still expected to continue to lead industrial growth in the future, once plans are pursued to diversify and integrate their production lines. The ultimate aim of developing these industries would be to reduce the heavy reliance on imports of certain essential consumer goods (such as textiles, engineering products, packing materials, and so on). They would also provide appropriate intermediate inputs and capital goods for further processing and intensive manufacturing operations, and for other economic sectors such as agriculture, construction, and transportation and communication, while developing export-oriented products to help reduce dependence on the crude-oil sector for the required foreign exchange.

Within the Gulf region, Iraq pioneered the idea of developing an oil-based industry. However, implementation of the major relevant projects of this industry was delayed, and after the events of 1980, utilisation of their production lines either dwindled or was halted. With the exception of its fertiliser and cement projects, all others, including the oil refineries, have had only modest capacities. Furthermore, its export-oriented industries have never produced any tangible effect on the economy.

Considering the risk involved in relation to marketing potentials because of keen competition in the world market from well-established production capacities, the initial investment required to build export-oriented industries (large economy of scale) may be very high. This is especially the case since those capacities, mainly in developed regions, underwent the necessary structural changes against deteriorating growth performance in the late 1970s and early 1980s. Although the developing world is still at the receiving end of those capacities, the Gulf region has, in the meantime, and sometimes in collaboration with worldwide established trading manufacturers, developed modest capacities in relevant oil-based industries while Iraq trails behind.

Iraq, which will in effect be a newcomer, will face accelerating structural changes and decelerated international trade. However, development of its manufacturing sector – since it is not starting from a low-level base – will depend significantly on the success of the rehabilitation programme: this will include measures to remove business restrictions and the negative aspects of a highly centralised economy. The positive outcomes of such a programme should ensure improved performance and help in alleviating the problems of funding further development of the sector, especially if ways and means can be found to address issues of heavy debts and war reparation, as well as scarce foreign exchange resources. This would also have to include effective negotiation with the World Trade Organisation for applying sufficiently mild provisions on the country's exports.

While there will always be room for further redeployment of petrochemicals and other related oil-based production capacities from industrialised regions, this requires a competitive status based on high-level manufacturing productivity, a wide range of products with diversified specifications according to international standards to meet the requirements of different end-users, and further downstream activities supported by locally institutionalised but internationally coordinated research and development activity. However, the major factors in achieving good results would be the efficiency of the marketing procedure, and the bargaining power used to infiltrate the export market. In this context, a well-conceived cooperation and/or coordination programme among the Arab countries, especially those of the Gulf region, ought seriously to be considered. Among other things, this would try to avoid duplication of investment and over-capacities in certain products. It would attempt to create a collective bargaining power, expand interregional trade, abolish trade barriers, develop a regional stock market, and establish joint ventures. It would also support the establishment of efficient regional industrial-services institutions. These would operate on a commercial basis and ensure maximum utilisation of the region's technical capacity, but would not function as intergovernmental agencies or simply for political purposes. If directed toward the above objectives, a well-defined cooperation programme, based on authentic consultative dialogue involving the private sector, might also enhance national economic development and promote industrialisation in all member states of the region.

The desirability of regional cooperation among the Arab states in general cannot be overemphasised. The creating and strengthening of regional, economic, and political blocs has gathered momentum worldwide, especially since the collapse of the Soviet Union. These blocs aim at maximising the benefits of the complementarities that exist between member states, in order to achieve improved productivity and competitiveness in the world market. This trend will make it more difficult for exporters from outside the bloc to penetrate its integrated market. Therefore, even when the exporting country has a competitive advantage, sustaining an export-oriented industry may not be easy without strategic alliances with other countries. It is in this context, rather than for political pan-Arab fervour, that cooperation among the Arab countries becomes essential.

For Iraq, particularly at this stage, cooperation with the Arab Gulf states may be important for attracting capital in the form of loans and investment. This would not only be for rehabilitating the manufacturing sector, but would also coincide with plans for diversifying and/or creating a broader industrial base, marked whenever feasible by an adequate level of specialisation. Should oil revenues decline in such a way that they were unable to meet the country's socio-economic infrastructure requirement, specialisation within the context of cooperation among those states would becomes preponderant. This argument becomes more forceful when one considers that the location factor for oil-based industries within the Gulf region may not continue to be advantageous, especially when world production patterns are heading toward specialised rather than performance commodities. In view of technology advances and structural changes, even in management, the emphasis is shifting to mergers and collaboration among large national and international specialised enterprises.

It can be deduced from the above that policy formulation in respect of promoting export-oriented industries to enhance their international competitiveness will have to take into account changing patterns in the world market and among prospective major trading partners. UN sanctions and the economic blockade may not last forever. Even so, Iraq's manufacturing industry may take a long time to reach a level comparable to what it might have been in 1980 if the first Gulf War had not taken place. The same can be said about crude-oil exports. However, rebuilding the oil-export market is not by itself as complicated as reconstructing manufacturing industries, and hence it may initially act as a catalyst for rehabilitating the latter. In the long run, carefully drawn oil-marketing schemes may help to promote the manufacturing export of certain commodities such as refined oil products, petrochemicals, and so on, but excessive reliance on oil revenue should be avoided, or at least scaled down and limited to importing selective manufacturing inputs. The same principle should be applied to excessive reliance on oil revenue for the importing of consumer goods, not to offset but rather to streamline the domestic consumption patterns that emerged in conjunction with local production during the oil-boom days of the 1970s.

Modern management and production methods are essential in broadening the industrial base. This has already been suffering from lack of diversity and overstaffing, coupled at times with harsh political treatment, and the imposition, or even forcing, of a low level of skills. In order to undo some of the problems, it will be essential to achieve high productivity in all areas and levels through upgrading and updating the technical, administrative and professional skills of the workforce. These skills include modern management and production methods; marketing skills and procurement services; information systems; computerisation and automated control-system applications; and innovation and development schemes. All these functions should be subjected to continuous and/or recurrent training and constantly upgraded teaching programmes. It should be evident that, in addition to the personal sense of responsibility and interest in the work, motivation and incentives can go a long way in obtaining positive results out of training.

The lack of access to information and proper communication will be a major constraint for the development of the manufacturing sector as well as others. Even before 1980, rigid control and a constant and domineering surveillance over all sorts of telecommunications use, publication, and printing, had been in force. Overreaction by successive governments to the issue of national security has created difficulties for the private as well as the public sectors in gaining access to all sorts of domestic and international information services. This problem has to be addressed within the context of augmenting or developing industrial-support institutions and industrial-related services that are just as important for small- and medium-scale industries (SMIs) as they are for export-oriented heavy and large industries.

The issues of continuous protectionism in relation to industry and of heavy subsidisation of public-sector enterprises need to be discussed and resolved, in conjunction with the criteria to be adopted for privatisation and in the light of the long-term strategies for developing the manufacturing sector. The prevailing living conditions, high rates of unemployment, and the results of the industrial survey mentioned earlier all need to be taken into consideration. In the same context, certain industrial support institutions may need to be restructured in order to function efficiently. Apart from improving their technical capacities, a new approach in respect of their managerial functions might also be in order. For instance, while national statistics and telecommunication institutions may continue to be official government departments, a relaxed procedure for accessing their use will be important in promoting industrial development, especially by the private sector. Other industrial-support institutions, such as those involved in the quality control and international standards certification of products, may be commercially managed but run as government-supported agencies supervised by a consumer-protection department or by the Government Bureau of Standards. In addition, NGOs such as the Federation of Industries, the Chamber of Commerce, and the professional and trade unions and scientific societies should all play an active role, directly or through keen consultation, in all matters related to industrial development.

Finally, industrial pollution is an important area, which needs to be addressed in conjunction with the rehabilitation and restructuring of the manufacturing sector. This area can be tackled to some degree on the enterprise level, as an issue of safety within the manufacturing facility, but can be dealt with more effectively at the sector level. It then becomes an environmental issue – and one among many of the crucial factors that Iraq will have to consider during all the stages of its future industrial development.

NOTES

1 The US$14 billion included investment in crude-oil refining and gas treatment and distribution systems, as well as gas-fired power plants and other-than-oil mining operations and mineral treatment plants. It did not include crude-oil exploration and production activities.

2 The objective of Law No. 46 1988 was to encourage industrial investment by the Arab private sector, offering limited openness. The Iraq Revolutionary Command Council's Decree no. 774 of 21 September 1988 was intended to encourage the private sector to invest in industry, due to declining government resources; the war and the inefficiency of the public sector had caused revenues to fall, thereby reducing the government's ability to generate profits for financing expansions.

3 In speaking of technology, the term 'advanced technology' refers to up-to-date, state-of-the art techniques used for specific production processes that incorporate sophisticated automation and quality-control units. 'Modern technology' refers to popular mechanised production facilities rather than to simple forms of tools and techniques. These latter are more likely to be a characteristic of 'appropriate technology', since this term refers to technically adapted or developed production

facilities that are suited to local conditions and levels of apprenticeship in developing or least-developed countries.

4 Government of Iraq, *Statistical Bulletin*, 1994.

5 UNIDO *Global Report,* 1988/1989.

6 UNSC Resolution 1051 in effect authorises the investigation, at any time, of any civil production facility whose components might be liable to produce inputs to the military industry, even if this means temporarily or permanently shutting off the main stream of production. As such, a wide range of industrial units may be out of commission.

PART III

ASPECTS OF THE POLITICAL ECONOMY

6

Consuming Interests: Market Failure and the Social Foundations of Iraqi Etatisme

Kiren Aziz Chaudhry[1]

1. Introduction

The end of the Iran–Iraq war terminated established methods of governance in Iraq, throwing the Ba'thist regime into a multi-faceted crisis. This was not just an economic crisis, centring on the imperatives of reconstruction in a context of international indebtedness and fiscal chaos, but was also a political and administrative one. The war had institutionalised a broad and stable coalition between the state and Iraqi consumers, held in place through a highly complex set of administrative relationships but maintained at its foundation through the state's direct control of domestic goods, prices, services and distributive systems. The market reforms that were introduced in 1989 were one possible response to the conditions in which Iraq found itself. They also undid the coalitional basis of the Ba'thist regime, creating a much deeper disjuncture in the state's capacity to rule.

Iraq's market reforms were unusual, not only in their breadth but also in the speed and apparent effortlessness with which they were implemented. Yet they were as short-lived as they were dramatic: by the end of 1989, a brief six months after the reforms had taken effect, the government reimposed price controls, rescinded many of the liberal policies and tried to recentralise control over the economy. Out of this failed effort to recentralise came an economic disaster of such dimensions that even the repressive apparatus of the Ba'th Party was unable to guarantee political stability. The remobilisation of the army in the winter of 1989 and the subsequent invasion of Kuwait followed.

From property rights to the price of tomatoes, Iraq today has almost nothing in common with the economy I studied in 1989. Since 1990, the Iraqi middle class has been wiped out; the economy has been completely recentralised at the apex and quite radically decentralised at the base; some extractive and distributive functions of the state have been relegated to recreated 'tribal' hierarchies;[2] and large segments of the bourgeoisie holding dinars have been reduced to penury.[3] For a time, until 1996, Iraq was in a state of hyperinflation, and today a middle-level civil servant lucky enough to have kept his job can buy two small chickens and a kilogram of sugar with a month's salary.

Indeed, the ruptures of the past ten years are so striking that one is forced to question the validity of the basic apparatus of conventional economic history, built as it is on the fundamental assumption of association, if not continuity. Iraq today is not one economy, but many, in which transnational networks, local power structures, and a variety of free and administered prices are enforced by a host of formal and informal authorities, ranging from the US Central Intelligence Agency to the United Nations to the village-level bosses, all of whom have dubious legitimacy. What relevance, then, do the economic reforms of the late 1980s have for Iraq's economy today?

Political systems reveal in their decline the skeletal structure of their longevity. Iraq's experience with economic liberalisation in the late 1980s gives us important insights into the political economy of Iraqi populism. For, contrary to popular scholarly opinion, which is as insulated from fact as Saddam Hussein's bunker is from smart bombs, Iraqi developmentalism, and the role of the Iraqi state in the economy was not an artefact of either the Ba'th Party or of the country's current leader. Through the Mandate period and on into the 1990s, often while pursuing other, possibly conflicting priorities, the Iraqi state's principal method of governing the economy has been historically stable: fix prices, stabilise consumption. In modern history, Iraq has never had a functioning market economy. The price liberalisation of the late 1980s therefore marked a radical and a politically unmanageable departure not just from the regime but in the way the state itself was socially constructed. Moreover, liberalisation occurred at a historical juncture in which the state's control over the economy had achieved its most direct form.

The aim of this chapter is to initiate a discussion of Iraqi statism as seen through the market reforms of the 1980s. The argument is simple and can be summarised at the outset. The role of the state in setting urban prices is the centrepiece of the state's economic role in modern Iraqi history. The shifting terms of this role were set in periods of intense class conflict and can be traced historically from the Mandate period through to the Iran–Iraq war. The main constituents of the Iraqi state, in its economic guise, are consumers: a group notorious for its inability to act collectively. In the 1970s and 1980s, the spread and sectoral location of state employment created large groups that would have been opposed to liberalisation, had they known what liberalisation meant.

Price liberalisation in a context marked by absolute scarcities in foreign exchange undercut the regime's traditional base of support. At the same time, the structure of the industrial sector as it emerged by the end of the Iran–Iraq war, and the stagnation of agriculture, together permitted the implementation of a rapid privatisation programme, which provided the distributive base for a new private-sector client group. These two facts made reforms easy in the beginning; they also meant that the subsequent dysfunctions of the market would radically affect the largest base of political support the regime had cultivated.

The liberalisations of the 1988–9 period thus marked a radical disjuncture in the fundamental mechanism through which the Baʿth maintained itself and in the basic coalitional base of the regime. While the failure of the reforms have a host of economic, institutional and political explanations, the key force propelling recentralisation less than a year after the reforms had been implemented is traced to the political threat attending the impact of liberalisation on consumers and the military.

This argument would be nonsensical if it were based on positing a direct one-to-one connection between the state's ability to stabilise prices and social support for the regime. Even hunger is relative: what people were not willing to suffer at the end of the war with Iran, for example, they have tolerated under the sanctions of an international system that must appear unjust, cruel and conspiratorial. On the ground, domestic actors are evaluated comparatively, not by recourse to some theoretical ideal. For reasons less wrenching than our silence about the obscenity of a global system of power that would starve a nation, it is difficult to write about Iraq. One can only imagine the complexity of the game theoretic model that would capture the valuative processes that come into play, or the span of history that seems relevant, or the analytical basis for apportioning blame and credit, that operates when one watches one's children starve.

In addition to recounting my interpretations of liberalisation, which have already appeared elsewhere, it is my aim to debunk two important myths, one about Ba'thist Iraq, the other about market reform. The first holds that, unlike all other regimes in the world the Iraqi Ba'th maintains itself without a serious base of domestic support, and that the regime 'hovers' above society, unaffected by social pressures. The other suggests that markets sprout, naturally and fully developed, when the state ceases to regulate. These two points, self-evident as they seem, have prevented even the most knowledgeable authorities on Iraqi politics from recognising the strength and durability of the social base of the Ba'thist regime in Iraq, and from appreciating how both contemporary and historical experiences of Iraqi society with market economies have been dramatically unsatisfactory. At some level, part of trying to understand what Saddam's Iraq is, and how it was constructed, must centre on the extent to which both the interventionist and the coercive powers of the state arose as substitutes for administrative and regulatory powers that have historically eluded the Iraqi state.

2. Constructing the social base of Iraqi etatisme

Iraqi etatisme grew and took its particular shape as a result of class-based struggles between urban labour, agricultural elites, and commercial groups pre-dating independence. The protagonists in these struggles changed, as did the particular mix of policies. Thus, during the Great Depression and the Second and World War, shortages and inflation were fought mainly with price controls, but in the immediate post-Mandate period a host of developmentalist interventions were undertaken, including the provision of credit and the construction of infrastructure. Overall, at each identifiable juncture, the state expanded its control over the economy, not to achieve predefined developmental goals or to fill gaps in expertise or investment capital left by a 'weak bourgeoisie', but because of its inability to regulate a national market characterised by shortages and by struggles over scarce resources.

Nothing demonstrates better the continuities in the sources of Iraqi etatisme than to recall that the first wide-scale intervention of the state in domestic markets occurred during the Mandate period, when, in theory, a capitalist alliance between foreign companies and local landlords, upheld by legal monopolies enforced by the British, was ascendant. Throughout the Mandate, and particularly during the economic disruptions caused by the global depression and later by the Second World War, when the Iraqi economy was intimately linked with global markets, British conglomerates and local merchants with monopolies in trade and agricultural produce cooperated to create widespread shortages through hoarding and other speculative behaviour which was extremely profitable.[4]

The early 'independence' period was marked by dramatic confrontations and devastatingly costly failed attempts to enforce price controls and curtail profiteering.[5] Comprehensive attempts to regulate domestic prices thus began early on, first in 1939 with Regulation 58, governing 'Life During International Crisis',[6] and then in 1942, with the 'Law Regulating Economic Life of Iraq'. The intention of the latter was to prevent monopolies and price manipulation and to maintain required supplies in basic goods. To achieve these aims, the government began to restrict exports and to import commodities to fill domestic needs. Import licences were required and prices of commodities were fixed.[7]

These regulations, born of Iraq's immersion in, not her exclusion from, the international economy, failed miserably. For a variety of reasons, including weak enforcement capacities, interference from the British authorities, the involvement of powerful foreign companies, and the cohesiveness of the local merchant group, the private sector was able to ignore the law. The results were devastating. Speculation, hoarding and profiteering abounded, contributing to inflation rates of between 200 and 300 per cent in the late 1930s and early 1940s;[8] rates which continued, on and off, through the early 1960s.[9]

The response of the government to these repeated administrative failures began with direct entry into trade to alleviate shortages and lower prices,[10] and reached its zenith with the creation of government monopolies in agricultural goods.[11] Not only does Iraqi statism pre-date the republican period, but the causes of state intervention in the period from the 1930s to the 1960s bear a remarkable resemblance to why the state recentralised economic control after the failed liberalisation of 1989.

Regulatory struggles, together with the pressing need to create a unified national market, were linked with the problem of regime maintenance and nation-building. So great was the economic hardship caused by the speculation and profiteering of the private sector during the 1930s and 1940s, and so widespread was the threat to the incumbency of the regime, that the government attempted, unsuccessfully, to legislate against the foreign monopolies and local merchants who, in every other way, controlled the state. While merchants, landed elites and foreign companies had consistently come into conflict with urban consumers in Iraq well before labour became a major political force, their struggles overlapped with the emergence of a growing and increasingly organised urban working class, which gave voice to the lower strata and put pressure on the state to craft a minimal level of consumer security. Thus, when the economic hardship of the Second World War precipitated riots and fomented labour unrest among the Iraqi employees of the large railway, oil and port authorities,[12] they were demanding price stability, not the end of class exploitation.

If the first conflict that prompted state intervention was waged between commercial elites and urban consumers in Mandate Iraq, the protagonists in the second were urban labour. Even before 1958, organised labour (and the Communist Party of Iraq with which it was allied) became increasingly vocal in struggles over economic policy. Although it had gone through several cycles of activism and quiescence, the potential for labour to organise had grown dramatically through the 1930s and 1940s. By the 1950s, labour constituted a cohesive and organised group that was expanding rapidly, especially in service and transportation sectors: between 1926 and 1954 the number of workers in industries with more than one hundred workers rose from 13,140 to 62,519 – by 375 per cent. These industries were concentrated in Baghdad and Basra, which had 43 per cent of the large industrial firms. Employment in government, especially the ports and railways, increased by 106 per cent between 1938 and 1958.[13]

The decisive conflict between organised labour and business occurred between 1958 and 1967, marking a juncture in which the economic arena became the venue for a political struggle over who would define the structure of the domestic legal and economic system. During 'Abd al-Karim Qasim's tenure immediately following the 1958 coup, private investors withheld business investments and created shortages in basic commodities destabilising a regime which was seen as being too close to the Communist Party. Business leaders, many of whom were Shi'i, feared the populist agendas of the new military leaders that came to dominate politics, not because they anticipated the bloodbaths of the 1970s but because they were threatened by the new government's interest in expanding competition, protecting consumers and undercutting the long-standing monopolistic pacts that bound domestic commercial elites to the foreign corporations that had thrived under colonial rule and after. Despite Qasim's apparent support of local industry, his populism was based on economic measures – such as limits on the profits for consumer goods, cuts in prices and rents and land reforms – that could hardly have won the support of the commercial classes.[14] In 1959 higher income taxes were imposed on the wealthy, the immunity of agricultural land from taxation was repealed, an inheritance tax was introduced, and labour unions were legalised.[15]

The intervention of the military in national politics was bound up in these struggles, but not in the way liberals would like to imagine. For, quite in contradiction to the widely cited support of the bourgeoisie for democracy, the commercial classes were deeply involved in supporting the brutality of the military when they believed it would be to their economic benefit. Qasim, in this regard, was a great disappointment, since it was he, and not the Ba'th, who started a tradition of military populism that combined the brutal repression of labour, regulation of the private sector and price stability for consumers.

The explicit beginning of the military–consumer coalition can be clearly traced to Qasim's reforms, as can the attendant trend of simultaneously suppressing labour and curbing the business class. Class struggles were mediated directly through the military, which alternated its attack on the left with attacks on acts of economic sabotage propagated by the private sector. In 1958 the formation of the Communist Party of Iraq dominated parallel armed force, mass demonstrations organised by the communists, and the CPI's success in the 1959–60 student elections[16] combined to fuel the conviction of the wealthy that stability could be guaranteed only through the intervention of the largely Sunni army.[17] To their disappointment, Qasim first passed laws to fix prices and redistribute income and only then took measures against his former allies in the CPI and labour. The actual process of de-institutionalising labour took place under a formally promulgated political liberalisation programme. Qasim first created and then granted a party licence to a bogus communist party, stripping the CPI of legal status.[18] Press censorship of communist newspapers, purges of members in the government, and the disbanding of youth organisations was carried out, and workers' organisations and unions were dissolved in 1961.[19] In the end, the Labour Union Federation Council, purified of communists, was granted legal status as the representative organisation for industrial and service-sector labour.

What Qasim and the Islamists started, the first Ba'th government continued during its nine-month tenure in 1963. The Ba'thists discovered and eliminated huge segments of the Communist Party rank and file and either imprisoned or killed labour organisers. Then, with labour and the communists in retreat, the regime of 'Abd al-Salam 'Aref, closely identified with the Sunni north, nationalised all agricultural lands, industry, banking, insurance and services which, after the consolidation of land reforms in 1970, resulted in the virtual elimination of the top layer of the largely Shi'i commercial and landed elite.[20] The triumph of Iraq's 'republican moment' was thus not supported by its entrepreneurial elites. As the state alternated between fighting the left and controlling the commercial classes, the political significance of the sectarian split between the military and the bourgeoisie rapidly became an independent and important influence on state policy and social reaction to it.

While the nationalisations of 1964 are often attributed to the Iraqi government's preparations for economic union with Egypt, there are more likely reasons for the seizure of private property. These can be found in the speculation, capital flight, and shortages that racked the economy between the creation of the Joint Presidential Council with Egypt in May, and the final nationalisations in mid-July. The behaviour of the private sector at this juncture added to already high levels of inflation.[21] From incremental and sporadic entry into commodity markets and banking, coupled with spates of unsuccessful regulation, the Iraqi state finally began to manage all aspects of the economy from imports and production to distribution and retailing. Subsequent regulation restricted the private sector to the confines of small industry, retailing, and transportation.

The dynamic aspects of the sectarian divisions that plagued Iraq's economy have been discussed elsewhere and need not be recounted. Here it is important simply to point out that the pursuit of economic policies for political purposes and vice versa is not unique to the second Ba'thist regime. In fact, the developmentalist coalition in Iraq was built well before the Ba'th came to be in charge, and it was constructed on the simultaneous elimination of capital's rampant entitlements and on the suppression of independent labour organisation.

From its inception, Iraqi populism, the political skeleton on which economic developmentalism was built, was always a pact between the state and urban consumers, undertaken at the cost of politicised segments of the labour force and the commercial classes. Unlike the Gulf monarchies, the Iraqi state did not create a permanent class of private-sector clients. Economic insecurity and the rotation of economic contracts in construction were key means by which the once all-too-powerful bourgeoisie was kept under state control.

3. Etatisme in the context of fiscal autonomy

The oil boom of the 1970s did not create, but rather it expanded, the government's role in direct ownership and control of industry, trade, agriculture, services and retailing. Iraq had pursued import substitution industrialisation since the early 1950s. However, with the oil boom the state appropriated the stewardship of the entire industrial sector, with the explicit intent of investing in large industrial establishments to meet domestic demand for light and durable consumer goods. Oil revenues were spent on guaranteeing supplies and increasing wages, on social infrastructure and consumer subsidies. The Iraqi state's core supporters were located in the army and the bureaucracy, but through its direct setting of wages and prices, the Iraqi government cultivated a broader level of economic dependency among labour and consumers, who were protected from inflation during the 1970s and then through eight years of war with Iran.

The military leadership, now firmly in the hands of Sunni Ba'thist officers, made explicit their attack on commercial and industrial groups in the 1970s. So acute was the Iraqi government's sense of threat from the Shi'i commercial classes and their co-sectarians who form a majority of the Iraqi population, that it took every opportunity to eliminate the Shi'i leadership in both the economic and the socio-religious realms through mass deportations, imprisonment and other draconian methods. In contrast, even though the independent power of the trade unions was destroyed, labour won significant gains including a minimum wage, severance pay, social security, housing and a number of other legal guarantees. As the populist base of the Ba'thist state broadened through the boom, power centralised at the apex.

The expansion of the state's economic role is not difficult to document. In the mid-1960s, the Iraqi government employed 52 per cent of the total industrial workforce. By 1987, state-sector industry employed a full 89 per cent of the industrial workforce.[22] Government shares in imports were only 5 per cent of the total in 1955. Between 1975 and 1983, they averaged 90 per cent of total imports. By 1987, when the reforms began in earnest, 96 per cent of the industrial workforce (excluding establishments of less than ten workers) was employed in state-owned factories that produced in excess of 84 per cent of total industrial output. Government control of foreign trade was complete and its part in retailing was not insubstantial. Over 50 per cent of agricultural land was under some form of state ownership. Similarly, the government owned banking, insurance, and other major services.

What was the structure of this economy? In sectoral terms, Iraq experienced all the effects of classic Dutch Disease: an oil-supported rise in the exchange rate of the Iraqi dinar put pressure on tradeables and encouraged the rise of the service economy. Yet, through the deliberate and purposive acts of the state, industrial investment actually grew rapidly. While the service sector grew, employing 40 per cent of the labour force,[23] and while agriculture declined in absolute and relative terms (see Figure 6.1), industry actually grew at unprecedented rates. Although some observers credit Iraq with a heavy industrialisation programme, it had in fact pursued import substitution industrialisation since the early 1950s. With the oil boom, the state appropriated the stewardship of the entire industrial sector, and investments in non-military heavy industry were made only with the intent of feeding import-substituting firms producing for the domestic market. Iraqi industrial investment averaged 23 per cent of development budgets in the 1960s; in the 1975–80 period, it averaged 32 per cent.[24]

With the onset of the war in 1980, industrial investments were shelved and the government began to borrow heavily to finance imports of basic goods and armaments. Dependence on imported food was dramatic, but not atypical for neighbouring oil exporters: by the end of the 1970s Iraq imported over 75 per cent of her food. During the war the state stopped allocating foreign exchange to local entrepreneurs and the domain of the private sector was confined to the construction sector alone.

TABLE 6.1
Public sector share in large manufacturing in Iraq

	% of firm	% workers	% wages	% inputs	% outputs
1956	0.5*	n.a.	n.a.	n.a.	n.a.
1960**	16	32	n.a.	n.a.	n.a.
1964	22	52	64	55	65
1970**	26	64	n.a.	n.a.	n.a.
1974	27	74	76	74	74
1984#	33 (31)	87 (82)	88 (83)	84 (74)	88 (78)
1987#	29 (27)	89 (83)	89 (82)	87 (78)	91 (84)
1989	17##	n.a.	n.a.	n.a.	n.a.

Notes: n.a. = not available. 'Large manufacturing' category before 1983 includes industries with ten or more workers. After 1983, only industries employing 30 or more workers are included. #Includes 'mixed sector'; percentages in brackets exclude mixed sector. ##Calculated by subtracting 64 privatised industries from 1987 figure. Does not account for new state sector industrial investment.
Sources (general): *Annual Statistical Yearbook*, 1988; and Joe Stork, 'The War in the Gulf', *MERIP*, No. 97, June 1981, p. 15.
Sources (specific): *Kathleen Langley, *The Industrialisation of Iraq*, Cambridge, Mass.: Harvard University Press, 1961, p. 281; and **Hanna Batatu, *The Old Social Classes and Revolutionary Movements of Iraq*, Princeton, N.J.: Princeton University Press, 1978, p. 1121.

TABLE 6.2
Public, private and mixed-sector imports (%)

	Government	Mixed sector	Private sector	Foreign
1955*	5	–	n.a.	n.a.
1975	92	0	8	–
1976	89	0	11	–
1977	89	2	9	–
1978	89	2	9	–
1979	91	1	8	–
1980	89	1	10	–
1981	89	2	9	–
1982	83	2	15	–
1983	90	2	8	–
1984	84	3	13	–
1985	86	3	11	–
1986	86	2	12	–
1987	85	4	11	–
1988	59 (total through state foreign exchange)		41**	–

Note: n.a. = not available; – = not applicable/nil. *Kathleen Langley, op. cit., p. 181; **Private sector imports through the 1988 law permitting use of foreign exchange held abroad for imports. Calculated from Ministry of Trade data for 80,843 import licences. No information distinguishing capital held abroad from foreign exchange provided by the government at official exchange rates was available. Up to 28 September 1989, 30,000 import licences had been issued for the private sector for a total of about ID one billion (US$3.2 billion at the official exchange rate).
Source: *Statistical Yearbook*, 1989; figures for 1975–80 from Robert Springborg, 'Infitah, agrarian transformation and elite consolidation in contemporary Iraq', *Middle East Journal*, vol. 50, no. 1, 1986, p. 49.

TABLE 6.3
Private-sector share of GDP for selected economic sectors (%)

Economic sectors	1980	1981	1982	1986	1987
Agriculture	53	49	47	47	49
Manufacturing	37	45	41	27	19
Construction	88	94	94	89	80
Transportation and communication	71	73	76	43	60
Distribution	n.a.	n.a.	n.a.	41	52
Social and personal services	n.a.	n.a.	n.a.	6	7
GDP at factor cost	n.a.	n.a.	n.a.	31	33

Source: Robert Springborg, 'Infitah, agrarian transformation and elite consolidation in contemporary Iraq', *Middle East Journal*, vol. 50, no. 1, 1986, p. 47; and *Statistical Yearbook*, 1988.

FIGURE 6.1
Iraq food production per capita

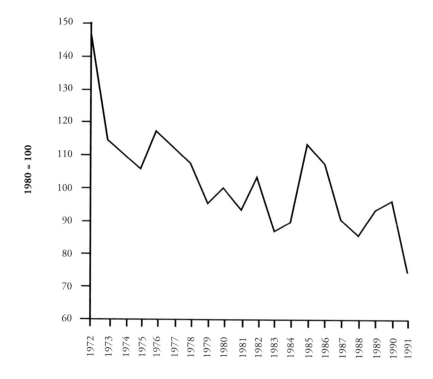

Source: World Bank. *World Tables.*

As a result of these tumultuous events, Iraq entered the recession with a very small private sector concentrated in the construction sector, which had long been subservient to, and existed at the sufferance of, the Party and state bureaucrats. The Ba'thist state upheld its populist pact with consumers even during the war with Iran, while systematically undercutting all independent forms of organisation in society. By the mid-1980s, the Ba'th had thoroughly penetrated all professional, occupational and social groups through corporatist institutions that served the multiple purposes of surveillance, control, distribution and repression. Labour and consumers, in contrast, had maintained their entitlements throughout the eight years of war, largely funded by foreign borrowing from the Gulf states and other international sources.

4. Privatisation and liberalisation in Iraq: 1983–9

Like other oil exporters, from Algeria to Iran, Iraq's first efforts at liberalisation were prompted by the 1983 decline in oil prices and were focused narrowly on agriculture. Early attempts to divest state-owned farms and agricultural projects in 1983 turned into a full-fledged privatisation programme in 1987, and started to gain momentum from the autumn of 1988 after the ceasefire with Iran was announced. Prompted by the steady decline in agricultural yield since the mid-1970s, which had generated a soaring imports bill and created widespread rural-to-urban migration,[25] the new agricultural policies reflected a radical shift from equity to efficiency.[26] In 1983 a new law allowed individuals to rent state farms for grain production, and introduced a variety of input and output subsidies to encourage vegetable and fruit production on privately held farms.[27]

On the eve of the reforms, the government controlled approximately 50 per cent of all agricultural lands. The growth of private business farming was relatively swift. Between 1959–74, only 5.9 million *donums* of the nationalised holdings been distributed; under Law No. 35 of 1983, 4.53 million *donums* had been privatised by 1989. By January 1989, according to government sources, the ownership structure of land had changed considerably: 53 per cent of land was privately owned; 46 per cent was rented from the state by farmers and private investors, and the remaining 1.0 per cent was state held.[28] Some of the land rented to new private entrepreneurs had previously been managed by state-owned agribusinesses,[29] while much of the rest was taken back from agrarian reform beneficiaries. Production subsidies to private agricultural producers were a departure from previous policies, which had held purchase prices down. In addition to renting lands and selling state farms and agribusinesses at extremely favourable prices, the government doubled production subsidies for wheat, rice, barley, corn and tobacco in 1989, bringing the total subsidies for grains alone to over ID300 million for that year.

Did these reforms change the historical bias against agricultural producers and in favour of urban consumers? The state purchase prices were raised in September 1988 as follows (all figures in Iraqi dinars per ton): wheat: 270 (from 170); rice: 500 (from 400); barley: 180 (from 120); corn: 550 (from 450). Whether or not the new prices raised the domestic price of these products above their international prices depends on whether the dinar was valued at official or at black-market rates. For example, the international price of wheat at the time was about US$170/ton. Using the official and black-market exchange rates, the domestic purchase price of wheat in 1989 would thus have been US$891 and US$90, respectively. Moreover, additional subsidies to consumers, administered in the form of the difference between purchase prices and government sale price to bakeries was US$80 per ton. In other words, at the restricted market exchange rates farmers were being squeezed and consumers were being subsidised.

A variety of very large poultry, dairy and fishing enterprises were sold to the private sector outright. By 1989, 19 of the state's 29 poultry farms, six of the large poultry-feed projects, six of the ten large dairy farms,[30] three of the four large government fisheries, and subsidiary agricultural services such as mills and bakeries, were sold to private investors.[31]

While agricultural reforms were the most dramatic, privatisation of state-owned industries during 1989 was also significant. Where Egypt's widely publicised *infitah* policy resulted in the privatisation of exactly two factories over a period of 15 years, the Iraqi government sold, in a single year, seventy large factories in construction materials and mineral extraction, food processing, and light manufacturing.[32] In addition, the government privatised parts of the service sector, such as small hotels, and leased gas stations to the private sector for a duration of three years.

In addition to divestment, the government adopted a number of policies to remove the existing barriers to large industrial investments that had been in place since the nationalisations of 1964. In 1983 the investment ceilings were raised to ID2.0 million and ID5.0 million for solely-owned and limited-share companies respectively. In 1988 the ceiling was lifted completely, and private investors were allowed for the first time to invest in any sector. The law against cross-sectoral investment was abolished, allowing for the development of large, vertically-integrated industrial, trading and agricultural conglomerates. Most importantly, previous tax laws, which claimed up to 75 per cent of profits and forced large private industries to pay 25 per cent of the remaining profits to social security funds for workers, were abolished. In 1987 the maximum tax on industrial profits was lowered to 35 per cent, and then, in 1989, all industries were given a ten-year tax holiday.

Changes in trade were equally dramatic. Imports had previously been completely controlled by the state, either directly or through strict licensing procedures. In 1988, all imports were liberalised for the private sector, provided that they were paid for with foreign exchange held outside the country. Export promotion policies were also initiated in the form of access to foreign currency at a one-to-one ratio (compared with the official rate of ID/3.3 US$ and the black-market rate of (3.3 ID/$), provided that export earnings covered initial investment by 120 per cent in the first two years.

As 1989 wore on, price liberalisation and declining wages created widespread economic deprivation. Trade liberalisation and export promotion policies were promulgated, but only the former had any effect. As a result, imports – which had previously been completely controlled by the state directly or rationed through strict licensing procedures – came under private control at the same time that prices were liberalised. Price controls on all but a handful of basic goods were lifted. With the cuts in government imports, access to state retail outlets was restricted to high-level bureaucrats and the army.

Important institutional changes accompanied the reforms, reflecting the regime's conscious decision to destroy potential sources of social resistance to the reforms *before* they organised. Thus, over 200 General Directors and their entire staffs were dismissed, thereby eliminating the group within the bureaucracy that would have been most opposed to the reforms. Changes that included worker incentives, greater management autonomy and production bonuses were introduced in the organisational structure of the remaining public manufacturing sector. As a result, production in the remaining state-owned factories increased by 27 per cent between 1987 and 1988 and labour productivity jumped by 24 per cent. The special treatment afforded to state construction companies, ostensibly set up to complete 'strategic' projects and compete with foreign contractors in the 1970s, was withdrawn, forcing them to compete with private contractors for government projects.

Observers of Iraq have rightly questioned the degree to which the reforms actually meant a smaller role for the state in the economy. To be sure, the reforms of the 1980s, dramatic as they appear, did not signal a fundamental change in the balance between public and private shares in the economy outside agriculture. Any boost in private agricultural and industrial investment[33] was more than matched by new state-sector investments in strategic and heavy industries. Despite tax breaks, price liberalisation and pro-business labour laws, total private industrial licences for 1989 amounted to only ID33 million for 117 new industrial projects. In contrast, government investment in oil and downstream petrochemical projects alone was US$3.0 billion for the same year. The economic changes of 1988 and 1989 at no point resulted in the state's share of industry falling below a historically stable 76 per cent. Indeed, even in construction, the traditional preserve of the private sector, the share of the public sector increased slightly, while state-financed private-sector imports actually declined in 1989.[34]

In addition, the government's role in pricing policy remained significant. Only ten basic food products retained fixed prices after the 1989 reforms, but the government twice resorted to temporary price fixing and emergency imports of foodstuffs in 1989, in response to excessive profiteering. At a more fundamental level, the state retained control of the 'feed' projects for agribusinesses. As a result, private dairy farms, ranches, fisheries and poultry farms remained dependent on the state for inputs of all kinds, including hatching eggs, imported livestock, feed components, fish, machinery, spare parts and so on. By leasing and not selling land, and by controlling inputs, the state retained control over basic commodities even after they had been privatised.

Thus, the most important aspect of the privatisation and liberalisation programme was not a relative decline in the role of the state in the economy, but rather was a decline in the kinds of activities that the government engaged in. The most important impact of the reforms lay in what they meant for the broadest base of the regime's beneficiaries, consumers and the military. To understand why this was so, it is important to first explore why the reforms were possible and why they failed.

5. The political economy of the market reforms

Part of the reason why the reforms proceeded so swiftly is purely political. Following a pattern identified with the reform programmes of bureaucratic authoritarian regimes in Latin America, the government began by neutralising the organisational structures where social opposition to the reforms was most likely to be located. As mentioned earlier, over 200 General Directorships were dissolved and their staff dismissed, thereby eliminating an important anti-reform group. The Ministry of Industry was merged with the Ministry of Military Industries, centralising control over strategic and heavy industries. The earlier dissolution of the labour unions by fiat in 1987, and the elimination of the category of 'worker' in the state sector, was one important aspect of this social engineering. Workers remaining in public-sector enterprises were turned into regular members of the civil service, while private sector employees were formally given the right to reconstitute labour unions on their own. The new 'right' to form unions only applied to private-sector establishments with over 50 workers, a category that covered only 8 per cent of the total industrial workforce. The membership in the General Union of Iraqi Labour, which had included agricultural, industrial and service workers in both the private and public sectors, plummeted from 1.75 million in 1988, to a total possible membership of only 7,794. The abolition of the minimum wage and the new liberalised labour import laws associated with the short-lived Arab Cooperation Council meant that the economic entitlements long exchanged for political quiescence were now obsolete.

As important as it was, political 'will' only works in material context: the structure of the Iraqi economy was critical to why the regime could implement its reforms as quickly as it did. Industrial strategies and state employment patterns illustrate the structural characteristics that permitted Iraq to implement the reforms; ironically, they also explain why the economic crisis of 1989–90 resulted in military aggression.

FIGURE 6.2
Industrial structure by type of industry

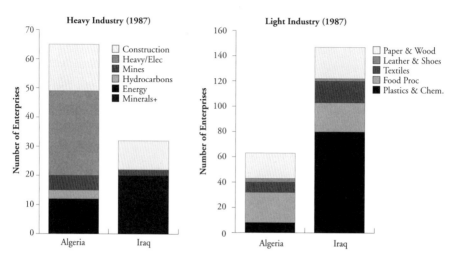

Notes: Includes 'National Enterprises' for Algeria and 'Socialist and Mixed' sectors for Iraq. +Non-metallic minerals only.
Sources: Government of Algeria, ONS, Bulletin #18, p. 48. Government of Iraq, *Statistical Yearbook*, 1988, p. 139.

FIGURE 6.3
Industrial work force by type of industry

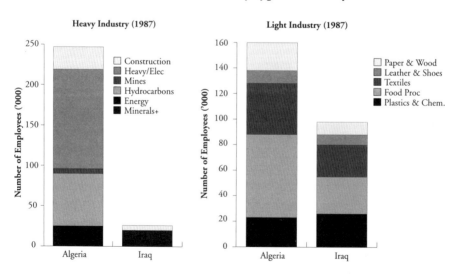

Notes: Includes 'National Enterprises' for Algeria and 'Socialist and Mixed' sectors for Iraq. +Non-metallic minerals only.
Sources: Government of Algeria, ONS, Bulletin #18, p. 48. Government of Iraq, *Statistical Yearbook*, 1988, p. 139.

Socialist countries typically squeeze agriculture to pursue heavy industrialisation, with the aim of achieving self-sufficiency. With the exception of strategic industries, Iraq's plans in this direction were thwarted in the 1980s. As a result, the country had an industrial structure based on import substitution, and composed of relatively small plants designed to produce for the domestic market. The structure of Iraqi industry was an important part of why the state could privatise easily. A quick glance at figures comparing Algerian and Iraqi industry dispels the notion that Iraq's heavy industry was highly developed. In contrast to cases like Algeria, Iraqi national capitalists were able to contemplate buying and running formerly state-owned industries.

Because industry was the most heavily protected sector, liberalisation affected it most dramatically. The small size of the import-substituting (IS) industries and their relatively large number meant that there were private investors able to take them over and that, in addition, the state could cultivate a relatively large number of new clients with a relatively small outlay of the national patrimony. Politically, the privatisation of those industries that produced for the domestic market also allowed the state to set up a conflictual relationship between consumers and industrialists that would insulate the state from responsibility for inflation or shortages. However, it would also set in place the regime's capacity to play capitalists and consumers off against one another – a time-honoured Ba'thist device.

Another important clue to why the reforms could be implemented so quickly is to be found in the structure of state-sector employment in Iraq. Again, the point can be illustrated in comparative terms by noting the following: by 1987 a full 44 per cent of Algerian workers were in plants of over 5,000 workers, and 23 of Algeria's largest industries had twice the number of workers as all of Iraq's state industrial sector.[35] Moreover, only 4 per cent of Algeria's labour force was employed in the private sector, compared with 11 per cent in Iraq.

State employment captures the most striking characteristic of the 'beneficiaries' of the Iraqi regime. Where Algerian industrial workers comprised 12 per cent of the active population and the Algerian civil service absorbed about 24 per cent, about 30 per cent of the urban population was unemployed and directly dependent on government subsidies for their livelihood. In Iraq, industrial workers made up only 3 per cent of the active population but a full 23 per cent of Iraq's workforce was in the military. The figure for Algeria is 2 per cent. The state's low reliance on industrial labour for political support was critical in its ability to privatise light industry, where the vast majority of industrial workers were employed.

The nature of Iraqi statism and the structure of domestic industry and services allowed the Iraqi government to privatise the bulk of its IS industries in one year, while simultaneously centralising control over heavy industries.[36] The privatisation of importing-substituting industry severed links between the economic ministries and the private sector, removing import quotas, preferential foreign exchange allocations for importing capital goods, and price controls. Iraq's IS industry structure meant that privatisation could serve as a shortcut to cultivating new clients in society.[37] State firms, sold at nominal prices to private-sector, Ba'th, and military elites generated new social support.

Compared with the difficulties faced by most countries in generating sufficient political will to undertake such massive restructuring programmes, the first stages of the reforms went smoothly. Initially the reforms appeared to boost production. In the industrial sector, cuts in labour costs and substitution of local for imported inputs appeared to be an 'efficient' result, but on the whole, levels of industrial production actually declined due to shortages in imported raw materials.[38]

However, despite these early successes, the Iraqi government's policies soon resulted in high levels of inflation (estimated by the Ministry of Finance to have been as high as 60 per cent in 1989), as well as unemployment, shortages in basic goods, growing and highly visible economic inequality, and the emergence of a brisk black market in foreign currencies.

There were deeper and more fundamental causes for the failure of the reforms. An important part of the problem was institutional: the Ba'thist state's formidable coercive powers did not empower it to regulate a market economy. The economic functions of the state had in the past determined the structure of the bureaucracy. Not unlike other liberalising states, the institutions that governed the economy could not be revamped overnight. To begin with, the administrative problems of monitoring the private sector were far more difficult than anticipated, particularly since cooperation was required between the remaining state-sector and private enterprises.

A formidable reorganisation of the bureaucracy became necessary simply, for example, to ensure that bread reached consumers at the fixed prices. The government subsidised producers and consumers alike for this basic commodity, but as the farms, mills, bakeries and retail outlets were divested, price fixing began to require substantial administrative resources. Direct links between either local producers and processing industries or between exporters and producers were non-existent: indeed, it was the deliberate policy of the government to force key transactions through the bureaucracy. Those agricultural enterprises and food pro-cessing units that were retained by the government for 'security reasons' were also responsible for monitoring inputs and prices, supplying spare parts and machinery, and performing a number of activities that involve the government directly in the day-to-day operation of the firms.

Similarly, the privatisation of agriculture transformed both the social organisation of rural Iraq (where agricultural producers had previously been forced to join state cooperatives and collective farms), and reshaped the administrative tasks of the state. In the 1970s and 1980s, the distribution of nationalised agricultural land had been restricted to those farmers who joined cooperatives, with the result that the cooperative movement grew apace in that period. With the reforms of 1983, land leases were de-linked from the requirement to join co-operatives, and lending to co-operatives through the Agricultural Co-operative Bank plummeted from ID21.5 million in 1985 to only ID91,000 in 1988.[39]

The extent to which the socialist and collective farming sector was held together by law alone was reflected in the spontaneous disbanding of agricultural cooperatives in the 1980s.[40] However artificial the cooperatives, they were nevertheless the only organisational entity in the rural areas entrusted with credit, inputs, transportation and processing. Thus, initial gains in agricultural yields were temporary, and the historical problems of the agricultural sector, as well as the exploitative nature of 'traditional' forms of credit provision, soon re-emerged, with predictable consequences for production.

Ironically, the very factors that guaranteed the initial ease with which these measures had been undertaken presaged their failure: while opposition from labour, consumers and the bureaucracy could be repressed, the market reforms required the confidence of local and Arab investors as well as that of the large contingent of Iraqis abroad, who were encouraged to return. Investors, though clearly the most favoured group, lacked confidence in the regime's long-term commitment to private enterprise and responded to this uncertainty by pricing goods at the highest possible level. Thus, attempts to attract serious long-term investments, such as the sale of the large government-owned luxury hotels, failed quite seriously due to lack of confidence on the part of both domestic and Arab investors in the regime's commitment to upholding property rights. Private-sector apprehensions about the government's desire to retain shares as well as representatives on the boards of directors reflected this lack of confidence.

Even in the mushrooming foreign-trade sector, importers of luxury goods responded to the new policy with caution, and with good reason: while the new imports policy allowed merchants to use external funds, the law that prescribed the death penalty for holding foreign currency abroad was not formally repealed. In addition, the government's promise to review the free imports policy after three years engendered a certain amount of apprehension among importers and produced hoarding among consumers. Contradictions in the trade liberalisation policies included the prohibition against the repatriation of profits to fund more imports. Designed to encourage fixed-capital investments, the law against capital repatriation, even for purposes of buying raw materials for domestic industry, was one of the most destructive aspects of the reform: a major cost of the trade liberalisation policies, as noted earlier, was the emergence of a brisk black market in currencies.

Unlike other command economies, Iraq had avoided the emergence of a black market throughout the war by enforcing the death penalty for dealing in foreign currency. The free imports policy enabled individuals to use currency held abroad to fund imports of any goods, but prohibited the repatriation of profits to import more. The policy failed to achieve the stated aim of forcing profits into local investment, resulting instead in the physical transport of dinars to Kuwait, where the high volatility of dinar circulation put increasing pressure on the value of the Iraqi currency.[41] While access to dollar-denominated oil revenues protected the government from the effects of the deteriorating value of the dinar, Iraqi industrialists and importers using the black market were forced to pass on costs to consumers. Shortages of raw materials added to the trend, as did the high cost of imports, most of which were financed through the burgeoning black market in currencies. While the government allowed exporters to use 100 per cent of foreign currency earnings for capital investments and up to 30 per cent for raw material imports, few were able to take advantage of this policy since the privatised factories were geared toward supplying domestic, not foreign markets. By the summer of 1990, a competitive regional market had been found only in dates and related products. At the same time that most food-processing and agricultural enterprises were turned over to the private sector, the government stopped providing raw materials for industry and drastically curtailed state imports of foodstuffs. For a variety of reasons, including severe foreign exchange shortages that prohibited capacity expansion, raw material imports and even access to spare parts, large-scale shortages in basic goods were created.

The costs of the transition in Iraq was clearly borne by fixed-income groups, consumers and labour, who, in the winter of 1989, fondly recalled the long years of war as times of certainty and plenty. Private sector elites were given wide privileges, but their reluctance to use their own foreign reserves and the government's tight control of oil revenues prevented them from meeting domestic demand. In short, while enjoying unprecedented profits and benefits, private investors suffered from the lack of foreign exchange for raw materials and capital goods. Furthermore, the institutional and social changes necessary for them to gain enough confidence to invest showed no signs of emerging. Like other capitalists, Iraqi investors perceive free repatriation of capital as the only true guarantee for investments.

Finally, the reforms failed completely in introducing domestic competition. Investors, wary of each other, of the legal system, and of the regime's commitment to upholding contracts, designed new investments with the exclusive aim of insulating themselves from other entrepreneurs by gaining control of all upstream and downstream activities in a particular product. This tendency was enhanced by the lack of new investment that would promote competition among producers. As a result, the sale of industries to the private sector in Iraq resulted not in the unleashing of competitive forces but in the transformation of public monopolies into private monopolies.

For many of the same reasons, foreign capital remained equally shy of Iraq. British and American businessmen clamoured for entry into the Iraqi market and were encouraged by what appeared to be signs of an overall softening of the Ba'thist's refusal to allow foreign companies free access to that market. However, while General Motors and Mercedes had begun construction on joint-venture plants with the government, non-Arab foreign investment remained strictly prohibited.

The Arab Investment Law of 1989, which gave Arab investors unprecedented benefits that included capital repatriation, guarantees against nationalisation and unrestricted labour import rights, failed to lure any investors from the rich Gulf states. Palestinian, Lebanese and Jordanian businessmen, however, because of the high levels of mobility they enjoy and their experience with making quick investments with great alacrity, used the favourable laws speedily to effect highly lucrative transactions in trade and services. Neither industrial investment, nor investments in the large state-owned hotels that were earmarked for privatisation in 1989, actually materialised.

There was a characteristically Machiavellian aspect to the government's choices, which made investors even more uneasy. By widely publicising its withdrawal from the economy and divesting itself of those industries that catered specifically to the needs of the average consumer, the government freed itself from direct responsibility for inflation, shortages, and the escalating black market in goods, currencies and services. Furthermore, by stepping out of its heavy role in mediating the relationship between consumers and labour on the one hand and producers on the other, the blame for economic hardships was placed squarely on the shoulders of the newly affluent private industrialists and businessmen. The new terms under which workers or service-sector employees with fixed incomes were to bargain with business were deliberately left undefined. It did not escape the attention of businessmen that the rapidly growing resentment of the consumers, bureaucrats and workers who had to bear the burden of these policies became a political facility which could be used with impunity by the state against the new entrepreneurs at any time.

Thus, while liberalisation and privatisation clearly benefited the new business group in the short term, the small private-sector elite that emerged in the late 1980s was hamstrung in several ways. While there is not enough evidence to prove or disprove the widely accepted view that this group forms an important social base of support for the regime, the contradictory economic policies pursued so far almost guarantee the failure of the private sector, laying open to question the extent to which it is a true and protected client of the regime. The regime's reluctance to define property rights in such a way as to produce investor confidence, coupled with its refusal to publicly revise its ideology and legitimate the growing inequalities in the cities, recalled the historical acrimony between private capital and the state on the one hand, and capitalists and the public on the other.

The regime had long managed without a permanent support base among economic elites, and the failure of the reforms to attract investment may have created the crisis. However, the true political cost of the reforms appears to have been unanticipated: it was the cumulative result of the reforms on the broadest base of domestic support for the Ba'th – consumers and the military.

At the onset of the reforms, the Iraqi workforce was fully employed. Labour shortages, particularly during the Iran–Iraq war, had resulted in the inward migration of approximately three million workers and in the entry of over 280,000 women into the civil services.[42] The General Union of Iraqi Labour, itself an organisational device for the political demobilisation of labour, had largely served as a top-down organ of the regime designed to discipline and to organise labour for political and economic purposes defined by the state. For reasons unrelated to anything in history, analysts of modern Iraq are unwilling to credit the idea that, as in all other militarised societies, demobilisation was an exceedingly disruptive event. Moreover, Iraq began demobilising its military as privatisation was pushing thousands of skilled industrial workers into the market and as migration from Arab Co-operation Council countries was opening up to attract more or less unlimited numbers of significantly cheaper workers. In short, the government began its programme by completely undercutting the bargaining power of labour, by removing minimum wages, and by opening up the domestic labour market to Arab labour at a time when some 200,000 members of the armed forces were returning to the civilian workforce.[43] As early as October 1989, civil order in Baghdad was beginning to break down as large segments of the army returned to find themselves without employment in a highly inflationary environment.

The military was merely one of the highly organised (and trained, and armed) groups within the broader category of consumers that had made up the mass base of the Ba'th from the onset of the oil boom.

Unlike labour and the bureaucracy, consumers lacked organisation as well as a formal representative body. A direct consequence of the government's privatisation and imports policy was that it allowed all prices, apart from those of some basic goods, to rise completely unchecked, thereby creating inflation rates of over 60 per cent by the end of 1989. These sudden changes were experienced by a population long accustomed to stability and certainty in consumer goods, and to relatively equitable income distribution. Previously, the state had set prices by selling imported and local goods directly in the market through retail outlets that were well stocked and available to all citizens. In 1989, however, access to these was restricted to bureaucrats and the army, and the goods stocked by government retail stores was narrowed to a few basic items, which were available under a quota system. By the winter of 1989, shortages were so severe that the Baghdad metropolitan area, usually the best supplied in the country, had black markets in eggs and chickens. The comments elicited from average Iraqis concerning rising inequality in Baghdad reminded this researcher that it was not for nothing that the main contenders for political power in Iraq had traditionally been the communists, not the liberals.

Economic liberalisation, as in other command economies, created instant millionaires and widespread downward mobility. Unlike the Soviet Union, Poland, Syria, Egypt and Algeria, the Iraqi regime felt no compulsion to solicit a public mandate for these policies or even publicly to revise its pronouncements on the redistributive aims of the regime. While recent attempts in much of the Gulf to expand the stock market have been based on the explicit desire of the government to attract small investors and to give the poor 'a stake in the system', no attempts were made to design policies that would expand ownership in Iraq. The total number of shareholders trading on the Iraqi stock market remained small, and shares in the lucrative 'mixed-sector' plants which had access to government foreign exchange and boasted annual profit rates of between 30 to 90 per cent have been used for political patronage purposes. Administrative sluggishness and legal restrictions on share trading curtail the entry of new investors and discourage small investors.[44]

Not surprisingly, the privatisation and liberalisation programmes produced highly inequitable results. Ownership of the newly privatised enterprises was highly concentrated. Thirteen of the 70 factories that were privatised were purchased by the Al Bunniyyah family. Not counting agricultural projects, this same family owns 36 of the very largest industries and over 45-million square metres of land. Many of the new importers, as well as the investors who bought and leased the large government holdings in agriculture and industry, were not traditional private-sector elites. Rather, most of them made their fortunes in contracting. Some of the smaller investors, particularly in agriculture, were retired army officers. Unlike the old commercial elite, the new private sector that emerged from the reforms had strong political, financial and kinship ties with the regime.

By the turn of the decade, broad-based consumer discontent was directed, conveniently, not against the state but against the new industrial and commercial elites. As the crisis deepened, the economic disaster became a political crisis, and in this context the government reimposed price controls and eventually rescinded many of the reforms. Under increasing international pressure and unable to control the value of the dinar on the market, the Ba'th decided to look elsewhere for financial relief.

It is important to emphasise that what the regime could not withstand in peacetime, both ideologically and materially, it easily weathered when it remobilised the army and was again at war.

6. Conclusion

Following development theory in general, it is a staple of Iraq studies that the Iraqi bourgeoisie was 'weak' and the state 'strong'. Added to this, the 'theorists' of the rentier state have promoted the impression that oil wealth by itself endows states with extraordinary capacities. Oil has become a catch-all variable that suffices, mantra-like, to explain everything from dictatorship to sectoral spending patterns. In fact, oil exporters have radically different economic structures that evolved out of historical struggles between social groups contending for ascendance. There is a need for scholars to attend to these differences through empirical investigation rather than 'theory'.

Iraqi developmentalism was constituted during periods of intensifying class conflict between workers and capital that occurred while the colonial economies were being restructured following independence. The stabilisation of these conflicts was achieved simultaneously by eliminating segments of the traditional commercial and land-holding classes and by demobilising labour. The political exclusion of labour and capital coincided with the cultivation of a broad constituency of support among consumers, supplied with fixed-price basic goods through direct state control over imports. State control over industry and agriculture meant that wages were centrally regulated, and wage/price ratios remained relatively stable over long periods of time. In political terms, the developmentalist state in Iraq, replete with its valorisation of industrial production, industrial workers and national self-sufficiency, actually rested on the acquiescence, if not the support, of a broad base of consumers, not producers.

The economic liberalisation policies of the 1980s undercut precisely this broad base of the developmentalist coalition through subjecting consumers to world prices in many goods previously subsidised by the state. What changed, in the process, was the underlying relationship constituted in the autarkic period between the state as supplier and society as consumer. In recrafting state–society relations along a supplier–consumer nexus, the developmentalist coalition destroyed class affinities and disorganised class-based associations in the two decades preceding the liberalisations of the 1980s. In political terms, national developmentalism was a response to class conflict, but the workings of national developmentalist systems produced social groupings that were not class based.

Rhetoric notwithstanding, the fact that consumers were the social base of the developmentalist state had a number of important implications for how social coalitions unravelled in response to economic liberalisation. Consumers typically face severe collective action problems, even in industrial democracies. While organised opposition to the reforms came from corporate groups already institutionally constituted in the pre-reform period, the onset of liberalisation generated not class conflict but consumer revolts that (a) did not occur within an institutionalised framework, and that (b) articulated very broadly based and general criticisms of economic policies. The preferences of constituted corporate groups in post-liberalisation Iraq fragmented within and between class lines, producing cleavages within industrial labour, the private business sector, and the state sector. These cleavages reflected company- and industry-level evaluations as to how a particular sub-sector would fare in the new liberalised economy. The very majority group that neoliberal reformers argued would benefit from liberalisation – the consumers – actually lost substantial entitlements and experienced increasing levels of insecurity as a result of the economic reforms of the 1980s.

In privatising, the Iraqi state was moving toward an unprecedented pattern of creating permanent private power. This was completely in contrast with its previous strategy of rapid rotation, in which the oil state constantly shifted its private interlocutors, thereby breeding a politics of insecurity that left private interests fragmented and dependent. I have shown elsewhere that, unlike the Gulf states where permanent classes of state-sponsored private-sector elites were created under the rubric of 'capitalism', the Ba'th used insecurity to promote loyalty, especially in the lucrative field of contracting; a sector larger than both agriculture and non-strategic industries put together.

I have also argued that the conflicts which emerged around liberalisation reflect the fundamental contours of the way the Ba'thist state was run. Liberalisation undercut the economic pact that had kept the Ba'thist regime in control, but it also reflected a perennial problem in the economic history of modern Iraq: the inability of the state to regulate a private economy. Given the attention that has been rightly lavished on the coercive powers of the Iraqi state, it is ironic that at no time in its history, including the pre-independence period, has it successfully regulated a market economy.

Under economic sanctions – and even with the institutional mutation of the state into a network of tribes and formal bureaux, the complete rearrangement of property rights, a ravaging of the middle classes, and the death of thousands by starvation and disease – the state has reconstructed its legitimacy by being the price-fixer. We can measure the ineffectiveness, if not the moral bankruptcy, of the sanctions by considering the extent to which the sanctions reinstalled the state in the position of stabilising prices at a time when the 'market' was telling the people of Iraq that they had no option but to starve. The conduct of the Gulf War allies since 1990 has been a masterful confirmation of the wildest conspiracy theories of the Ba'th that are deployed domestically to justify all manner of policies.

In the macabre, violent world that is underwritten and sanctified by the rhetoric of international human rights so shamelessly deployed by international organisations and the United States government, Saddam Hussain recently justified punitive measures taken against exploitative traders in terms that combined Stalin and Marx:

The poor should know exactly what their interests are and should not side with the rich . . . some of the poor even side with the traders who exploit them. The same poor before the executions of the traders were crying and shouting against the traders asking: 'Where is the state to protect us?'[45]

As in the 1930s, today's Iraq gives us a stark and disturbing illustration of the central paradox of modern political life: the state is at once the only prophylactic against unbridled capitalism and at the same time is the wellspring of the most brutal and pervasive oppression. This continuity, and the consequences that flowed from it, are critical to understanding both the present and future of Iraq's political economy.

NOTES

1 This work draws on heavily on data previously published in articles that appeared in *Comparative Politics*, *Middle East Report*, and *Business and Politics*. Fieldwork in Iraq was conducted in 1989. The author has not visited Iraq since 1991.

2 For a descriptive account, see Amatzia Baram, 'Neo-tribalism in Iraq: Saddam Hussain's tribal policies 1991–96', *International Journal of Middle Eastern Studies*, vol. 29, 1997, pp. 1–31.

3 See *Middle East Economic Digest* [*MEED*], 21 May, 1993.

4 Merchants would often create shortages by getting licenses but not using them, thereby precipitating shortages which yielded windfall profits: cf. *Economic Report*, July 1951, PRO EQ 1013/8, v. 91644. After the departure of the Jewish merchants, their Basrawi replacements prevented confiscated Jewish merchandise from being unloaded, as huge profits were being made due to scarcities. *Basra Report*, May 1951, PRO EQ 1013/6, v. 91644.

5 See, for example, the account of the merchants' resistance to the Ministry of Agriculture's announcement of fixed commissions on agricultural machinery, in *Middle East Report*, February 1954, PRO VQ 1101/4, No. 2, v. 111005. The powerful Jewish community opposed the Banking Control Ordinance of 1950. Of their opposition, the British wrote, 'it is generally believed that it was in order to prevent them from trying to block it in parliament that the Minister of Finance decided to put it into force by special ordinance'. The law placed a minimal capital requirement on moneylenders and required licensing. See *Control of Banking Ordinance*, Law no. 34 of 1950, PRO EQ 1117/1, v. 82434, 1950. After the legislation went through, there was a run on the banks because of the regulating of the moneychangers. See 'Report from Sir Henry Mack', Baghdad, 23 March 1950, PRO EQ 1103/1, v. 82422.

6 This law created a Central Supplies Committee, restricts exports, and is expanded in Ordinance 63 of 1939 to prevent re-export of local or imported products. See E 6579/78/93, vol. 23203, 1939.

7 The 'Law Regulating the Economic Life of Iraq' of April 1942 is reproduced in PRO FO, vol. 31360, E 3292. It makes transit trade illegal. In addition, 'Regulations for the Occupation of Immovable Property' allows the government to occupy any property necessary during the war; see Law 27 of 1942, PRO FO v. 31360, E 4250/17/93. Furthermore, the 1942 law authorised the government to take possession of commodities and sell them at fixed prices, or to take over and operate mills, factories and transportation facilities. The law mandated the use of compulsory labour service on fixed wages.

8 See *Economic Report*, 22 October 1942, PRO FO E 6607/31/93, vol. 31361.

9 Hanna Batatu, *The Old Social Classes and Revolutionary Movements of Iraq*, Princeton, N.J.: Princeton University Press, 1978, pp. 470–1.

10 For a detailed discussion of government policy, see 'Review of Commercial conditions in Iraq', 22 December 1944, PRO E 7888/7888/93, v. 40109. See also *Basra Report*, December 1952, PRO Q 1015/1, v. 104664, and *Economic Monthly Report*, January 1951, v. 91644.

11 See *Basra Report*, January 1953, PRO Q 1015/2, v. 104664.

12 See Kiren Aziz Chaudhry, 'On the way to the market: economic liberalization and the Iraqi invasion of Kuwait', *Middle East Report*, vol. 21, no. 3, May/June 1991.

13 Batatu, op. cit., p. 482.

14 See Fran Hazelton, 'Iraq to 1963', in *Saddam's Iraq*, London, CARDRI, 1989, pp. 1–29.

15 Batatu, op. cit., pp. 838, 839 and 841.

16 Ibid., p. 839.

17 Ibid., pp. 859–60.

18 Ibid., pp. 937–40.

19 Ibid., pp. 944–6.

20 See U. Zaher, 'Political Developments in Iraq, 1963–1980', in *Saddam's Iraq*, London, CARDRI, 1989, pp. 30–53.

21 Ibid.

22 Iraqi figures calculated from Government of Iraq, *Annual Statistical Yearbook, 1988*, and Joe Stork, 'The war in the Gulf', *MERIP*, No. 97, June 1981, p. 15. Algerian figures from Mahfoud Bennoune, *The Making of Contemporary Algeria, 1830–1987: Colonial Upheavals and Post-Independence Development*, New York: Cambridge University Press, 1988, p. 159, and Government of Algeria, *Statistical Yearbook 1991*, p. 54.

23 See Figure 6.3 for Algerian data. The Iraqi estimate for 1976 is from United Nations, Economic Council for West Asia (ECWA), 'Industrial Development in Iraq – Problems and Prospects', Beirut, EWA/UNIDO, 1979, p. 28.

24 Phebe Marr, 'Iraq: Its Revolutionary Experience Under the Ba'th', in *Ideology and Power in the Middle East*, Durham, NC, Duke University Press, 1988, f.n. 28, p. 475.

25 Agricultural policy and production trends are summarized in Robert Springborg, 'Infitah, agrarian transformation and elite consolidation in contemporary Iraq', *Middle East Journal*, vol. 40, no. 1, 1986, pp. 33–52.

26 It is perhaps slightly ironic that after the disruptive land reforms of the 1960s and 1970s, the missionary zeal of the rural development programmes eclipsed to the point where lands leased under the current programme are legally prohibited from sinking below 62.5 acres.

27 See Robert Springborg, 'Infitah, agrarian transformation and elite consolidation in contemporary Iraq', *Middle East Journal*, vol. 40, no. 1, 1986, pp. 33–52.

28 Government of Iraq, *Statistical Yearbook for 1988*, p. 130.

29 Government-owned grain farms that were rented are huge and in many cases were leased by a single person. Two of these, for example, were 29,000 and 25,300 *donums* each. Between 35 and 40 of the smaller state farms remain under government control.

30 All of these enterprises were very large. For example, the capacity of each poultry farm was between 100 million and 270 million eggs. The capacity of the dairy farms was not less than 800 milk cows each. In at least three cases, the dairy farms were in mint condition, as they had not yet begun production.

31 The government retained two of the 12 mills and 14 of the 20 bakeries, and planned to sell all but one of the bakeries by the end of 1990.

32 Of these, 66 factories were sold by auction to individual investors and four were transferred to the 'mixed sector', in which shares are held by the Industrial Bank, individuals and other mixed-sector companies.

33 Licences granted for private factories to be set up with the use of foreign exchange held abroad went from seven in 1987 to 63 in 1988 to 117 in 1989. However, total investments in these factories was small, comprising only ID17m in 1988 and ID33m in 1989. At the official exchange rate, the 1989 figure comes to about US$99 million. Most of the planned private investment is in plastics (60 per cent), the rest in textiles and construction (15 per cent), food processing (15 per cent) and metal industries (10 per cent).

34 In 1988, total private-sector imports were ID1.1 billion for 80,843 licences, while in 1989 only 30,000 licences were issued for ID1 billion.

35 For a more detailed discussion and data, see Kiren Chaudhry, 'Prices, politics, institutions: oil exporters in the international economy', *Business and Politics*, vol. 1, no. 3, November 1999.

36 I have discussed the Iraqi reforms at length in 'Economic liberalization and the lineages of the rentier state', *Comparative Politics*, October 1994; see also my 'Economic Liberalization in Oil Exporters: Saudi Arabia and Iraq', in Eliya Harik and Dennis Sullivan (eds), *Privatization and Economic Liberalization in the Middle East*, Indiana University Press, 1992.

37 This point is made, for different countries, in much of the literature on economic liberalization and privatization. See, for example, David Stark, 'Privatization in Hungary: from plan to market or from plan to clan?', *East European Politics and Societies*, vol. 4, no. 3, pp. 351–92; Thomas Callaghy and Ernest James Wilson III, 'Africa: Policy, Reality, or Ritual?', in Raymond Vernon, (ed), *The Promise of Privatization*, Council on Foreign Relations, 1988; Tony Killick and Simon Commander, 'State divestiture as a policy instrument in developing countries', *World Development*, vol. 16, no. 12, 1988, pp. 2465–79; Nicolas Van De Walle, 'Privatization in developing countries: a review of the issues', *World Development*, Vol. 17, no. 5, 1989, pp. 601–15; and Pan Yotopoulos, 'The (RIP) tide of privatization: lessons from Chile', *World Development*, Vol. 17, no. 5, 1989, pp. 683–702.

38 An example might clear up this otherwise confusing assessment. The Dialah Tomato Canning Factory, recently sold to an experienced industrial family, used to produce 7,350 tons of canned tomatoes from concentrate imported from

Czechoslovakia. When it was government owned, the factory worked all year and employed 286 workers. Concentrate imports were cut completely in 1989, forcing the factory to use only domestic fresh tomatoes. Working for only three months and employing only two hundred workers, under private ownership the factory produced 3,000 tons of canned tomatoes in 1989. While retail prices were ID2.56/kg in 1988, in 1989 the retail price was 3.66/kg, (a price increase of 43 per cent).

39 Government of Iraq, *Statistical Yearbook for 1988*, p. 128.

40 From 1,635 co-operatives with 23,109 members in 1975, the number of co-operatives plummeted to 713 by the end of 1988. Collective farms declined from 79 to seven, and specialised co-operatives shrank from 173 in 1975 to 52 in 1988. See Robin Theobald and Sa'ad Jawad, 'Problems of Rural Development in an Oil-rich Economy: Iraq 1958–1975', in Tim Niblock (ed.), *Iraq: The Contemporary State*, London: Croom Helm, 1982, p. 204; and Government of Iraq, *Statistical Yearbook for 1988*, p. 125.

41 Prior to the ban on Arabs purchasing real estate in Iraq, the dinars were widely used by Kuwaitis to purchase urban land in and around Baghdad.

42 Government of Iraq, *Statistical Yearbook, 1988*, Table 2/16.

43 The resulting violence against foreign workers, mainly Egyptians, prompted the government to lower the limit on legal remittances to ID10 per month, precipitating an exodus of foreign labour.

44 In 1989, there were 59,348 shareholders in the 19 mixed-sector industries and the three private joint-stock companies in Iraq. The nominal capital of these industries is only ID242 million. The liberalisation policies apparently had little effect on the stock market. As the table below shows, shares grew from 1985 to 1988, but the number of transactions has remained small. Owing both to low trading and also to policies prohibiting large shareholders from selling their shares within one year of purchase, formal prices have stayed almost constant, although informal trading prices often exceed the registered share prices.

Year	Number of transactions	Number of shares traded
1985	334	482,207
1986	408	644,441
1987	423	1,142,829
1988	465	762,965
1989 (till August)	310	428,199

45 Quoted in *MEED*, 28 August 1992, pp. 15–16.

7

Humanitarian Needs and International Assistance in Iraq after the Gulf War

Sarah Graham-Brown[1]

1. Introduction

After the 1991 Gulf War and the implementation of United Nations Security Council Resolution 687, the economic embargo on both Iraqi exports and sales of goods and services to that country remained in place, with certain humanitarian exceptions relating to food and medicine. Few people expected this regime of economic sanctions to continue for more than a year or so. However, because sanctions were in fact prolonged, the dramatic decline in living standards for the Iraqi population created pressure to find ways of relieving the most acute forms of suffering.

In April 1991, the Iraqi government began a reconstruction drive to repair damage caused during the war. A three-month emergency campaign was organised immediately after the uprisings to restore basic services and infrastructure. Initial repairs – to roads, some telecommunications, electricity and water services – were completed more rapidly that most aid officials had anticipated.[2]

However, there were longer-term difficulties. Civilian infrastructure and health facilities did actually improve to some extent from the post-war low in early 1991, but always remained far below pre-war levels, due to shortages of spare parts, shortage of vehicles and equipment, and problems with the supply of basic healthcare inputs, medicines and paper. A considerable amount was achieved using cannibalised parts, but cumulatively these problems caused a further deterioration in services after a couple of years. In September 1991, for example, the international study team found water-treatment plants were operating on average at about 50 per cent capacity. In 1996, the level appeared about the same – no further improvements had been achieved. The water and sanitation situation remained serious in most of the southern governorates.[3] Thus, the problem of declining living and environmental conditions has remained.

Reports published in 1991, especially that by Prince Sadruddin Aga Khan, the UN Secretary General's Executive Delegate, suggested that massive investments would be required to restore Iraq's infrastructure and services to their previous condition.[4] Under the sanctions regime, Iraq itself was unlikely to be able to pay for this, and the international community had no intention of doing so. The humanitarian aid programme administered by UN agencies and NGOs was able to meet only a fraction of need, and it quickly became clear to aid agencies working in Iraq after the war that their efforts were little more than a 'sticking plaster'.

Until 1990, Iraq, which was categorised as an 'upper-middle-income' country by the World Bank, had not been a recipient of humanitarian assistance. There were UN programmes in the country – run mainly by UNDP, UNICEF, WHO and FAO – but these were mainly technical assistance and some primary healthcare programmes, carried out in association with the government. Since 1991, the international community has constantly sought ways of using Iraq's own funds to pay for humanitarian needs. The first part of this chapter looks at attempts to find such a mechanism, and their outcome.

2. The 'food-for-oil' resolutions

The first attempt to resolve this impasse was a Security Council initiative (UNSC Resolutions 706 and 712 of 1991) that would have allowed Iraq to sell limited amounts of oil (US$1.6 billion every six months) under conditions controlled by the UN. 'Food-for-oil', the journalistic shorthand used for these resolutions, is a somewhat misleading description. The proceeds of the sales were not to be used solely to pay for humanitarian goods. Thirty per cent of each tranche was to be contributed to the UN Compensation Fund, established under UNSC Resolution 687 to deal with individual, corporate and government claims relating to damage and injuries during the Gulf War. Further appropriations were to be made for UN expenses in Iraq, particularly for the work of UNSCOM, the UN Special Commission that was dealing with Iraq's weapons of mass destruction.

This left considerably less funding available for humanitarian needs than had originally been proposed by Sadruddin Aga Khan.[5] In his July

1991 report, he had argued that to restore a level of services described as 'greatly reduced' from the level that had prevailed in the pre-war period – including water, sanitation, power generation, health services, agriculture and basic repairs to the oil industry – would require a sum of approximately US$6.85 billion over one year. He had suggested an initial sale of US$2.65 billion over four months.[6]

After several rounds of negotiations during 1992, it became clear that Iraq was unlikely to accept Resolutions 706 and 712. At the same time, it was becoming increasingly difficult to raise funds from donor states for the UN humanitarian programme in Iraq. The Security Council therefore sought a new source of funding by encouraging member states holding frozen Iraqi assets to place them in a UN escrow account (under Resolution 778, passed in October 1992). This mechanism, put into place unilaterally without Iraqi consent, did not solve the overall problem of funding the UN aid programme, but it did make a significant contribution to its survival between 1993 and 1996 (see below).[7]

During 1994, another scheme was proposed to realise funds from Iraqi oil through an agreement to flush out and sell the oil remaining in the Iraq–Turkey pipeline (closed down since 1990). However, this did not materialise. Part of the proceeds from the sale of the oil in the pipeline would have been used for humanitarian aid, channelled through the UN escrow account.

The continuing impasse over sanctions as well as reports from humanitarian agencies of a sharp decline in health and welfare conditions during 1994, especially after the reduction in September of the monthly state ration allocation,[8] led to a new attempt by the Security Council to implement a 'food-for-oil' resolution. For the USA and the UK, it seems that part of the incentive to adopt such a measure was the imperative to retain overall economic sanctions. The increasing evidence of civilian suffering was creating greater pressure to move towards lifting sanctions, wholly or in part.

Resolution 986, adopted in April 1995 followed in outline UNSC Resolutions 706 and 712 (1991) which Iraq had rejected.[9] However some modifications were adopted, especially with regard to the humanitarian aspect, partly in response to changed circumstances and also to accommodate Iraqi objections to the formulas in Resolutions 706 and 712.

First, the ceiling on sales over a six-month period was raised from US$1.6 billion to US$2 billion, though the proportional deductions for the Compensation Commission and for the expenses of UN operations in Iraq (including UNSCOM) remain the same.

Second, a portion of this total (US$130 to US$150m every 90 days) was assigned directly to the three northern governorates, with the option that the UN as well as the government of Iraq (GOI) could be involved in purchasing humanitarian goods. However, the resolution also reiterated the political formula that 'nothing in this resolution should be construed as infringing the sovereignty and territorial integrity of Iraq'. In August and September 1991, when Resolutions 706 and 712 were adopted, the question of separate allocations for the north did not arise, since the central government did not withdraw from northern Iraq until the end of October 1991.

Third, Iraq was permitted to export oil through its Mina al-Bakr terminal in the south as well as through the Turkish pipeline. Iraqi objections to the exclusive use of the Turkish pipeline had been a major sticking point in previous negotiations.

Fourth, provisions were made for 'observers' appointed by the UN Department of Humanitarian Affairs (UNDHA) to 'observe' the distribution of humanitarian goods on the basis of the government of Iraq's distribution plan, and their impact on the humanitarian situation. In Resolutions 706 and 712, UN agents from the various UN organisations (such as UNICEF, WHO and WFP) were to 'monitor' in-country distribution. This was another area in which Iraq had raised objections in the earlier negotiations.

Iraq finally accepted Resolution 986 in May 1996. Its acceptance, after a year of resistance, appears to have been in part because of economic difficulties. However, many observers consider the main reason was that, after the defection of former Defence and Military Industries Minister Hussein Kamil in mid-1995, it became clear to the Iraqi government that no Security Council members considered Iraq had complied with the conditions relating to weapons of mass destruction in Resolution 687. Such compliance would have allowed the implementation of Paragraph 22 of that resolution to lift entirely the embargo on Iraqi oil exports. The latter had been the Iraqi government's major goal since it had abandoned negotiations on the implementation of Resolutions 706 and 712.

3. The humanitarian assistance programme 1991–6

The rest of this chapter examines the nature of the international humanitarian assistance programme and its impact in different parts of Iraq between the dramatic events of 1991 and the initiation of the 'food-for-oil' programme under UN Security Council Resolution 986 (1995), which came into effect in December 1996.

First, from the earliest days of the Gulf emergency, humanitarian priorities were subordinate to political priorities in the eyes of most Security Council members. Many commentators argue that in northern Iraq humanitarian aid became a substitute for a political solution to the crisis that developed with the uprising after the Gulf War. Operation 'Provide Comfort' and the 'safe haven' were intended as short-term responses to a major refugee crisis. In particular, they were responses to Turkey's refusal to accept those refugees who had fled to its borders. Unforeseen developments since 1991 – especially, the Iraqi government's withdrawal of troops from the three northern governorates in October 1991, its imposition of an internal economic embargo, and withdrawal of rations and services from the north – impelled the international community to provide ongoing relief aid; something which was not envisaged in 1991.

The fact that the UN was overseeing the implementation of sanctions as well as providing humanitarian assistance meant that it was inevitably seen as having a political agenda. Decisions about where and to whom aid was given were regarded by many in the Middle East region and outside it, as well as in Iraq, as being guided by political rather than strictly humanitarian considerations.

Finally, the recipient authorities – the government of Iraq and the Kurdish authorities since 1992 – developed their own agendas with regard to aid. For the Iraqi government the main priority was to control the way in which aid was distributed. In the Kurdish areas, aid funds certainly provided a measure of rehabilitation but also became part both of the political competition between various contestants for power and of the open conflict between the Kurdistan Democratic Party (KDP) and the Patriotic Union of Kurdistan (PUK) between 1994 and 1997.

Between 1991 and 1996, an estimated US$1.2 billion worth of humanitarian relief assistance was channelled through the UN system, through the international Federation of Red Cross and Red Crescent Societies, and through NGOs.[10]

However, the overall figures are deceptive because proportionately far more funds were expended during the period of acute emergency in 1991. Of the funds channelled through the UN, UNHCR alone spent US$230m in 1991,[11] while the European Union spent ECU111m in 1991, most of which went to Iraqi Kurdistan. Between 1992 and 1997, the EU's total expenditure via the European Community Humanitarian Office (ECHO) was estimated at ECU 103m. In 1991 the USA, which did not channel its assistance funds through the UN, spent US$581m on Operation Provide Comfort and associated costs in 1991,[12] half the total of US$1.1 billion which it spent in Iraqi Kurdistan in 1991–4.

A further factor which has to be taken into account is that, throughout the period up to the end of 1996, expenditure on security in northern Iraq by the UN – primarily for the UN Guards contingent (UNGCI) – was included within its assistance budgets, and comprised up to 10 per cent of the annual budget.

From 1992 onwards, political developments altered the nature of the aid programme and the willingness of donors to fund it. The demands of other more recent emergencies in former Yugoslavia and Somalia were also consuming emergency funds, while the refusal of Iraq to sell limited amounts of oil under UN control to fund humanitarian needs (Resolutions 706 and 712) further diminished interest in assisting government-controlled Iraq.

If the declining availability of funds to support humanitarian aid in Iraq was indicative of failing international interest and political will, it was compounded by the UN's increasing difficulties in operating freely in government-controlled areas of Iraq. UNSC Resolution 688, on which leading Security Council members based their criticisms of Iraqi government behaviour in regard to the humanitarian aid programme, insisted that Iraq 'allow immediate access by international humanitarian organisations to all parts of Iraq and make available all necessary facilities for their operations'.[13] However, the implementation by the UN of the humanitarian programme was governed, not directly by UN Resolutions, but by a series of Memoranda of Understanding (MoU) with the Iraqi government, so that the latter has in fact had significant influence on how programmes have been implemented.[14]

In the course of 1991, the UN backed away from confrontations with the Iraqi government on questions of access – first with regard to the establishment of a UN sub-office in Kirkuk, and then in respect of establishing a 'humanitarian centre' in the southern marshes. Such centres were provided for under the April 1991 MoU, which was renewed in November 1991, but only a few were in fact established. Leading Security Council members and UN officials expressed concern at the Iraqi government's actions, but the lack of any resolute insistence that the terms of the MoU as well as Resolution 688 were fulfilled must have signalled to the Iraq government the absence of a serious commitment to maintaining full access for humanitarian purposes.

Consequently, by 1992 when there was a major problem over renewing the MoU, the government of Iraq must have been clearly aware that the issue of humanitarian access was a low priority for the Security Council, and in fact it encountered relatively little difficulty in imposing its wish to restrict UN access. This was reflected in UN acquiescence when the Iraqi government insisted on the withdrawal of international staff from the south before the new MoU was signed, and its restrictive terms were eventually accepted in the new agreement signed in October 1992. This agreement allowed UN staff to be based only in Baghdad. Travel outside the capital had to be notified in advance to the Foreign Ministry, and staff were to be accompanied by counterparts from the relevant Iraqi ministry. NGOs working in Iraq were also expected to sign the MoU. The only warning statement issued by the Security Council in early September was ignored by the Iraqi government. This has to be compared with the energetic pursuit of Iraqi compliance over weapons of mass destruction on the part of UNSCOM, leading to frequent threats of military action from leading Security Council members.

During the summer of 1992 a number of NGOs withdrew from Iraq when visas were not renewed or when they found the difficulties of working in government-controlled areas became too great. After October 1992, the majority of NGOs that continued to work in Iraq decided that rather than signing the new MoU, they would work only in the Kurdish-controlled north (Iraqi Kurdistan), entering and leaving the region through Turkey. Thus by 1994/5, there were some 50 NGOs working in the north, with funding mostly from their own governments and from ECHO, but in government-controlled areas there were only four NGOs with a permanent presence.

Despite the fact that the three northern governorates were no longer under Iraqi government control after the end of 1991, all UN humanitarian operations in Iraq, including those in Iraqi Kurdistan, continued to be run from Baghdad. In addition, all UN staff had to get visas from Baghdad and enter and leave Iraqi Kurdistan via Iraqi government-controlled areas, although many of the supplies for the north under both UN and NGO programmes came through Turkey.

While the Iraqis had imposed an economic embargo on the Kurdish-controlled areas, UN economic sanctions remain in force over the whole of Iraq. The UN Security Council has not been willing to consider lifting the UN sanctions on Iraqi Kurdistan, mainly because this would be seen as acknowledging its separate status. It has also been argued that the 'leakage of goods' into government-controlled areas would weaken the effect of sanctions there.

These developments resulted in an overall decline in the funds available to the UN programme. From early 1993 until late 1995, additional funds were generated by the procedure under UNSC Resolution 778 by which the USA used funds from its frozen Iraqi assets (up to a ceiling of US$200m) to 'match' voluntary donor contributions to the UN humanitarian programme if they were channelled via the UN escrow account. However, steep declines continued in the total funds available.[15]

Furthermore, the Kurdish-controlled areas received an estimated two-thirds of total assistance, both in relief supplies and rehabilitation. The perception among donors was that the UN humanitarian programme in government-controlled Iraq was unduly influenced by the Iraqi government and that effective monitoring of aid projects was therefore impossible. Hence many donors either refused to fund work in government-controlled areas (GCAs) or provided only very limited amounts of money. This situation led to an imbalance in funding between the GCAs and the north. Given the relative size of the populations – about 3.5 million in the north and some 17 to 18 million in GCAs – the GCAs received far less funding in per capita terms than the Kurdish-controlled region. Many donors earmarked their funding to a particular region – usually the north – or to a particular UN agency.

The record of the World Food Programme, which received on average almost half the funds provided by donors to the UN humanitarian programme, illustrates the skewing of contributions to the north and also the effects of the shrinking availability of funds. Its total number of beneficiaries in 1992/3 was about 1.26 million people, of whom

750,000 were in the north and 510,000 in the GCAs. In 1995/6, diminished funding of the programme meant that the total number of beneficiaries was reduced to approximately 400,000, of whom 300,000 in the north received reduced rations monthly, and 100,000 in the GCAs (refugees, hospital in-patients and persons in social welfare institutions) received a regular full ration. To those on the Ministry of Labour and Social Affairs 'destitutes' list, the WFP failed for the first time since 1991 to deliver any rations for four months.[16]

In 1996/7 the number of target beneficiaries was raised to 2.15 million. For the first time, the majority were in government areas, including more than half a million people from poor urban families (mainly female-headed households with children under five). This was in addition to the Ministry of Social Affairs and Labour's 'destitute' category, covering 450,000 people.[17] However, towards the end of the year the agency was reporting 'alarming' stock shortages and lack of donor response.

The EU's Humanitarian Office estimated its average distribution of its funding in Iraq between 1992 and 1996 as 65 per cent of the total to the north and 35 per cent to the GCAs. In 1997 this was adjusted to 53 per cent and 47 per cent respectively.

The US government has been one of the main donors of bilateral assistance in Iraq. However, apart from providing matching frozen assets to voluntary contributions to the escrow account, much of which is used in government areas, its direct assistance was targeted on the Kurdish-controlled areas. There, its bilateral contributions have been important, particularly in supporting winter programmes to import food and fuel, but it only channelled a relatively small proportion of this to the UN humanitarian programme, mainly to the World Food Programme.[18]

Until fiscal year 1995, US funding came mostly from Department of Defense funds. Only a very small proportion of funding came from USAID's budget. Much of the US assistance was goods in kind purchased in Turkey and transported to northern Iraq. The sums varied between approximately US$30m and US$50m a year from 1993 to 1995. According to the US Office of Foreign Disaster Assistance (OFDA), 'The US does not fund humanitarian programmes in central and southern Iraq because of the inability to monitor humanitarian aid and prevent interference by the GOI [Government of Iraq].'[19]

Of the other permanent Security Council members, France in recent years has only made modest contributions to UNGCI and UN humanitarian agencies – in addition to contributing to the work of French NGOs, mainly via the EU and the French Red Cross. The Russian Federation made periodic donations in kind, and the Chinese offered little or nothing. In contrast, a number of European states, including the Netherlands, Sweden, Germany and the UK contributed substantially to the overall aid programme and to UNFCI, in addition to contributions via the EU. The British Overseas Development Admin- istration (ODA) – now the Department for International Development (DFID) – a major bilateral donor, also continued to give the major part of its aid to the Kurdish-controlled region: during the years 1992/3 to1996/7, only 16 per cent of its direct bilateral assistance went to the 'centre and south'.[20] Turkey made several major in-kind donations for both the north and government areas after 1994.

ECHO, OFDA and ODA funds, as well as those from other European states, were increasingly channelled directly to NGOs in the north. By 1995/6, NGO-funding components of the overall aid programme in Iraq made up US$62m, while direct funding of UN agencies was only US$85m.

Despite UN statements from 1993 onwards that the programme should put greater emphasis on rehabilitation, a great deal of the funding continued to be spent on relief inputs, including food aid and fuel, especially to the north in winter. In 1993/4, 60 per cent of UN expend- iture from April–December 1993 still went on relief and on the UN Guards. Health, across the whole country, made up 13 per cent; and agriculture – mainly in the north – 14 per cent. The proportions in proposed expenditure for 1996/7 were little changed: food assistance and nutrition plus UNGCI funding still made up some almost 48 per cent of the total, with agricultural rehabilitation making up 15.5 per cent, water and sanitation 14.7 per cent, and health 15.8 per cent. Education, an area of acknowledged crisis in areas under both government and Kurdish control, received 1.7 per cent.[21] NGOs working in the north were more successful in focusing on rehabilitation, but their efforts were often weakened by limited coordination and planning of their efforts, and a series of security problems which affected both their staff and their beneficiaries (see below).

The donors complained increasingly of the UN aid programme's ineffectiveness, but did little to improve it. The message was very much that it was a sideshow. A report on UN coordination in 1992 noted 'a contradiction between the heavy bilateralism of donor-governments and their oft-stated insistence that the UN exercise a coordinating role'.[22] This observation was borne out by subsequent developments. The difficulties were compounded by the fact that UNDHA, as coordinator of the humanitarian effort, lacked strong political influence with other UN agencies, the Security Council and donor governments.

Both the uncertain political situation and lack of secure funding encouraged 'short-termism' among UN agencies, and frustrated any attempts at strategic planning. The short-range thinking usually associated with emergencies still tended to prevail. Although the notion of a chronic emergency has now entered the UN's vocabulary, its implications, in the case of Iraq, for planning and funding were not tackled either by the UN or by donors. Annual plans were made but often these did not seem to have been based on any realistic assessment of what could be achieved: rather, they were a bargaining chip with donors and a way of dealing with the Iraqi government.

Furthermore, there was no independent or systematic means of monitoring the overall impact of sanctions on the population. As a result, the aid available was not necessarily targeted in the most effective way. Both the Iraqi government and the rival Kurdish political parties also played their part in trying to direct aid to fit their political goals rather than the welfare of the population at large.

Humanitarian assistance in government-controlled Iraq

In addition to the issue of access to those in need, already dealt with above, there was also the difficulty of defining and choosing beneficiaries. In GCAs, the UN was largely dependent on Iraqi government and Iraqi Red Crescent lists of needy people. However, it is unclear how far these covered the whole population, particularly with regard to displaced people, and especially to those from the southern marshes or to Kurds and Turkomans forced out of their homes in Mosul and Kirkuk.

The government retained a large measure of control over what kind of assistance it was possible to provide. It basically wished to see aid organisations as a channel for imports. In this respect, aid agencies were useful, since humanitarian goods which they ordered usually passed through the UN Sanctions Committee processes with few difficulties. However, the delays in receiving goods were still long and the quantities were relatively small compared with need.

The government also had considerable control over where both the UN and NGOs could operate. NGOs that remained in Iraq to carry out assistance programmes rather than simply to deliver periodic consignments of goods were constrained as to where and with whom they could work.

Most of the work done was in relief and providing 'sticking plasters' for a crumbling infrastructure. Food distribution remains a major item in aid – through WFP, and also through the International Federation of Red Cross and Red Crescent Societies, which resumed a programme of food relief in cooperation with the Iraqi Red Crescent in 1994. Medical imports by UNICEF, WHO and several NGOs were small in scale compared with the overall level of need.

UNICEF and the International Committee of the Red Cross (ICRC) made a significant contribution to keeping water and sewerage systems running, though by no means at full capacity. This is a critical area because water pollution and environmental contamination by sewage are major health hazards and are thought to be a significant factor in the rises in child morbidity and mortality since 1990. However, their work has mostly been basic maintenance rather than rehabilitation of the system. The Iraqi system depends heavily on expensive imported equipment, and maintenance work is hampered by the long lead times required to obtain parts from abroad because of sanctions procedures.[23]

In other sectors where sanctions may have lasting effects – for example in education and agriculture – the aid effort has been too small and marginal to make a difference. Because of very limited funding from donors for these agencies, FAO's contribution to agricultural rehabilitation was small, as was UNICEF's and UNESCO's contribution to education.

The kind of goals frequently set by NGOs in rehabilitation – of sustainability and increasing self-sufficiency of communities – are not realisable in the GCAs, both for political reasons and because of the scale of the problems.

Humanitarian assistance in the Kurdish-controlled north

A 1991 report by the US-based Lawyers Committee for Human Rights warned, in relation to northern Iraq, that

> the international community must fully appreciate the enduring commitment necessary to undertake arrangements such as those that have been made for displaced persons and returned refugees in northern Iraq. There are no easy or inexpensive solutions to the needs of refugees and displaced persons for new permanent homes.[24]

Regrettably, much of the assistance provided after 1991 continued to go into relief rather than to rehabilitation or development. While the north received a considerably higher proportion of aid per capita, it suffered from the embargo imposed by the Iraqi government, as well as from economic sanctions. The Iraqi internal embargo on the Kurdish-controlled areas, which remained in place until late 1996, meant that the majority of aid funds were consumed by basic winter food and fuel needs for the north, supplied through the UN humanitarian programme and US bilateral aid.

NGOs (both international and Kurdish) that have worked in the Kurdish-controlled region generally attempted to focus on rehabilitation of housing, infrastructure, and agriculture devastated by the Ba'thist regime's Anfal campaign (1988 and earlier) as well as after the 1991 uprisings. Their main achievement was to have restored in some measure the rural economy of Iraqi Kurdistan. In future, however, the maintenance of this rural economy, especially small-scale farming in the mountain valleys, will depend on the attitudes and plans of whatever authority controls the area. Also, this rehabilitation has affected only a small proportion of the population which lives in villages and engages in small-scale agricultural production.

As in government-controlled areas during the mid-1990s, it was the poorer sections of the urban population that experienced the most difficult conditions. For this part of the population, with the exception of displaced persons living in towns, the humanitarian assistance programme had a much more limited impact.

In the north, these difficulties were compounded by the virtual collapse of financial administration between 1995 and 1997, when salaries of public employees were only paid erratically, with teachers, nurses and administrative staff sometimes going months without pay. It seems to have been the experience of areas under both parties' control that only very meagre sums were available for public services, while some party and tribal leaders as well as merchants were enriching themselves, and the parties were spending money on sustaining their conflict.

Humanitarian programmes were also severely constrained by political developments after 1992. The ambiguous and 'temporary' status of the Kurdish-controlled region created serious difficulties in establishing medium- to long-term rehabilitation programmes in a climate beset by political uncertainty, security concerns and, from 1994 to 1997, by internal conflict. Aid agencies faced problems not only in dealing with the different Iraqi Kurdish factions, but also in increasingly difficult relations with the Turkish authorities, whose role in the region grew in scope in the mid-1990s.

Furthermore, Turkish military activities inside the borders of northern Iraq created a precarious security situation for people living in rebuilt villages in those regions, putting in jeopardy the work of both villagers and aid agencies to restore the rural economy. Access to Iraqi Kurdistan via Turkey, initially easy, became more and more difficult after 1994, until Turkey closed its border to most aid personnel, including those from donor organisations such as ECHO, in late 1996. Aid agencies were then obliged to seek visas from Syria.

Finally, the Iraqi army's incursion into Arbil in September 1996, and the subsequent growth in influence of the Iraqi government in parts of northern Iraq, further increased the potential difficulties for international NGOs and their local staffs, since the Iraq government labelled them as 'agents' of external powers. Most personnel, both international and local, associated with US-funded programmes were evacuated, which caused a good deal of unease among other local aid-agency staff. These difficulties were compounded by the absence of any clear long-term strategies articulated by the international community in regard to the future of this region.

4. Postscript: the impact of UNSC resolution 986, 1997–9

The system instituted under UN Security Council Resolution 986 which allowed Iraq to sell US$2 billion of oil every six months began operating in December 1996. Thirty per cent of each tranche went to the UN

Compensation Commission, with further deductions to meet UN expenses in Iraq. The remaining 66 per cent was divided between funds for the purchase of humanitarian goods for areas under government control (the so-called '53 per cent account') and for those under Kurdish control administered directly by the UN (the '13 per cent account').

Humanitarian goods purchased with the proceeds of oil sales only began to arrive in Iraq in March/April 1997. The resolution was renewed in June 1997 (Resolution 1111) and in December 1997 (Resolution 1143). Although this was dubbed a 'food-for-oil' deal, it did allow the import of other humanitarian goods as well as food. However, the Iraqi government, which decides on the distribution plan for humanitarian goods, has used the funds mainly to buy food and medicines. By the end of 1997, concerns were being expressed that the serious problems caused by Iraq's failing infrastructure – particularly electricity, water and sanitation services – were not being alleviated by the programme.

The actions envisaged by Resolution 1153 (1998) included more than doubling the amount of oil to be sold (to US$5.52 billion every six months), making some changes to administrative procedures, focusing on the urgent need to restore the infrastructure, including the electricity system in areas under both government and Kurdish control, and targeting humanitarian aid to the most vulnerable groups inside Iraq. However, the sharp fall in oil prices in early 1998, and the physical limits on Iraq's ability to lift oil because of infrastructural problems in the industry, meant that the increased sum expected for humanitarian goods was not achieved until late 1999, when oil prices rose rapidly from the low levels of 1998.

A key issue for the framers of the food-for-oil programme in 1995 was to ensure that strict controls were maintained over Iraq's access to funds (as opposed to imported goods), and its ability to use oil revenues for purposes other than humanitarian needs. Despite the redefinition of these needs to include more attention to the infrastructure, there were other problems inherent in the programme. First, no internal purchases were permitted in either the GCAs or in Iraqi Kurdistan. Although inputs to agriculture are included in distribution plans, there have been persistent concerns about the impact of importing food instead of purchasing local agricultural produce when it is available. Furthermore, the six-monthly planning horizon further encourages the short-termism that has characterised much humanitarian assistance in Iraq since 1991.

The programme was certainly not intended to address the long-term crisis of poverty, declining standards of health and education, and infrastructural collapse.

Since early 1997 the external aid programme has dwindled dramatically, even in the north which previously received the lion's share of external assistance. The UN ceased its appeals for international funds, and relatively few aid agencies maintain programmes in the country. The DHA's functions were handed over to the Office of the Iraq Programme in New York and to the UN Office of the Humanitarian Coordinator in Iraq based in Baghdad. In government-controlled areas, its role is mainly to 'observe' the flow of goods under the food-for-oil programme, through a Baghdad-based team of observers that travels round the country. However, it is the government that makes the main decisions on what to order and how to distribute it. Only in the north does the UN play an active role in distributing aid, with the two separate local authorities preparing the six-monthly distribution plans.

In the north, the advent of the food-for-oil programme led to several significant political changes. The Iraqi government's internal embargo was gradually eased and the increased flow of goods into the country appears to have modified the intense economic rivalry between the KDP and PUK. Nonetheless, the two regional party administrations remain separate and no integration is imminent, though a measure of cooperation has developed between them.

Some international NGOs continue to work in the north, but their numbers are much reduced, mainly because donors ceased to fund their work. The era in which international NGOs had a key role in determining the direction of work in Iraqi Kurdistan came to an end in 1996. The closure of access via Turkey combined with the enhanced role of the UN in administering the food-for-oil programme sharply diminished their influence. Kurdish NGOs also continue to operate, but the situation still does not favour an independent NGO sector, particularly because of the competition generated for contracts between rival political groups and leading players in the economy.

Humanitarian needs in Iraq have continued to be used as a political football, with evidence of suffering used to gain political advantage. A good example was the interpretations given to the findings of UNICEF's 1999 report on child and maternal mortality, which give the first country-wide indication of the state of key indicators of health and well-being since the nationwide survey by the International Study Team (IST) in August/September 1991.[25] The IST survey had shown an estimated threefold increase in deaths of children under five in the eight months from the beginning of 1991. Those figures reflected the conditions created by the war and refugee flight, as well as the initial impact of economic sanctions, first imposed in August 1990. In the intervening period, infant and child mortality statistics have been based only on samples from particular areas, from which a national picture was extrapolated.

UNICEF's 1999 survey, which was based on a household survey covering 24,000 households in government-controlled central and southern areas of Iraq, indicates that mortality rates for children under-five have more than doubled in a decade: from 56 per thousand in 1984–9, to 91.5 for 1989–94 and 130.6 from 1994–9. In the north, a survey of 16,000 households showed a different pattern. The under-five mortality rate for 1984–9 was much higher than in the rest of the country, and the rise in 1989–94 to 90 per thousand was less dramatic. Finally, in contrast to the rest of the country, the rate fell to 72 per thousand between 1994–9.

The Iraqi government used the report to blame sanctions for all the problems revealed, without taking any responsibility for its own role. On the other hand, the US State Department claimed that the UNICEF report showed that

> Iraqi mismanagement – if not also deliberate policy – not sanctions, is responsible for malnutrition and deaths. In Northern Iraq, where the UN administers humanitarian assistance, child mortality rates have fallen below pre-Gulf War levels. Rates rose in the period before food-for-oil, but with the introduction of the program the trend reversed, and now those Iraqi children are better off than before the war.[26]

However, UNICEF dismissed the idea that the difference in the current child mortality rates can be attributed to the differing implementation of the food-for-oil programme. Food-for-oil imports of goods only began in March/April 1997. 'Therefore it is too soon to measure any significant impact of the food-for-oil program on child mortality over the five-year period of 1994 to 1999 as is reported in these surveys.'[27]

The differing mortality figures for the north were undoubtedly influenced by the distinctive history of the northern region since the 1980s. The higher mortality figures for the north in the late 1980s probably reflect the severe disruption caused by the forced clearance of Kurdish rural areas by Iraqi forces in the Anfal campaigns. The refugee crisis in 1991, when almost two million people fled to the country's borders, certainly involved a significant number of child deaths, but in the years that followed, the three northern governorates received significantly more aid that other parts of the country.

From 1994 to 1997 the continuing decline in economic and health conditions in the GCAs is clearly reflected in the rise in infant and child mortality levels shown by the UNICEF study. UN reports to the Security Council in 1999, after two years of food-for-oil, suggested that child mortality rates were levelling off and nutritional levels were stabilising from the high levels reached in the 1990s.

US officials have focused on the reluctance of the Iraqi government to implement measures to target special assistance to vulnerable groups, especially mothers and children, which UN agencies have been advocating for several years. The slow distribution of goods in several sectors, particularly of medical supplies, and the government's refusal to order special nutritional items, has raised questions about its priorities.

However, the blockages are not all on the Iraqi side. In the Sanctions Committee, scrutiny of orders continues to result in rejections or long delays on many items. On 19 August 1999, the UN Secretary General noted in his report to the Security Council on the operation of the food-for-oil programme that 'there has been a significant increase in the number of holds being placed on applications, with serious implications for the implementation of the humanitarian programme'.[28] The total of applications on hold was valued at US$500 million, with the highest concentrations in telecommunications, electricity, water and sanitation sectors. The lack of transparency in Sanctions Committee decision making means that it is difficult to find out the reasons for these delays.

It is certainly possible to blame the Iraqi government for both incompetence and malfeasance in the use of scarce resources. Nonetheless, the overall scarcity of resources compared with the previous decade is the result of prolonged economic sanctions which, even by 1993, had led to a drop in GDP in real terms to less than one-fifth of its value in 1979. The UN Security Council Panel on Humanitarian

Issues concluded in March 1999, 'Even if not all suffering in Iraq can be imputed to external factors, especially sanctions, the Iraqi people would not be undergoing such deprivations in the absence of the prolonged measures imposed by the Security Council and the effects of war.'[29]

5. Conclusion

Much of the criticism of aid programmes in Iraq has focused on the cumbersome bureaucracy and inefficiency of UN agencies. While these factors are certainly elements in the outcome, arguably greater weight should be given to the underlying political context, which has not favoured the creation of an efficient or effective aid programme.

The hostility between the government of Iraq and donor states left UN humanitarian agencies stuck in the middle, blamed by both sides for lack of effectiveness. The lack of effectiveness has been real enough, but the situation has encouraged it. The highly politicised attitudes of states which have both implemented the embargo against Iraq and, in the years up to 1997, provided some measure of humanitarian assistance, has often been problematic. NGOs have had more freedom of action, and have therefore largely escaped these criticisms. However, political pressures, both from the international community and from within Iraq, as well as the assessment of need, have shaped their programmes and limited what it has been possible to achieve.

Both the UN and NGOs have had to operate in a context of acute and prolonged political conflict. There has been ongoing civil conflict in parts of Iraq, and a continuing conflict between Iraq and the international community. Both these factors have placed aid agencies in a difficult position. The existence of an economic embargo on Iraq by the international community and a further embargo imposed by Iraq between 1992 and 1996 on the three northern governorates restricted the scope of programmes. Large amounts of money continued to be spent on relief of the 'symptoms' of these embargoes – a source of considerable frustration to those attempting to promote a greater emphasis on rehabilitation. The latter has also been hampered by continuing civil conflict. Therefore, for all but a small minority of the Iraqi population, the large scale of expenditure has not been translated into productive activity or greater self-sufficiency.

NOTES

1 The research for the chapter was supported by a grant for Research and Writing from the John D. and Catherine T. MacArthur Foundation.

2 The International Study Team which visited Iraq between 25 August and 4 September 1991 found that electrical generation had been restored to about 68 per cent of 1990 peak load and about 37 per cent of installed capacity, with 75 per cent of transmission lines operable (International Study Team, *Electrical Facilities Survey*, October 1991, p. 1). The team found water-treatment plants were operating at between 30 and 70 per cent of capacity, with chlorine supplies being rationed. Waste-water systems in Baghdad and the south were operating at between nought to 70 per cent capacity; that is, some plants were not operating at all, mainly due to lack of spare parts and electricity supply (International Study Team, *Water and Waste Water Systems Survey*, October 1991, p. 1).

3 Centre for Economic and Social Rights, *Unsanctioned Suffering: a Human Rights Assessment of UN Sanctions on Iraq*, New York, May 1996, p. 26.

4 See United Nations, *Report to the Secretary-General on Humanitarian Needs in Iraq*, by a mission led by Sadruddin Aga Khan, Executive Delegate of the Secretary General 15 July 1991; and the [UN] Secretary General's *Recommendations to the Security Council on implementing Resolution 706*, with Annex I: *Report of the Executive Delegate* [Sadruddin Aga Khan] dated 28 August 1991 on estimates of humanitarian requirements.

5 *Gulf Newsletter* No. 1, November 1991 (Refugee Council, London), p. 2.

6 UN, *Report to the Secretary General on Humanitarian Needs in Iraq*, op. cit., 15 July 1991, p. 8.

7 For details of the working of the UN escrow account (The United Nations Trust Fund for Humanitarian Assistance to Iraq) and contributions to it, see UN Department of Humanitarian Affairs, United Nations Consolidated Inter-Agency Humanitarian Programme in Iraq, *Extension of the Cooperation Programme for 1995–1996* (March 1995), Annex I 'United Nations SCR 778 Escrow Account', pp. 93–4.

8 For a description of the rationing system see Peter Boone, Haris Gazdar and Athar Hussain, 'Sanctions against Iraq; Costs of Failure: Report prepared for the Center of Economic and Social Rights on the impact of United Nations-imposed sanctions on the economic well-being of the civilian population of Iraq', London, 1997. For a more critical analysis of access to rations, see United Nations General Assembly, *The Situation of Human Rights in Iraq* (Report of the UN Special Rapporteur on Iraq), 51st session, Agenda Item, 110(c), A/51/496, 15 October 1996, para. 71–3.

9 For the texts of these resolutions, see *The United Nations and the Iraq–Kuwait Conflict 1990–1996*, The United Nations Blue Book Series, Vol. IX, Department of Public Information, UN, New York, 1996, Documents 72, 81 and 207.

10 UN Department of Humanitarian Affairs United Nations Consolidated Inter-Agency Humanitarian Programme in Iraq, *Extension of the Cooperation Programme for the Period 1 April 1996–31 March 1997*, 3 April 1996, p. 5.

11 UNHCR, *Information Bulletin* No. 8, on operations in the Persian Gulf Region, 27 January 1992.

12   Commission of the European Community, *Humanitarian Aid from the European Community: Emergency Aid, Food Aid, Refugee Aid*, DE 70, Brussels, January 1992.

13   Paragraph 3 of Resolution 688 in *The United Nations and the Iraq–Kuwait Conflict 1990–1996*, op. cit., Document 37.

14   For the texts of the three MoUs, signed respectively in April 1991, November 1991 and October 1992, see *The United Nations and the Iraq–Kuwait Conflict 1990–1996*, op. cit., Documents 44, 93, 135.

15   See also note 6.

The following table shows the decline in overall funding for assistance programmes in Iraq, the decline in the UN programme's share of that funding, and the large discrepancies between the amounts projected in the annual or biannual appeals, and the amounts actually received.

	Total appeal	Funds to UN US$ million	Funds to other organisations
January–June 1992	143.2	89.1	–
July 1992–March 1993	201.7	134.7	–
April 1993–March 1994	467.1	95.2	77.9
April 1994–March 1995	288.5	61.2	53.4
April 1995–March 1996	138.8	39.4	34.3

Source: *The United Nations and the Iraq–Kuwait Conflict 1990–1996*, op. cit., p. 59.

16   Department of Humanitarian Affairs, UN Inter-Agency Humanitarian Programme in Iraq, *Implementation Report* (1 July 1992–13 March 1993), Geneva, June 1993; United Nations Consolidated Inter-Agency Humanitarian Programme in Iraq, *Extension of the Cooperation Programme for the Period 1 April 1996–31 March 1997*, 3 April 1996, p. 5.

17   UN Department of Humanitarian Affairs, United Nations Consolidated Inter-Agency Humanitarian Programme in Iraq, *Extension of the Cooperation Programme for the Period 1 April 1996–31 March 1997*, 3 April 1996, p. 20.

18   For example, in 1995–6 the US provided 18.51 per cent of total humanitarian assistance to Iraq, but only 8.26 per cent of UN programme funding. UN Consolidated Inter-Agency Humanitarian Programme in Iraq, *Extension of the Cooperation Programme for the Period 1 April 1996–31 March 1997*.

19   US Agency for International Development, Office of US Foreign Disaster Assistance (OFDA), *Situation Report: Northern Iraq – Displaced Persons*, no. 1 (FY 1995) (9 December 1994); no. 3 (FY 1995) (26 April 1995), p. 2. In 1996, after Congressional allocations of humanitarian aid funds through the Department of Defense had ended, US funding for WFP (US$7.3 million) announced in October 1996 came 'with the caveat that the money should be used to assist the Kurds in northern Iraq' (US Department of State, *Daily Press Briefing*, 30 October 1996).

20   Calculated from ODA, 'Emergency Aid to Iraq 1991–6', 28 May 1997 (London).

21   *Gulf Newsletter*, no. 9, March 1994, p. 2; derived from United Nations Consolidated Inter-Agency Humanitarian Programme in Iraq, *Extension of the*

Cooperation Programme for the Period 1 April 1996–31 March 1997, 3 April 1996, Table II, p. 9.

22 Larry Minear et. al., *United Nations Coordination of the International Humanitarian Response to the Gulf Crisis, 1990–1992,* The Thomas J. Watson Jr Institute for International Studies, Brown University, Occasional Paper no. 13, Providence RI, 1992.

23 For a description of the practical difficulties of ordering and importing equipment for water and sanitation plants, see ICRC, *Water in Iraq,* Special Brochure, Geneva, July 1996.

24 Lawyers Committee for Human Rights, *Asylum under Attack: a Report on the Protection of Iraqi Refugees and Displaced Persons One Year After the Humanitarian Emergency in Iraq,* New York, April 1992, p. 5.

25 UNICEF/Government of Iraq, Ministry of Health, *Child and Maternal Mortality Survey 1999: Preliminary Report,* Iraq, July 1999.

26 US Department of State, 'Saddam Hussain's Iraq', 13 September 1999.

27 UNICEF, 'Questions and Answers for the Iraq Child Mortality Surveys', Baghdad, 16 August 1999.

28 UN Security Council, *Report of Secretary-General pursuant to Paragraph 6 of Security Council Resolution 1242 (1999),* S/1999/896, 19 August 1999.

29 Quoted in UNICEF, 'Questions and Answers', op. cit.

8

Refugee Camp or Free Trade Zone?
The Economy of Iraqi Kurdistan since 1991

Michiel Leezenberg

1. Introduction

Our knowledge of the economy of Iraq as a whole still displays considerable gaps, but the situation is even worse when it comes to possible divergences in the economic development of different parts of the country. Of course it makes perfect sense to treat the Iraqi economy as a nationally integrated whole, as earlier studies have done.[1] Nevertheless, a more systematic attention to regional variation is not only interesting in its own right, but – at the very least since the 1991 uprising – inevitable. Since the uprising and the subsequent withdrawal of government personnel, Iraqi Kurdistan has been de facto independent from the rest of the country, politically and to a considerable extent economically. Its economy displays a number of significant differences in comparison with that of government-held territory. It has, for all practical purposes, a separate currency, with distinct inflation and price developments. The Kurdish region has also received a greater per capita amount of humanitarian aid than government-held territory. Yet it has suffered as much as the rest of the country under the UN sanctions imposed on Iraq in 1990; and has, moreover, been subjected to an internal blockade affecting salary payments, petroleum supplies and food distribution.[2]

The dismal economic situation of Iraqi Kurdistan is largely determined by outside forces, and is a main cause of the continuing political instability and social chaos in the region. However, despite the massive input of foreign aid, and the presence of a large number of foreign aid workers with an obvious interest in acquiring a greater knowledge of the quantitative and qualitative features of the local economy, very little has been done towards providing a general economic overview. There remains an extreme dearth of reliable information on such basic variables as employment, gross domestic product, and so on. This very fact, is, however, significant, in that following the 1991 withdrawal of the Iraqi state bureaucracy, a lively informal economy developed, in which data collecting, planning and centralised allocation of resources played at best a minor role.

This chapter is a largely qualitative account of the most important structural aspects of the economy of Iraqi Kurdistan against their social and political backgrounds. Neither economic nor political factors have been decisive by themselves. Thus, though the emphasis is on economics, it is more or less impossible to obtain a comprehensible economic picture without taking both the local and the international political context into account. Conversely, a firmer grasp of the economics of the region is essential for a more adequate understanding of political and social developments. Attempts to explain the relation between the central government and the Kurdish insurgents and, latterly, the persistent armed conflict between the main Iraqi Kurdish parties, have so far kept largely to the political level, and have tended to focus on such factors as personal ambitions, tribal rivalries, and the like. Consideration of the economic interests involved can help clarify the motivations for political behaviour that might otherwise be hard to explain rationally.[3]

2. Background: structural features of the regional economy

Many of the difficulties facing the economy of Iraqi Kurdistan grew out of the economic policies pursued by the Iraqi government from the early 1970s, when state revenues rose dramatically because of the nationalisation of the oil industry and the subsequent sharp increase in oil prices. Urbanisation in the Kurdish areas proceeded more slowly than in other parts of Iraq, but the majority of the population had, nonetheless, become urbanised by 1980.[4] The reasons for this rapid urbanisation were partly socio-economic and partly political in nature.

The Kurdish region in Iraq saw considerably less industrialisation than other parts of the country. According to 1987 government statistics,[5] the autonomous region of Kurdistan possessed a mere 4.9 per cent of large industry and 11.1 per cent of smaller plants in the country. In 1990, there were only nine major factories in Iraqi Kurdistan, notably for the production of cement and cigarettes. Iraqi industry was highly centralised and concentrated around Baghdad, and the Kurdish region remained dependent on the centre.

In the 1980s, the private sector rapidly expanded all over Iraq, but it was mostly engaged in construction and infrastructure projects commissioned by the government. In fact, a generous form of 'state capitalism'[6] developed in the 1980s; although the state maintained basic political control, those in the private sector who were politically loyal to the regime benefited. The lucrative market for government-commissioned construction projects also left ample room for corruption. Although the education level of the local population was relatively high, skilled work relied heavily upon foreign technical personnel, notably from Eastern Europe and East Asia. At the same time, up to two million unskilled labourers, mostly originating from Egypt, Sudan, and South Asian countries, replaced the domestic labour force of Iraqi citizens away on front duty.

The government's relative neglect of the agricultural sector led to a stagnation in agricultural productivity,[7] which was only exacerbated by successive land reforms. The reform laws of 1958 and 1970 had been only partially implemented, and tended to lead to a fragmentation and loss of productivity of fertile lands. In the Kurdish region, implementation had been hampered by the intermittent insurgency from 1961 onwards. Law No. 90 of 1975 aimed at increasing the number of agricultural cooperatives, and was at least partly politically motivated, since it applied only to the north of Iraq, and was aimed at breaking the power of the Kurdish landowners. However, as an attempt at collectivisation, it ended in failure, and was largely undone by the privatisation wave (*infitah*) of the 1980s. Law No. 35 of 1983, for example, allowed for the private leasing of state lands for private profit, on terms especially favourable to business groups with connections to the regime.[8] In the late 1980s, more than half of the arable land was in private hands,[9] the remainder being for the most part leased from the government by private individuals.

Domestic political considerations had a dramatic impact on the regional economy, especially on the agricultural sector. After the collapse of the Kurdish front in 1975, many villages had already been forcibly evacuated. When guerrilla warfare resumed on a large scale during the 1980s, the Iraqi government created the so-called 'National Defence Battalions', consisting of Kurdish irregulars to protect the rural areas against Kurdish guerrillas, and imposed an economic blockade on the villages in territory outside its control.[10] Little is known, however, about economic developments in this period. The *mustashar*, the leaders of the irregular troops, many of whom were landowners with a traditional tribal power base, received huge sums of money from the government, and often increased these amounts by presenting highly inflated figures of the number of men under arms. Some of this money was reinvested in local capital-intensive agriculture, such as chicken farming; some *mustashar* thus transformed themselves from tribal leaders into businessmen. For Kurdish civilians, it was economically and politically advantageous to enlist in the National Defence Battalions, as this exempted them from front duty in the war with Iran. Many of them never did any active service, and could continue their regular employment.[11]

By contrast, the economic blockade in the rural areas held by Kurdish insurgents created enormous hardship; basically, the rural population was forced into a subsistence economy, with few possibilities for marketing goods, let alone for further development. The Patriotic Union of Kurdistan (PUK) apparently introduced limited land reform in the areas under its control: as a reaction to the *mustashar* system, it had village councils (*anjuman*) installed in the regions under its control, for example in the Qaradagh area.[12] In the course of the 1980s, the government's successive counter-insurgency measures, which culminated in the 1988 Anfal operations, destroyed a considerable part of the Kurdish agricultural sector. In these operations alone, over one thousand villages were evacuated and dynamited. Possibly as many as 100,000 Kurdish non-combatants were killed by the security forces, and many more were forcibly resettled in poorly-equipped relocation camps (*mujamma'at*).[13]

From 1970, some 4,000 Kurdish villages[14] were destroyed and declared out of bounds for civilians. Much of the rural population was forcibly urbanised and rendered dependent on government distribution of food and fuel, often handed out by local *mustashars*. Kurdistan's traditionally self-sufficient agricultural sector was thus incorporated into the far-reaching welfare and food distribution system of the repressive Iraqi state – a system that to some extent remained in place after the 1991 uprising – and Iraqi Kurdistan's productive rural community was transformed into a consumerist society.

Following the UN Security Council's imposition of an economic embargo on Iraq on 6 August 1990, the Iraqi government hastily tried to encourage the rehabilitation of the domestic agricultural sector. To this end, it rented land to former *mustashar* leaders in the North, thus carrying the privatisation of the 1980s one step further. The extent, if any, of agricultural rehabilitation by the time of the uprising is unclear; however, the large-scale destruction of the agricultural infrastructure and the conflicting claims to the same lands based on different reform laws or on different political loyalties were later seen to be serious structural problems. Despite the privatisations of the 1980s, the urban economy remained strongly dependent on the state; or rather, the private sector itself had flourished largely on the basis of capital provided by the state. After the uprising, the main challenge for the Kurdish administration was to take over, or to transform, this highly unproductive and expensive form of 'state capitalism'.

Even before the second Gulf War, however, the Iraqi welfare state itself was already in the process of undergoing a radical – and as yet inadequately studied – transformation. After the ceasefire with Iran in August 1988, the Iraqi government dramatically accelerated the pace of its economic liberalisation and privatisation programme. Price controls on basic commodities were relaxed, the bureaucracy was trimmed, and the private industry received a ten-year tax holiday. All of this led to a steady increase in economic inequality. There have been very few studies of the 'shock therapy', which coincided with the massive demobilisation of the Iraqi army.[15] It does appear to have caused, as elsewhere, considerable economic chaos and social instability, but these domestic developments were soon overshadowed by the 1990 Gulf crisis, and by the ensuing war and its dramatic aftermath.

3. The 1991 uprising: state collapse or radical liberalisation?

The uprising in the spring of 1991 and the subsequent refugee crisis along the Turkish and Iranian borders brought a massive international aid effort to the region. The total amount of immediate relief aid is difficult to establish, but must have run into the hundreds of millions of dollars. This massive operation will not be discussed here, but some long-term effects of the activities of foreign humanitarian agencies will be described below.[16] In October 1991, the peace talks between the Kurdish parties and Baghdad collapsed and all government security personnel and civil servants were withdrawn. Following renewed fighting in several Kurdish cities, an area of some 40,000 square kilometres came under control of the parties of the Iraqi Kurdistan Front (IKF), although the oil-rich areas of Kirkuk and Khanaqin, as well as the largely Kurdish

Jabal Sinjar, remained in government hands. The population of the Kurdish-controlled area exceeded 3 million, and included many displaced persons and refugees returning from camps in Iran and Turkey (see Table 8.1). Most lived in cities or in the quasi-urban *mujamma'at*, and consequently were directly or indirectly dependent on state support. According to UN sources, there were still 586,000 rural and 140,000 urban displaced persons in March 1993.[17]

TABLE 8.1
Iraqi Kurdistan: population statistics as of October 1992

Governorate	Established	Returnees/Displaced	Total
Duhok	565,000	17,000	582,000
Shaikhan/'Aqra districts	253,000	unknown	253,000
Arbil	unknown	unknown	1,100,000
Sulaimaniyya	1,204,000	70,000	1,274,000
Total			3,209,000

Notes: The source of this OFDA data is not clear. The figures given are notably higher than those of the 1988 Iraqi census,[18] and may derive from regional government estimates or a privately conducted population survey. The population of those areas of Kirkuk governorate under Kurdish control, notably the Germian plains, is not included in the OFDA report, but may be roughly estimated at some 320,000 persons.[19] After October 1991, the Kurdish-held parts of Shaikhan and 'Aqra districts, which had hitherto belonged to Nineveh governorate, came under the administrative control of Duhok governorate. The OFDA estimates the total urban population at 1,488,000 persons, or 46.4 per cent of the population; but this figure is certainly too low, as it excludes the inhabitants of smaller towns and *mujamma't*.
Source: OFDA (Office of US Foreign Disaster Assistance), *Report on Joint Emergency Winter Humanitarian Needs Assessment Mission to Northern Iraq, October 9–16, 1992*, pp. 77–81. [Editor's note: According to the results of the supplementary Population Census conducted in 1989 in the northern governorates, total population in these governorates was recorded as 2,284,000. Adding the estimated figure for the Shaikhan and 'Aqra districts in Table 8.1, and accounting for natural population growth, the total would be around 2.7 million as of late 1992, as opposed to the OFDA figure of 3.2 million reported above.]

On 23 October 1991, Baghdad imposed an economic blockade on the areas under Kurdish control. Salary payments to the remaining civil servants (an estimated 45 per cent of the urban work force) were discontinued, shipments of food and petroleum products were gradually cut back, and some ID135 million of savings in local banks were confiscated. The amounts of essential foodstuffs allocated for redistribution were reduced to less than half, and rations were further reduced in 1992, when medicines were also included in the embargo. Overall, Iraqi government expenditure on the North diminished by one half from October 1991 and by a further half to June 1992.[20] In July 1992, delivery of the supply of subsidised petroleum products stopped altogether. Since much of the regional economy (from industry and transport to cooking and heating) depended on a cheap and abundant supply of such products, this was particularly disruptive.

At the time, many local and foreign observers regarded the internal blockade as a backfiring attempt to force the Kurds into compliance, which (although it contributed to serious hardship) in fact increased their economic independence. And while it did not produce any immediate results, it certainly helped create the economic chaos that seriously hampered the establishment of an effective regional administration. Free market prices rocketed with the discontinuation of petroleum products and an intensive smuggling activity developed across the demarcation line. With these decreased supplies, individuals close to the Baghdad government could increase the profit margin of 'trade' with the North.[21] More importantly, humanitarian aid organisations subsequently resorted to buying petrol products and foodstuffs for the North from the Iraqi government, thereby providing it with an important source of hard currency (as will be further discussed below).

To fill the administrative vacuum, the IKF parties held regional elections in May 1992. A Kurdish Regional Government was formed a month later, consisting mostly of officials associated with the two Kurdish parties that had emerged victorious in the elections – the Kurdistan Democratic Party (KDP) and the Patriotic Union of Kurdistan (PUK).[22] These regional government elections proceeded according to the 1974 law that had provided for an autonomous region in Northern Iraq, with its own parliament and government. Nonetheless, the international community was reluctant to support the new entity, or to facilitate any long-term economic development (for example, by lifting the UN sanctions for the Kurdish-held territory), fearing that this might promote the emergence of an independent Kurdish state. The lack of international support, combined with a restricted budget and other structural difficulties, meant that the regional government was unable to take over the hitherto dominant role of the Iraqi state apparatus and to tackle the enormous economic problems. It had the daunting task of paying and feeding the largely unproductive and state-dependent population, and of reconstructing the infrastructure, which had been badly damaged by Iraqi government policies and by allied bombing. At the same time, it had a limited and variable budget at its disposal, most of which derived from the cross-border trading of oil products. This, however, was a highly uncertain source of income, since Turkey repeatedly closed the border with Iraq.

The regional government also lacked adequately trained personnel. Many of the ministries that had been taken over from the Ba'thist government of the Autonomous Region were seriously overstaffed, while others, like the Ministry for Humanitarian Aid (which acted as an unofficial ministry of foreign affairs), were newly created and consequently lacked qualified staff. The sharing of responsibility among several ministries (notably Humanitarian Aid, Infrastructure, and Reconstruction and Development) also hindered the coordination of economic reconstruction activities. Although there was talk in 1992 of setting up a development board involving these three ministries, no plans had materialised by 1994 when the regional government effectively ceased to function.

Re-emerging KDP-PUK rivalries further diminished the capability of the regional government. The fifty-fifty division of parliamentary seats was therefore extended to all ministries, but since each minister from one party had a deputy minister with almost equal powers under him, the government's potential for effective decision making, let alone action, was greatly reduced. The regional parliament, just as much as the government, became rapidly deadlocked in inter-party disputes, and was thus unable to address important economic and social issues like new land reforms and women's rights. Initially, there was some debate as to whether development should be primarily state controlled or left to market mechanisms and private initiative,[23] but no clear-cut policies were formulated, much less implemented.

The end of civil regional government was signalled by the outbreak of large-scale internecine fights between the KDP and PUK in May 1994. In 1996, both parties set up new cabinets in the areas under their control, both claiming to be the legitimate regional authorities; but these had even less potential for independent decision making than the government formed in 1992. The resulting state of affairs could be described as a caricature of a market economy, in which the state apparatus played no guiding role because it was weak, subservient to private interests, or simply absent.

4. Monetary developments

Before the Gulf War, the official exchange rate was 3.2 US dollars to the Iraqi dinar, though the black-market value of hard currency in the late

1980s was already some ten times higher. The imposition of the UN embargo led to a sharp rise in inflation. Unable to raise revenues, the Iraqi government introduced new, locally printed bank notes of 10, 25 and 100 Iraqi dinar in 1992, but in the Kurdish-held region, most people preferred to stick to the old foreign-printed bank notes, known as the 'Swiss' print. The latter inspired much more confidence than the newly printed currency, and the exchange rates of the old and new Iraqi dinars quickly came apart.[24]

In May 1993, the government announced the withdrawal of the old ID25 notes, and people living in government-held territory were given a few days to exchange old notes for new ones. However, because of border closures with Jordan and with the North during this time, substantial losses in individual and business savings were incurred in the North, where most people had not deposited their money in banks. Total losses in the Kurdish region alone may have amounted to some US$20 million.[25] After the October 1991 administrative pullout and the imposition of an economic blockade, the withdrawal of currency notes also effectively cut off the North from government-held territory, in monetary terms. The Kurdish regional government was unable to take any countermeasures. At one point it considered the option of adopting the Turkish lira as the new common tender. Turkish financial officials were cautious, but not in principle opposed to the idea. Some feared that this would lead to an increased Turkish money supply and thus raise inflation; but a Turkish Central Bank spokesman pointed out that the macroeconomic effects for Turkey would be minimal since the total money supply in Iraqi Kurdistan did not exceed the equivalent of US$50 million.[26] However, due either to Kurdish reluctance to make a move with such obvious political implications or to Turkish or Iraqi pressure, the plan did not materialise.

The regional government's inability to act in the financial crisis reflected a general malaise in the banking system. In October 1991, the assets of the regional branches of all Iraqi banks had been frozen, thereby paralysing the entire banking system in the North. A lack of trust meant that people did not deposit their savings (insofar as they had any) in the regional banks. A Kurdish Central Bank of sorts had been established after the uprising, but since it was mostly a deposit for customs revenues, its activities were limited. It supplied funds for salary payments and to the IKF parties, but did not perform other banking functions. Calls to establish a Kurdish National, or Regional Bank, or even a bank for reconstruction and development,[27] were ignored.

The regional Ministry of Finance was unable to get to grips with the financial system. Though set up in July 1992, it was still not functioning properly by the autumn, and the Iraqi Kurdistan Front remained the main supplier of funds, most of its limited budget being drained by payments to civil servants, which required ID45 million per month.[28] To the extent that they had ever got off the ground, the Finance Ministry and the regional bank more or less ceased to function after the outbreak of fighting between the KDP and the PUK in May 1994. The KDP refused to pass on the customs revenues it collected at the Habur border gate to the bank, which in turn accused the PUK prime minister, Kosrat Rasul 'Ali, of embezzling bank funds. After the uprising, the main Kurdish parties funded their *peshmergas* and other clientele directly from their customs revenues. Individuals trying to obtain loans had to turn to foreign sources of capital, as specific banking services, such as those formerly provided by the Iraqi Agricultural Co-operative Bank, were largely taken over by foreign NGOs.

In the absence of any centralised monetary policy, the local black market became the main locus for monetary transactions. The local exchange markets were capricious, and often dependent on political or mass-psychological rather than on economic factors; there were also slight variations between the exchange rates in different Kurdish cities. In December 1998, the local exchange rate was roughly 23 'Swiss' print Iraqi dinars to the US dollar. Between 1991 and 1996, the exchange rate was quite unstable, mostly corresponding to capricious political developments. For example, when KDP–PUK fights broke out in May 1994, the 'Swiss'-printed dinar depreciated to ID83 per dollar; but with the start of peace talks between PUK and KDP and the simultaneous reopening of talks between Iraq and the UN Security Council, it appreciated to ID69 and continued its rise, reaching ID47 in November 1994. Likewise, when the food-for-oil deal between Iraq and the UN, based on UNSC Resolution 986, was announced in May 1996, the dinar's exchange rate improved and food prices fell sharply.[29] This monetary instability made long-term planning rather difficult for local contractors and the smaller NGOs.

Although commodity prices were equally (and for the same reasons) unstable, inflation was on the whole remarkably low (see Table 8.2). Price increases between the pre-war period and late 1992 lay between 1,000 and 10,000 per cent, but between late 1992 and late 1995, they remained in the order of 100 per cent. This rise in prices, while by no means negligible, hardly qualified as hyperinflation. Iraqi Kurdistan was unlike the rest of Iraq in this respect, and was also different from other contemporary civil war zones such as former Yugoslavia.[30] Moreover, inflation in the Kurdish-controlled region was far less than in the rest of Iraq: thus, according to Ishow,[31] the market exchange rate of the Iraqi dinar in government territory plummeted from 30 to 2,000 to the dollar between the summer of 1992 and March 1995. There were several reasons for this remarkable divergence.[32] First was probably the continuing use of the 'Swiss-print' bank notes, which were in limited supply. The central government, by contrast, regularly printed new bank notes in order to finance its expenditures, thus fanning inflation in the region under its control. Moreover, the numerous foreign aid agencies, which had far easier access to the Kurdish-held area than to the territory under government control, steadily injected hard currencies into the local economy. However, despite the constant influx of foreign currency and goods, many basic commodities were simply too expensive for the average family. Thus, as Table 8.2 indicates, prices of meat and fish, which continued to be out of reach for the poorer strata of the population, apparently remained almost unchanged between 1992 and 1995.[33]

TABLE 8.2
Food prices in Iraqi Kurdistan (ID)*

Commodity	Pre-war	October 1992	Late 1995
Sugar (1 kg)	0.2ID	9.0ID	20.0ID
Flour (1kg)	0.125ID	5.5ID	10.0ID
Rice (1 kg)	0.2ID	16.5ID	25.0ID
Cooking oil (1 ltr)	0.7ID	17.0ID	30.0ID
Meat (1 kg)	6.0ID	45.0ID	35.0ID
Petrol (1 ltr)	0.07ID	8.0ID	6.0–20.0ID
Average civil-servant wages	200.0ID	200.0ID	200.0ID

Note: *Iraqi dinar here refers to the old currency which continued to circulate in the Kurdish-held area.
Sources: OFDA (Office of US Foreign Disaster Assistance), 'Joint Emergency Winter Humanitarian Needs Assessment Mission to Northern Iraq, 9–16 October, 1992', Report, 1992, p. 76; Netherlands–Kurdistan Society, Iraqi Kurdistan 1991–1996: Political Crisis and Humanitarian Aid, Amsterdam, 1996, p. 19.

5. Humanitarian aid

After the uprising transnational agencies, such as the humanitarian organisations of the United Nations and numerous foreign non-governmental relief organisations (NGOs), were among the main economic actors in the region; indeed, their impact on the economy was far greater than that of the regional government. Even excluding the emergency relief operation in the spring of 1991, the total amount of foreign humanitarian aid spent on Iraqi Kurdistan is thought to have run in the hundreds of millions, if not billions, of dollars.[34] However, despite this enormous investment, the usual cycle leading from immediate relief through reconstruction to development was barely followed. There was relatively little long-term planning for aid projects, and the local population remained heavily dependent on foreign handouts of food and fuel. In this way, the UN embargo and a lack of political will to promote an economically stable and self-supporting entity combined to thwart a more durable economic development.

In April 1991, the United Nations signed a Memorandum of Understanding (MoU) with the Iraqi government for carrying out a relief operation in Iraq as a whole. Significantly, the MoU was not reached on the basis of UNSC Resolution 688,[35] and gave Iraq considerable influence over NGOs working within the UN framework. Moreover, many of the relief supplies had to be purchased from the Iraqi government in dinars at the official exchange rate. The MoU was periodically renewed until April 1993.[36] The UN directed a disproportionate amount of its humanitarian effort towards the Kurdish region. Thus, in September 1992 it announced a 'winterisation' programme for the North, to provide relief from the hardship caused by Baghdad's blockade of foodstuffs and petrol.[37] The plan targeted 750,000 people in the North, out of a total of 1.3 million for Iraq as a whole (i.e., 58 per cent), as vulnerable and qualifying for relief aid.

Even by the UN's own admission, supplies were insufficient, as it had recognised the needs of 2.25 million people in the North, and could not even provide the target population with more than a fraction of its needs. Moreover, the rations would usually be reduced towards the end of summer, when stocks tended to run low.[38] The North received a relatively large amount of aid, less because it was economically worse off than the rest of the country (in fact, conditions in the South of Iraq were in some respects much more serious) than because the government distribution programme had been discontinued there. In other words, the UN took over part of the Iraqi government's responsibilities: it mostly provided relief aid (food supplies and fuel for heating), but also provided for the reconstruction of the infrastructure, through projects for the repair of schools, water facilities, etc. In consequence, humanitarian aid had a far greater influence on the economy in the North than elsewhere.

UN operations in Iraq were mostly funded by donations from its member states: frequently far less money was raised than was required, but the amounts involved were still considerable. Thus, the UN received US$146 million out of a budgeted US$288.5 million for the period between April 1994 and March 1995.[39] From early on, there were complaints about the inefficiency of the costly UN operation. Not only did UN agencies buy Iraqi goods at the official exchange rate, but high overhead costs also drained funds – a well-known phenomenon from other UN operations. Furthermore, many UN projects depended on concrete pledges by member states; these pledges declined steadily until they were replaced, so to speak, by the continuous flow of income that resulted from the food-for-oil agreement.

Apart from the UN, the United States and the European Union were the main sponsors of humanitarian aid. The American Overseas Fund for Disaster Aid (OFDA) was also involved, partly with village reconstruction through the Military Command Centre (MCC) in Zakho, and partly with providing funding for NGOs. In 1991 and 1992, the United States reportedly supplied a total of US$654 million in relief aid. However, according to Graham-Brown (this volume), most US expenditure came from Department of Defense Funds, and OFDA spent US$30–50 million annually in the region between 1993 and 1995.[40] The financial injection given by the European Union, particularly through ECHO (its humanitarian office), was also significant. The EU provided some 500 million ECU, or 60 per cent of the international relief effort, during the spring 1991 refugee crisis; in 1993 it contributed 21.5 million ECU to relief and reconstruction projects, ranging from the purchase of petrol products to village reconstruction and repair of water and sanitation systems.[41] Many foreign NGOs active in the region were funded, at least in part, by the above-mentioned transnational agencies. Sometimes this was done through escrow accounts involving unfrozen Iraqi assets, which effectively doubled the funds they had raised elsewhere. However, a number did not keep to the terms of the MoU, preferring instead to purchase their supplies and materials at black-market rates, either in Turkey or locally: this gave them far greater effective purchasing power.

A large proportion of the foreign humanitarian aid was directed to immediate relief aid rather than towards sponsoring long-term development, but following the establishment of a relatively stable area under Kurdish control, emphasis shifted from relief aid (which would keep the population dependent on external supplies) towards rehabilitation aid and income-generating projects. The potential for success of this approach was, however, severely constrained by long-term economic factors and international political considerations.

Agriculture
Agriculture seems to have been the one sector of the Iraqi economy that actually improved after the imposition of UN sanctions in 1990. In the North, agriculture benefited from substantial foreign aid, and there was high local demand after the central government discontinued the distribution of (largely imported) heavily subsidised wheat in 1991. However, even after the withdrawal of this wheat, the government continued its nation-wide practice of buying up wheat harvests at inflated prices, even across the demarcation lines with the Kurdish-held area. In the tense political atmosphere, such moves were immediately interpreted by Kurdish sources as attempts at economic sabotage through siphoning off already scarce cereals from a region suffering badly under the double embargo. The fact that government forces repeatedly set fire to wheat fields close to the demarcation line lends credence to this view.[42]

The bulk of the reconstruction efforts by foreign NGOs was directed towards villages and the agricultural sector. Important successes have been recorded here: according to the regional Ministry of Reconstruction some 2,850 villages had been rebuilt by early 1995, either by NGOs or with their financial support,[43] although reconstruction of schools and other infrastructural facilities in rural areas lagged behind the rebuilding of villages. Agricultural rehabilitation seemed an obvious strategy since it could engage much of the workforce and could relieve the pressure on the urban infrastructure. Even so, it faced a number of structural difficulties. First, the prospect of an economically viable and self-sufficient agricultural sector, let alone one able to compete on the international market, had been seriously jeopardised by Iraqi government policies of the preceding decades. In the absence of an adequate domestic transportation infrastructure, and with the high cost of petrol, it was often unprofitable to bring locally-grown goods to the local markets. Secondly, political insecurity and the lack of infrastructure prevented many farmers from returning to their villages from the *mujamma'at* where they had been resettled; many of those who did return maintained their households in the relocation camps.[44]

Other factors prevented the region's economic potential from being realised. Only 43 per cent of all arable land was under cultivation in 1992;[45] a more extensive cultivation was partly blocked by emerging land conflicts. After the uprising, many formerly deported peasants had returned, only to discover that their villages and the surrounding lands had meanwhile been seized by neighbouring tribes or local landowners that had remained loyal to the Iraqi government. Serious land conflicts arose in many areas, and it was mostly the landowners, many of them with a tribal or *mustashar* background, who emerged victorious, although in other areas peasants organised themselves effectively and kept the landowners at bay. In general, however, the prevalence of large-scale mechanised commercial farming strengthened the position of the land-owning elite. In 1997 it was reported that in Duhok governorate an estimated 85 per cent of the land was farmed for large-scale commercial production, and the remaining 15 per cent by subsistence farmers.[46] The regional government and parliament were too weak and divided to promulgate and implement a new land-redistribution law. The competition for supremacy between the two main Kurdish parties led them to seek the support of influential men in the rural areas, which gave the latter better opportunities to assert their power.

Natural conditions further reduced agricultural productivity. In 1995, agricultural production dropped by 10 per cent because of a lack of pesticides and fertiliser;[47] in 1996 and 1997, drought badly affected the harvests. At times, the scarcity of water led to conflicts over irrigation systems. Nor were the region's rich natural resources being put to good use. Environmental pollution took on new forms after 1991: the lack of diesel drove many to search for firewood, and the degree of deforestation was accelerated. The known oil reserves at Tawog, North of Duhok, Taqtaq and Qoratu, and smaller reserves near Makhmour and Khormal were left untouched, despite the great local need for petrol products.[48] According to experts, a small refinery could have been set up at a fraction of the annual cost of fuel purchases but for political reasons the aid agencies were unwilling to do this. However, in 1996 PUK personnel set up a small unit at Taqtaq near Koy Sanjaq for extracting oil, and apparently sold most of its output.[49]

The urban economy

Of the more than three million inhabitants of Iraqi Kurdistan, some two million may be estimated to live in urban structures, whether cities proper or *mujamma'at*.[50] Although urban centres thus contain the bulk of the population, most humanitarian aid has been directed towards the villages. In the cities (except for the *mujamma'at*), aid has been aimed at relief rather than rehabilitation, and there have been few concerted efforts to reduce the urban dependence on aid. According to UN sources, the number of urban destitutes had actually *increased*: by mid-1995, 590,000 persons (140,000 internally displaced persons, 250,000 in *mujamma'at* and 200,000 living on pensions or civil-servant wages) were directly dependent on external food support.[51]

Urban unemployment has been estimated at 70 to 80 per cent. Even those who have employment are barely self-supporting: average wages of civil servants amount to ID200, which is quite inadequate to cover basic living costs. Civil servants' wages have been paid only irregularly by the IKF, government, and finally the two main parties, and many town dwellers have resorted to moonlighting, working as taxi drivers or street vendors to make ends meet. Likewise, a lively trade has developed in second-hand furniture and clothing, and increasing numbers of children work as petty vendors.[52]

The radical impoverishment of the urban middle class is mainly, though not entirely, the result of the UN sanctions: statistical estimates of the development of GDP from 1950 to 1993 suggested that real per capita income, which had peaked in the early 1980s, had already dropped back to mid-1960s' levels by 1989, before the onset of the Gulf crisis.[53]

The urban infrastructure has deteriorated badly, though in some of the poorer urban quarters and many *mujamma'at*, it was never very well developed to begin with. Many urban quarters lacked adequate water supplies and sewage systems, and electricity was typically run at half power; in Duhok governorate, it was cut off completely between 1993 and 1995. While major efforts have been made to restore such basic infrastructural facilities, it has proved impossible to create a productive urban economy. The rehabilitation of industry was marred by the lack of raw materials and spare parts as a consequence of the UN embargo, while development of trade was equally badly affected. In short, humanitarian aid, no matter how massive, by itself was unable to prevent the continuing pauperisation of the urban population.

Problems of coordination

A major constraint on the international humanitarian effort was the lack of coordination between the respective government ministries, between the NGOs and the regional government, and among the various aid projects. Construction aid was fragmented and uncoordinated and lacked any master plan for development. Such fragmentation was understandable in the early days, but even after a regional government was established, a more centralised and coordinated approach failed to materialise, and this lack carried with it the risk of lopsided development, jealousy between different regions or parties, bottlenecks in reconstruction, and so on. In many reconstructed villages, for example, no new schools were built; or when they were, no teachers or educational materials would be available.

A further constraint was the close cooperation between foreign and local NGOs at the expense of the regional government. Few foreign NGOs worked directly with the regional government; in part, this reluctance to cooperate with the state apparatus is a general feature of present-day development aid.[54] In Iraqi Kurdistan, however, it also reflected an active wish to prevent political developments that the donor states considered undesirable, as well as an idea that pairing with local NGOs would better allow for market competition. In fact, since the local NGOs were typically affiliated to one or another of the major political parties, this approach had the unintended side effect of locally undermining the government's already weak position vis-à-vis the parties. Each major foreign NGO had a budget comparable to, or bigger than, that of the regional government. This widespread principle of affiliation meant, in effect, that local NGOs developed into private contractors sponsored by foreign capital, and could afford to bypass the regional government.

By 1993 there were already clear warnings and protests against this lack of coordination, and in particular against the form of parallel government that was being created.[55] Attempts to counter this trend failed to materialise, and the power of the elected bodies as distinct from the party leaders and politburos was further eroded.[56]

In short, the UN sanctions seriously inhibited the development of a healthy and productive economy in the region, although domestic long-term factors had dimmed the prospects of such development from the start. In Iraqi Kurdistan, economic stagnation was the result mainly of international policy decisions, or rather indecision: in 1991, for instance, the international community set up a 'safe haven' of an exclusively humanitarian nature without ever seriously addressing the question of the region's long-term fate. The international involvement in Iraqi Kurdistan has been termed 'humanitarian interventionism',[57] since it brought a purely humanitarian agenda into a major political conflict that left the underlying political problems unresolved, and actually tended to exacerbate the plight of the local population.

More generally, the activities in the region reflected a worldwide tendency to shift from development aid to relief aid, and from cooperation between states to support of private economic agents.[58] By providing relief supplies, the foreign agencies took over part of the Iraqi government's role in food distribution, and in doing this they relieved the local parties in power of their duty to feed the population. Humanitarian aid tended, mostly unintentionally, to undermine the strength and legitimacy of the state-related institutions, and to encourage a privatisation of the economy.[59]

Even so, although the dismal state of the rural and urban economy of Iraqi Kurdistan may suggest a heavy dependence on these formal, more or less 'official' sectors, the economy of Iraqi Kurdistan has in fact developed a distinct and rather unexpected dynamism: new, informal flows of finances, goods and persons have emerged, following rather different laws.

6. The informal economy

Following the imposition of the embargo in 1990, various local agents reaped considerable profits from the sudden scarcity of hitherto abundant goods. The power vacuum in the North, caused by the 1991 withdrawal of government personnel, further increased the potential for lucrative cross-border trade.[60] In the absence of an adequately functioning state apparatus to define the legal economy, it is perhaps hard to qualify this trade as either 'legal' or 'illegal'. However, given that it thrived largely on the deliberately created scarcity of the traded commodities (possibly even seeking to maintain this scarcity), and that it influenced prices and economic competition by using or threatening non-legitimate violence, there can be little doubt of its criminalised nature.[61]

Smuggling across the international borders has always been a characteristic feature of the local economy, but after 1990 it acquired a new scale and importance. The most obvious instance of this trade was the large-scale transport of Iraqi petrol or diesel to Turkey, which by 1992 was already becoming a major source of income for the regional government and the Kurdish parties, not to mention Iraqi state circles. Along with this 'fuel trade', the smuggling[62] to Iran of cigarettes, alcohol and other luxury goods rapidly gained importance.

After the establishment of a regional government, much of this revenue went to the regional Central Bank, while the rest was divided evenly between the KDP and PUK, much as it had previously been divided among the seven parties in the Iraqi Kurdistan Front. The profits to be gained from this rapidly burgeoning international trade were so large that in 1995 Talabani, the PUK leader, proclaimed a great future for Iraqi Kurdistan. He saw it becoming a prosperous free-trade zone for the entire Middle East, as Hong Kong and Singapore had become in South-east Asia. More relevant parallels for developments in Iraqi Kurdistan's economy would, however, seem to lie in Lebanon and Afghanistan, where comparable war situations created considerable economic opportunities for armed militias who tried to monopolise and exploit the most lucrative economic activities in the areas under their control for private profit.

Because of the informal nature of this economic activity, no precise data is available on the extent of such trade, but certainly by local standards, the amounts of money circulating in this business were enormous. Several well-informed sources reckoned that daily revenues collected at the Ibrahim Khalil (Habur) border checkpoint alone were running at up to one million dollars per day. Initially this estimate seems scarcely credible but on closer inspection it is understandable, given the highly profitable trade that was developing in the three scarce and much-demanded commodities: petrol/diesel, luxury goods such as cigarettes and alcohol, and cross-border permits.[63]

The smuggling of petrol and diesel products was by no means the only source of income, but it remained an important one. Every day, vast numbers of Turkish trucks carrying, on average, 4,000 litres of petrol, would cross the border; levies on these shipments were paid in (old) Iraqi dinars.[64] In 1994, two months' revenue was estimated at ID252 million; in spring 1997, one aid worker estimated that, given daily border crossings by several thousand trucks carrying up to 6,000 litres charged at one Iraqi dinar per litre, petrol duty revenues exceeded ID20 million *per day*.

Revenues from customs levies on luxury goods such as alcohol and cigarettes (not to mention the private profits on this trade) are even more difficult to establish than those of oil products, but can be assumed to be of a similar magnitude. In 1992, for instance, customs revenues on the Iranian border were reported to be only slightly lower than at the Turkish border – indicating that the trade in luxury goods must have been comparable in scale to that of the petrol/diesel trade, which was barely significant across the Iranian border. The PUK, which had long controlled most of the Iranian cross-border trade, claimed to have lost ID50 million or US$50,000 in customs revenues when Iran closed the border for two weeks. If this claim is correct, its daily revenues amounted to some US$3,500 per day – considerably smaller than the sum collected at the Turkish border by the KDP.[65] The PUK's actual income may, however, have been far higher.

The lucrative nature of the cross-border trade also helps to explain the character of the Kurdish internecine fights that broke out in 1994. Some of these were about controlling the principal lines of transport between Turkey and Iran; thus, the Hamilton road, one of the main trade routes to Iran, saw some of the fiercest clashes. The warlike situation also created new opportunities for extracting profits from commodities and people: the two hostile parties set up checkpoints between their respective territories, at which they charged additional levies on goods, and demanded transit fees from people crossing the demarcation line.

Cross-border trade revenues flowed mainly towards the local parties rather than to the regional government but also – and perhaps more importantly – spread to individuals high up in the party hierarchies. The locally accumulated capital was mostly deposited in bank accounts abroad, and was rarely invested in the region. An exception was the construction boom in cities like Zakho and Duhok, to which much of the local population came to make a living, and where small groups became rich from the cross-border trade. The revenues acquired by these privileged individuals were unlikely to be used for developing a productive economy.

Finally, substantial numbers of people moved across the borders, becoming instruments for private gain more than for the public good. After 1991, many fled from the region, typically to Western Europe. Since only the rich could afford the journey abroad, most of those who left were members of the educated urban middle class, especially engineers, higher-education personnel, and medical doctors, creating a brain drain that would undoubtedly have serious long-term consequences. This mass exodus, combined with dramatic impoverishment, effectively decimated the urban middle classes. At times the departure of the skilled labour force was even actively promoted by foreign countries: thus, in late 1996, following the takeover of Arbil by the KDP and the Iraqi army, 5,000 Kurdish employees or American-sponsored organisations were evacuated in two successive waves by the United States. Only a tiny minority of these could even remotely have qualified as CIA personnel; while many of them were employed by foreign aid organisations that had no links to the USA. This withdrawal, too, deprived the regional economy of some US$25 to US$40 million in local NGO personnel salaries.[66]

In the movement of people in and out of the region, financial and human flows became inextricably intertwined, as was clearly shown by the customs regulations applied to those wishing to enter or leave the enclave. Officially, only eight people with Iraqi passports (or no passport at all) were permitted to cross the border with Turkey each day, but actual numbers were considerably higher. In 1996 and 1997 a weekly average of 100 asylum seekers arrived in the Netherlands from Iraqi Kurdistan, most of whom had come through the Habur gate and Turkey; and comparable, or even larger, figures were recorded in other Western European countries. Numbers of persons wishing to leave Iraqi Kurdistan through the Habur border gate were reliably estimated at between 100 and 200 daily during this period. Depending on their KDP contacts, they often had to pay extortionate exit fees varying from US$100 to several thousand dollars per individual, and daily revenues from exit visas probably ranged between US$10,000, and US$200,000, or more. Those with foreign (i.e., Western European) passports paid an entrance fee of US$50 per person in hard currency (but no exit fees). In short, daily entrance and exit fees alone yielded revenues that might well have run to hundreds of thousands of dollars.

The large and rapidly expanding expatriate Iraqi Kurdish community in Western Europe quickly became a major source of cash in the region, where an amount of US$100 – more or less equivalent to a civil servant's average annual salary – represented a substantial contribution to meeting an entire family's need for food and fuel. With their newly acquired passports, many refugees could travel back to the region in relative safety, and most of these expatriates carried considerable sums of money with them, in cash, for the support of close relatives or on behalf of acquaintances in Europe. Reports suggest that individual travellers carried thousands of dollars at a time. In addition to these personalised money transfers, a number of informal 'money transfer desks' operated in Western European cities. Kurds living in Europe would transfer a given amount of money to the organisation's European bank account. The recipient in Kurdistan, notified of the transaction details by satellite fax or telephone, then collected the amount, minus a fee, from the organisation's local office in Iraqi Kurdistan.

It is clear that after the outbreak of internecine fighting, the Kurdish parties, especially the KDP (or at least some circles within the party leadership), commanded vast, if fluctuating, sources of income, although the financial picture looked somewhat bleaker on the PUK side. Following its conquest of cities like Arbil and Rawanduz, the PUK had controlled the main access routes to Iran, where most of the smuggled luxury goods were heading, and by mid-1996, its revenues were reportedly comparable to those of the KDP at Habur. However, after being ousted from Arbil in August 1996, the PUK was deprived of its main political asset – control over the regional capital – as well as many of its border checkpoints with Iran, notably the one along the Hamilton road running from Rawanduz to Haji Umran. Once the KDP had full command of the Hamilton road, it no longer needed to transport any goods through PUK territory. This military setback caused the PUK considerable difficulties; payment of salaries in its territory became a particularly serious problem. The KDP offered to pay salaries of civil servants in PUK territory, provided they were willing to come to Arbil to collect their money.

Restrictions on the flow of currency, commodities and persons were thus primarily of a political nature. After the outbreak of internecine fighting in 1994, additional blockades were constructed, and further levies were demanded along the front line between the territories of the two parties. In early September 1996, following the combined action by KDP troops and the Iraqi army to remove the PUK from Arbil, Saddam Hussain called an end to the internal blockade that had been in force for almost five years. When the PUK eventually recaptured much of the territory it had lost to the KDP, the blockade was only partially reimposed. Nowadays, people can move freely past the Kurdish and Iraqi checkpoints without the risk of having petrol or goods confiscated, although, according to one observer, flows of goods and people became more difficult between Arbil and Sulaimaniyya than between either Duhok, Arbil or Sulaimaniyya and government-controlled territory.

It can therefore be argued that the Iraqi welfare state did, in a sense, continue after the uprising, but with rather different sources of funding. It was no longer the state that supplied resources, but foreign NGOs and relatives abroad, and the main beneficiaries of this situation were the political parties, or rather, party-linked individuals. A prime economic activity in the region became what Bozarslan has called 'predation',[67] both of infrastructural materials and other state-related resources, and against the civilian population. Such predatory acts occurred frequently, especially in the immediate aftermath of the uprising, but continued and intensified following the security breakdown in 1994. Many materials for the Bakhama Dam project were sold off to Iran; other favourite items for predation included electricity cables, agricultural and irrigation machinery, and cars.

7. Conclusion: prospects and limitations

Under allied protection and with foreign humanitarian aid, an economy of a rather new kind developed in Iraqi Kurdistan. Given the UN embargo, the Iraqi blockade, and the infighting among the main Kurdish parties, individual agents backed by party or tribal power were able to turn the penury of the population at large into a source of private profit. It is unclear how best to characterise this economy. Despite the warlike situation that pertained in the region from 1991, and especially following the outbreak of internecine fights in May 1994, it cannot be described as a 'war economy' in the classical sense, since this involves strengthened state intervention and increased production, notably in the war industry.[68] Almost the exact opposite occurred in Iraqi Kurdistan: the state collapsed, an unchecked informal economy developed, and production continued to give way to consumption. Despite, or partly because of, the enormous amounts of money spent on relief and rehabilitation aid, the local power vacuum was primarily of benefit to local groups that were effectively able to appropriate a monopoly on violence for themselves.

One might even say that those reaping the benefits of the war between the parties have had an active economic interest in maintaining the ongoing predicament. Certainly only a small group has profited from these developments, while the urban middle class has found itself profoundly impoverished. A new class of entrepreneurs has emerged, overlapping only partly with the group of state-loyal individuals who had enriched themselves in the 1980s. However, many former *mustashars* have been able to maintain or increase their economic and political strength.

The situation in Iraqi Kurdistan is in many ways structurally similar to that pertaining in government-held territory. Both areas have suffered badly under the UN embargo, which has led to mounting social anarchy and large-scale predation by armed groups. Nevertheless, there are noticeable differences. However much the state has been 'privatised' for the benefit of Saddam Hussein's close relatives, the state apparatus in government-held territory has continued more or less to function. This has been due to the government's ability to formulate and implement economic and other policies, such as realising a reasonably efficient food rationing system.[69] In the Kurdish region, the distribution of essential goods was largely taken over by foreign aid agencies; certainly much more so than in the rest of the country.

The UN sanctions are likely to last for some time to come, and no major structural changes of the economy are to be expected. In the long run, things are much more unpredictable. The crucial variable here is, of course, the question as to whether or not the UN sanctions, for which international support is crumbling and which are increasingly difficult to monitor effectively, will be lifted. In this respect, conclusion of the food-for-oil deal between Iraq and the UN on the basis of UNSC Resolution 986 was significant, since it further institutionalised the population's dependence on humanitarian aid. Rather than providing for the rehabilitation of a productive, much less a self-supporting economy, the deal foresaw the resumption of regular food distributions to the entire population. Out of every billion dollars in revenues, US$150 million was earmarked for the North. The Kurdish parties had hoped to acquire administrative and financial control of food distributions in their region, but the Iraqi government successfully claimed supervision for the whole country; moreover, the food supplies for the North were to be stored in government-held territory. Although the actual implementation of the plan took considerable time to materialise, regular shipments of basic commodities had started to arrive by spring 1997.

With the implementation of the food-for-oil deal the Iraqi government regained a considerable part of its former economic influence over the region. An important step was taken towards the rehabilitation of the repressive welfare state that was Iraq in the 1980s when most of the responsibility for feeding the population shifted from foreign organisations back to the Iraqi state apparatus.

The stability of the informal economy that developed in the North has depended too much on capricious political factors to allow for any confident predictions, although the KDP secured an economically and politically quite advantageous position vis-à-vis the PUK and the neighbouring states. Despite tremendous humanitarian efforts, any eventual rehabilitation of a productive agriculture and industry in Iraqi Kurdistan (or, for that matter, in Iraq as a whole) seems, at present, to be unlikely.

It is to be feared that small groups of individuals, backed by force of arms, will continue to enrich themselves, and that the predatory economy will be further institutionalised while the bulk of the population remains poor, weak and dependent on external support. To put it cynically, most of the capital circulating in Iraqi Kurdistan, much as in the days of the Ba'thist government, ultimately still originates in external transactions. Nowadays, though, this capital is no longer acquired through the sale of oil, but rather by a privatised, politicised, and at times even criminalised form of trade in humanitarian aid projects, smuggled goods, and refugees.

Postscript, 1998

Events that occurred after this chapter was written in the summer of 1997 do not require any serious revision of its main argument, but do merit some discussion. First, the region regained a measure of stability and quiet with the cessation of hostilities between the KDP and the PUK in November 1997, and the initiation of peace talks in January 1998, which culminated in the September 1998 Washington agreement.

Whether the peace deal will hold is at best conjectural at the time of writing; nevertheless, both parties were stepping up infrastructural works in the regions under their respective control, and appeared to be increasingly active for the benefit of the population.

Secondly, despite the improvements in living conditions, the structural features of the regional economy have remained unchanged. The smuggling of petroleum products and other items continued as a prime source of income, despite serious efforts during summer 1998 by the Turkish Ministry of Economic Affairs to regulate the cross-border traffic, and to levy petroleum taxes that would benefit the state, rather than a restricted group of individuals. The smuggling of people continued unabated, and only in late 1997 did the European Union, the prime destination of most refugees, show any signs of alarm. The interception off the Italian shore in November and December 1997 of several boatloads of refugees, many of whom were Kurds from Iraq and Turkey, led to frantic meetings on how to counter this trade in humans. To those willing to look more closely, these events have revealed the high degree of international organisation of this traffic, and the collusion between organised crime and elements of the state apparatus in some of the countries involved.

Thirdly, implementation of the food-for-oil deal considerably relieved the suffering of the population, though it has shown itself to be no long-term solution for Iraq's problems.[70] The urban population in particular has a much better chance of receiving adequate food supplies, and, to some extent, the implementation of UNSC Resolution 986 has created new employment opportunities, with the UN commissioning construction projects for roads, schools, and the like. In the summer of 1998, tourists from Baghdad were able for the first time since the imposition of the UN embargo to visit the mountainous North – a clear indication that some Iraqis at least could afford leisure expenditure again. For the agricultural sector, however, the implementation of the food-for-oil deal was little short of a disaster, and confirmed all the warnings of aid workers. Farmers cannot compete with the heavily subsidised products of the programme, and have seen their profits melting away. They can do little more than smuggle their harvest to Iran, even though this brings in scarcely any more profit. In other words, implementation of the agreement risks undermining the one relatively flourishing sector in post-war Iraq.

The short-term relief of suffering and continuation of the population's dependence on welfare handouts were clearly not the same as working towards the rehabilitation, or establishment, of a healthy and productive economy.

NOTES

1 See, for example, M. Farouk-Sluglett and P. Sluglett, *Iraq Since 1958: From Revolution to Dictatorship*, London: I.B. Tauris, 1990, esp. Ch. 7; and A. Alnasrawi, *The Economy of Iraq: Oil Wars, Destruction of Development and Prospects, 1950–2010*, Westport, Conn.: Greenwood Press, 1994.

2 Here 'Iraqi Kurdistan' is defined neither as the predominantly Kurdish geographical area nor as the legal entity called the 'autonomous region of Iraqi Kurdistan' in either the 1970 agreement or the 1974 autonomy law. Rather, the term refers to the politically defined region under de facto Kurdish control since 1991. Little or nothing is known about the economic specifics of the (mostly) Kurdish areas under government control, in particular the oil-rich areas surrounding Kirkuk and Khanaqin, and the mountainous Jabal Sinjar.

3 For an overview of political and social developments, see M. Leezenberg, 'Irakisch–Kurdistan seit dem Zweiten Golfkrieg', in C. Borck et al. (eds), *Ethnizität, Nationalismus, Religion und Politik in Kurdistan*, Munster: LIT Verlag, 1997. A brief account of economic influences on the behaviour of the Kurdish political parties is given in M. Leezenberg, 'Barzani GmbH gegen Talabani AG: Militarismus und Kriegsgewinnlerei in Kurdistan-Irak', *INAMO*, nr. 8, 1996, pp. 4–7.

4 Cf. Phebe Marr, *The Modern History of Iraq*, Boulder CO: Westview Press, 1985, p. 285.

5 Quoted by S. Hafeed, 'The embargo on Kurdistan: its influence on the economic and social development', in F. Hussein, M. Leezenberg and P. Muller (eds), *The Reconstruction and Economic Development of Iraqi Kurdistan: Challenges and Perspectives*, Amsterdam: Netherlands-Kurdistan Society, 1993, p. 40.

6 Cf. Farouk-Sluglett and Sluglett, op. cit., 1990, pp. 230ff.

7 Cf. Marr, op. cit., 1985, p. 249.

8 Cf. R. Springborg, 'Infitah, agrarian transformation and elite consolidation in contemporary Iraq', *Middle East Journal*, 1986, 40, pp. 33–52; also Farouk-Sluglett and Sluglett, op. cit., 1990, pp. 243–5.

9 Hafeed, in Hussein et al., op. cit., 1993, p. 40.

10 Cf. Human Rights Watch/Middle East, *Iraq's Crime of Genocide*, New Haven: Yale University Press, 1995, Ch. 1.

11 D. McDowall, *A Modern History of the Kurds*, London: I.B. Tauris, 1996, pp. 354–7.

12 For a detailed description of the rural situation, see A.Fischer-Tahir, 'Die landrechtliche Situation in der Karadagh-Region/Kurdistan (Irak)', unpublished ms., 1993.

13 Cf. Human Rights Watch/Middle East, op. cit., 1995, for a detailed account of the Anfal operations and their extensive genocide.
14 This is the generally accepted estimate. See McDowall, op. cit., 1996, p. 360. [Editor's comment: human settlement figures are not available for Iraq. An official housing census carried out in 1956 reported a total of 11,404 villages and towns in the whole of the country (see Government of Iraq, Principal Statistical Office, *Taqreer a'n ta'dad al-masakin fi al-iraq* [Report on the Housing Census of Iraq for 1956], Baghdad: Al-Najah Press, n.d.). The total figure included 3,137 villages with an average of eight dwellings each. The Census used local definitions for what constituted a village. On the basis of these figures, the total number of Kurdish villages in Iraq was probably in the region of 4,000 to 5,000 in 1956, including the villages in the smallest category defined above. Given the large-scale rural–urban migration of subsequent decades, caused not least by the armed conflict in the region, it is unlikely that the number of villages will have increased. Using Kurdish sources, Makiya estimates the number of villages destroyed in the Anfal campaign to have been 1,276. (See K. Makiya, *Cruelty and Silence*, W. W. Norton, New York, 1993, p. 166). Many of these villages are likely to have also suffered destruction in earlier government campaigns. Makiya's figure for villages destroyed during 1969–1979 is 1,201 (ibid. p. 167). It is clear that a substantial proportion of the Kurdish countryside has been decimated by war and repression.]
15 However, see Kiren Aziz Chaudhry 'On the way to the market: economic liberalisation and Iraq's invasion of Kuwait', *Middle East Report*, May–June 1991, pp. 14–23; also Chaudhry's contribution in this volume.
16 According to some observers, the large-scale international supply of food aid was probably superfluous once the refugees had returned to their homes, since the Iraqi food-distribution system had largely continued to function.
17 'United Nations Assistance in Iraq', presentation before the Hearing of the Committee on Migration, Refugees, and Demography, Geneva, 11 June 1993.
18 Figures reproduced in Hafeed, in Hussein et al. (eds), op. cit., 1993, p. 38.
19 Cf. Dutch Consortium, *A Report on the Relief Operation in Iraqi Kurdistan, April 1991–April 1993*, Rotterdam: Memisa, 1993, p. 7.
20 Hafeed (in Hussein et al., op. cit.,1993) provides a detailed economic overview of the gradual imposition of the blockade up until the summer of 1992.
21 Cf. H. Bozarslan, 'Kurdistan: économie de guerre, économie dans la guerre', in F. Jean, and J.-C. Rufin (eds), *Economie des guerres civiles*, Paris: Hachette, 1996, pp. 117–18.
22 For more information on the elections, including a discussion of voting results and the question of fairness, see R. Hoff, M. Leezenberg and P. Muller, *Elections in Iraqi Kurdistan: An Experiment in Democracy*, Brussels: Pax Christi International, 1992.
23 Cf. contribution by Ja'far, in Hussein et al. (eds), op. cit., 1993.
24 In late 1992, for example, university salaries were paid out partly in old and partly in new bank-notes, but many recipients would change the latter for old currency as quickly as they could. In the Kurdish region, the old bank-notes were at this time already worth up to ten times as much as the new ones.
25 Cf. L. Schmidt, *Wie teuer ist die Freiheit? Reportagen aus der Selbstverwalteten kurdischen Region 1991-93*, Koln, ISP Verlag, 1994, pp. 196–9). Ishow interprets the withdrawal of ID25-notes as primarily a move to counter the hyperinflation in

government-held territory; abroad, it had the effect of further reducing confidence in the Iraqi dinar, and had adverse effects even on smuggling activities. See H. Ishow, *L'Irak: Paysanneries, politiques agraires et industrielles au XXe siècle. Contribution à la réflexion sur le developpement*, Paris: Publisud, 1996, p. 261.

26 *Wall Street Journal*, 25 May 1993; and cf. *Turkish Daily News*, 28 May 1993.

27 See the contributions by M. M. A. Ahmed and J. Fuad in A. de Boer and M. Leezenberg (eds), *Iraqi Kurdistan: The Need for Durable Development*, Amsterdam: Netherlands-Kurdistan Society, 1993.

28 Cf. C. Kutschera in *The Middle East*, November 1992.

29 Incidentally, price levels of basic commodities also sank dramatically, as local markets were suddenly flooded with local agricultural products. Many farmers hastily brought their products to the local markets, on the assumption (mistaken, as it turned out) that the deal would quickly materialise into shipments of subsidised goods.

30 Cf. P. Kopp, 'Embargo et criminalisation de l'économie', in Jean and Rufin (eds), op. cit., 1996.

31 Ishow, op. cit., 1996, p. 272.

32 See Bozarslan, op. cit., 1996, p. 110. The low inflation is explained by regional dynamism and the openness of Iraqi Kurdistan towards the economies of neighbouring states, by the massive injection of foreign aid, and by the relative scarcity of the 'Swiss' print dinars (of which, obviously, no new notes could be printed). However, as we will show later in the chapter, there are important restrictions on this alleged local openness. Bozarslan perceives these as contrasting with the dirigiste and nepotistic character of the Iraqi economy, which is monopolised by a small circle of people close to the highest echelons of the regime.

33 Some local sources explained that low inflation was the result of systematic Iraqi efforts to withdraw old dinars from the local economy; thus, Turkish truck drivers were alleged to have to pay for the petrol they acquired on government-held territory in old currency, which they could obtain in the North. Independent observers have not, however, confirmed these allegations.

34 It is practically impossible to make a reliable estimate of the total relief and rehabilitation input, given the fragmented nature of available information (most projects being of a small-scale and local nature) and the difficulty of obtaining access to the balance sheets of most NGOs. (For some statistics see Graham-Brown in this volume.)

35 The UN's humanitarian intervention was totally independent of, if not in direct contradiction with, its more political involvement (such as in the crisis leading up to the war and in the subsequent arms-inspection programme). For example, the internal blockade clearly violated Resolution 688 but the MoU simply accepted it as a *fait accompli*.

36 Strictly speaking, the Iraqi government's subsequent refusal to extend it made the continuing presence in the North of foreign NGO personnel illegal.

37 From October 1991 private individuals could not take any foodstuffs across the demarcation line.

38 See the following two UN publications: *United Nations Consolidated Inter-Agency Humanitarian Programme in Iraq: Extension of the Cooperation Programme for 1995–1996*, Document DHA/95/74, Geneva, March 1995; and *United Nations Consolidated Inter-Agency Humanitarian Cooperation Programme for Iraq:*

Mid-Term Review, Geneva, September 1995. Cf. also D. Keen, *The Kurds: How Safe is their Haven Now?*, London, Save the Children Fund, 1993, Ch. 3.

39 United Nations, op. cit., March 1995.

40 Elsewhere, much higher figures of OFDA spending are reported. Cf. Bozarslan, op. cit., 1996, pp. 120–1.

41 See A. de Boer and M. Leezenberg (eds), op. cit., 1993, pp. 13, 92–3.

42 Evidence for government burning, other than possibly for military purposes at the lines of control, is rather contradictory, and the rationale for crop burning when the government was willing to pay inflated prices for the wheat it required is not clear. Kurdish sources spoke of sabotage, and doubtless the reasons were political as well as economic. Eyewitnesses maintained the crop burning was aimed at destroying economic potential, but it could equally have been aimed at intimidating the peasant population.

43 United Nations, *United Nations . . . Mid-Term Review*, op. cit., September 1995, p. 56.

44 Cf. A. Fischer-Tahir, 'Deportation and Resettlement in Iraqi Kurdistan', paper presented at the international conference on 'The Kurds and the City', Sèvres, France, September 1996.

45 D. Keen, op. cit., 1993, p. 38.

46 Cf. B. Stapleton, 'Security Council Resolution 986: A Disaster Waiting to be Contained?', paper presented at a seminar on Iraq, June 1997, Christian Aid, London, 1997.

47 United Nations, *United Nations . . . Mid-Term Review*, op. cit., 1995.

48 D. Keen, op. cit., 1993: Appendix I; and cf. contribution by al-Jaff, in Hussein et al. (eds), op. cit., 1993.

49 Keen, op. cit., 1993.

50 Needless to say, this is only a rough estimate, obtained by adding the estimates of the properly urban population as 1,488,000 and those of the remaining rural displaced persons at 586,000. See, respectively, OFDA (Office of US Foreign Disaster Assistance), 'Joint Emergency Winter Humanitarian Needs Assessment Mission to Northern Iraq, October 9–16, 1992', Report, 1992, pp. 77–82; and United Nations Assistance in Iraq, presentation before the Hearing of the Committee on Migration, Refugees, and Demography, Geneva, 11 June 1993.

51 United Nations, *UN Consolidated Inter-Agency Humanitarian Programme in Iraq: Extension . . .*, op. cit., DHA/95/74, March 1995.

52 Cf. Gazdar and Hussain in this volume.

53 Abbas Alnasrawi, op. cit., 1994, p. 152.

54 Cf. F. Jean, 'Aide humanitaire et économie de guerre', in F. Jean and J.-C. Rufin (eds), op. cit., 1996.

55 See the various presentations in A. de Boer and M. Leezenberg (eds), op. cit., 1993.

56 M. Leezenberg, in C. Borck et al. (eds), op. cit., 1997.

57 R. Ofteringer and R. Backer, 'A republic of statelessness', *Middle East Report*, 187–188, 1995, pp. 40–5.

58 Cf. F. Jean, in Jean and Rufin (eds), op. cit., 1996.

59 Again, this development is not unique to Iraqi Kurdistan: cf. F. Jean, ibid., 1996, pp. 556–60.

60 Much of this information on various kinds of cross-border trade was gathered from a senior Kurdish official who preferred to remain anonymous. His remarks were largely consistent with my own observations and extrapolations, and with the fragmentary information available from other sources (press and other reports, observations by foreign aid workers, and so on).

61 Cf. Bozarslan, op. cit., 1996.

62 The term 'smuggling' is used advisedly – although the diesel trade was carried on openly through official checkpoints, the trade was semi-clandestine, privatised and in violation of UN sanctions.

63 Bozarslan estimates Habur border revenues much more conservatively at around US$100,000 per day, op. cit., 1996, p. 119.

64 *The Middle East*, November 1992; and see Schmidt (op. cit., 1994, pp. 182–3) for some observations on cigarette smuggling to and from Turkey.

65 The precise effects on this cross-border trade of the food-for-oil deal between Iraq and the UN are not known, but up to 1997 the amounts involved did not seem to have decreased seriously.

66 See Stapleton, op. cit., 1997.

67 Cf. Bozarslan, op. cit., 1996.

68 Cf. Kopp, op. cit., 1996, pp. 430–2.

69 Cf. Gazdar and Hussain, op. cit.

70 For a highly critical assessment of the early phases of the operation, see Stapleton, op. cit., 1997.

9

Iraq's Agrarian System:
Issues of Policy and Performance

Kamil Mahdi

1. Introduction

During the hundred or so years that led up to the events of 1958, Iraq experienced prolonged processes of sedentarisation, tribal disintegration, agricultural expansion, market incorporation, and land-title settlement. The outcome was a large but lacklustre market-oriented agricultural sector that was technologically backward and wasteful of scarce resources. The country's agrarian system was also highly iniquitous socially as well as unsustainable politically. This opened the way for a radical land-distribution programme, and subsequently led to a period of heavy state intervention, including a programme of state-enforced cooperation and even of collectivisation into the late 1970s. By the 1980s, and as a process of state compression was set in motion, these attempts at direct control of production began to give way to a policy of fostering capitalist farming. The state therefore found itself obliged substantially to disband the public agricultural sector, to withdraw from the provision of farming services, and to reduce the range of its policy intervention measures.

However, these are areas of policy that, politically, are heavily contested, and are thus subject to circumstances and to the pressures of interest groups. A coherent agrarian structure conducive to the application of a set of compatible and consistent policies has not been sought, and the agricultural sector has continued to be exposed by the government to its immediate political priorities in turbulent conditions. While it has abandoned its earlier etatist ideological drive, the government has not outlined clear parameters, nor has it developed and followed appropriate and flexible policies for the agricultural sector. Thus, the agrarian system continues to be unstable; it experiences drastic shifts that alter the ownership and landholding structures and that have a profound impact upon agricultural relations and upon production, employment and investment decisions.

This chapter charts the evolution of the agrarian system, beginning with the traditional system of large private estates that was established at the time of the accelerated demise of the tribal-communal system of agriculture. It highlights arguments made by the author in previous work concerning the performance of the traditional system, its utilisation of agricultural resources, its equity and its environmental stability, and then uses these earlier experiences as the background to an analysis of subsequent developments in Iraq's agrarian system and agricultural sector. The chapter's main argument is that the historic growth of Iraqi agriculture, which was based upon a phenomenal expansion of the cultivated area, was accompanied by a neglect of investment and of infrastructure development. Instead, the burden of adjustment of production activities and resource utilisation was often placed upon local communities – these operated in conjunction with large private landowners who were the prime beneficiaries of agricultural expansion.

Despite the ecological disruption that accompanied the implementation of agrarian reform, the agricultural economy recovered during the latter part of the 1960s. The chapter contrasts this recovery with subsequent stagnation and considers the impact of the agrarian system and agricultural policy on the performance of the agricultural sector. Policies and attempts to address the question of agricultural stagnation have usually involved radical shifts in the system of land tenure and in relations between the main social forces and policy institutions involved.

The unsettled agrarian issue offers both opportunities and dangers. With Iraq being at such a radical economic and social juncture, its economy must now face fundamentally different conditions from those it confronted prior to the two Gulf Wars, and the agricultural sector will have to play a crucial role in necessary adjustments. Furthermore, political circumstances now dictate a greater need for an independent capability for the domestic provision of food. The issue of food security has gained additional importance, given the USA's continuing policy of blockade and crippling sanctions against the civilian economy.

2. Background

A vigorous process of land settlement and expansion of cultivation took hold in Iraq during the second half of the nineteenth century. The 'regime of the tribes'[1] came under pressure from the combined forces of a rejuvenated central government and urban-based capital, both of which had been revitalised by the external shocks of military, transport and communications technologies. The growth of international trade provided an outlet for rising output and for surplus acquired by the new private landholders in the form of rent.

Settlement of private individual land title accelerated from around 1870, continued on a large scale into the 1950s, and it is still in progress to this day in project areas and in the controversial marshland reclamation. Issues of settlement of land title, the tenure conditions and the associated agricultural production relations were the subjects of field studies and analyses, and the iniquitous and socially divisive nature of the high concentration of agricultural land prior to the 1958 land reform was also noted. The social and political consequences of the agrarian system of the time, including the resurgence of widespread poverty, were widely remarked upon, but a few studies in addition considered the economic efficiency and growth aspects of the system of large estates and share-cropping farmers. Post-reform systems and policies have received less attention and critical appraisal, and they are composite and more complex structures.

The agrarian system and the expansion of cultivation prior to land reform

The system of individual private estates that had emerged after the break-up of tribal landholdings during the second half of the nineteenth and the first half of the twentieth centuries facilitated agricultural expansion and the settlement of what had formerly been grazing land. It had provided the social agents and the organisation necessary for that expansion by offering a framework for resource management and a minimal necessary protection against creeping soil salinity in the irrigated Mesopotamian plain. While it stabilised the delicate ecological conditions, the system was unable to offer any prospects other than low productivity agriculture and gradual decline as the process of expansion ran its course.

During the 1950s, when development expenditure was increased, the prevalent political economy dictated that projects would ignore the land system and address water provision and flood protection. Consultant blueprints and the programmes treated environmental and resource aspects solely as technical and engineering matters that were unrelated to economic institutions or to any social impact.[2] Hence, emphasis was placed upon resource development that reinforced the existing agrarian system and served the interests of the landlords. This was the rationale of the major water-storage and flood-control projects that were initially given emphasis alongside the continued neglect of irrigation and drainage networks. The main impact of these water projects was to postpone pressures for restructuring and for more intensive agriculture that would require investment in the land as well as technological and social changes.

High levels of investment in agricultural land were incompatible with the agrarian system prevailing in Iraq prior to the land reforms that began in 1958. In the centre and south of the country, the provision of water and the organisation of traditional irrigation methods were the critical functions of the system. Landowners held vast estates of irrigated land with water rights. Extensive cultivation with a biennial fallow system was the prevailing mode of land utilisation. Barley, the main winter crop, was originally produced for export under share-cropping arrangements that left the cultivators with low shares and an extremely poor living.[3]

In the northern rainfed zone, low and uncertain rainfall meant that in many areas the expansion of the 1950s was ecologically unstable, and in the long term unsustainable. In other parts of the rainfed zone, backwardness and geographical isolation were compounded by the land-tenure system, and this resulted in low productivity, low investment, insecurity and poverty.

Rigidities and constraints on the traditional system

The agrarian system rapidly transformed grazing land into land which supported the practice of cereal monoculture without the patterns of varied and mixed farming and the development of livestock, vegetable and fruit production. In effect, landlord and tenant competed over the degree of emphasis placed upon crop and livestock activities, with a resulting inefficiency in both. Similarly, industrial crops, legumes, fodder crops and other activities were discouraged by the predominance of large commercial operations which also hindered the farmers through a variety of social and institutional ties. Under the agrarian system, share-cropping was established with traditionally determined crop shares. The land-settlement process was associated with a range of technical and managerial constraints and cultural practices that also governed agrarian relations.

The tribal origin of the land system and traditional share tenancy remained pertinent, even though agriculture was commercialised and farm surpluses could supply a growing urban demand as well as international markets. Tribal affiliation was embodied in the Lazma Land Law of 1932. Legislation also restricted the mobility of sharecroppers who had been indebted to landlords.

In effect, land and labour markets were rigid and severely restricted. Land concentration in isolated rural environments also meant local monopolies of land and monopsonies in labour markets. The social/historical context and the legal framework of agriculture affected the nature of the relationships within agriculture. Wage labour and cash rents remained the exception rather than the rule in agricultural contracts, despite the fact that three out of four rural families owned no land at all, and despite a very heavy concentration of land ownership among those with land.[4]

Under the traditional system, indebtedness was widespread, and landlord–tenant relationships and landlord–merchant relationships were articulated with issues of rental contracts, seed and supply provisions, credit, and marketing of produce, all of which were interrelated. This was reflected in the Law of the Rights and Duties of Cultivators of 1933, which tied farmers to the landlord's estate, until the farmer was able to repay his debt.

The predominance of the institution of share tenancy had a profound impact upon the structure of Iraqi agricultural production, the organisation of farming activities, and the pattern of land use. A subsistence-food-crop production monoculture extended rapidly over large areas, even as a marketable surplus, destined in substantial proportions to international markets, was growing. Early efforts to establish plantation agriculture to produce cash crops, such as cotton, for export were not successful. Later production of cash crops was usually undertaken by middle farmers or enterprising agricultural capitalists, usually as part of a rotation which included food crops as wage goods. The production of cash crops, unlike that of main staple foods, was highly sensitive to price and was occasionally speculative and volatile, as occurred with the production of cotton in the early 1950s. Without the development of agricultural input supply and extension services, and without developing a broadly based and varied agricultural sector, the trend towards land concentration could not be changed.

In effect, a highly polarised income and wealth distribution, low average incomes, and dependence on a few staple crops, constrained the development of product and input markets and restricted responses to market stimuli in Iraq's traditional landlord system.

The inflexibility of the system and the predominance of landlord power in the countryside had the effect of accelerating the decline that had set in after the period of expansion; the latter had occurred during the 1940s in the irrigation zone and in the latter part of the 1950s in the rainfed zone. The limits to expansion that were reached were due to salinisation and seasonal water constraints in the irrigation zone, and to expansion into marginal land in the rainfed zone. They coincided with a new era of sharply rising oil income leading to inflationary pressures, real currency appreciation, and decline in the competitiveness of internationally tradable staple crops. Thereafter, barley exports declined and subsequently ceased, and wheat and rice imports grew rapidly.

This was the nature of the trend prior to land reform in 1958, but in discussions of the impact of land reform it has usually been ignored.[5]

Technology and the environment under the traditional system

In considering the performance of the traditional agrarian system of Iraq, it is important to mention two further characteristics. First, there was the system's ability to attract and absorb significant and lumpy investment and modern technological goods introduced by the private sector, such as irrigation pumps (from the 1920s) and tractors (from the 1940s). This frequently involved urban-based entrepreneurs who, in collaboration with their rural counterparts, contributed to the production process within the framework of the existing agrarian system which was able to accommodate both the technology and the social agents. It was thus possible to achieve a more rapid expansion of cultivation, to generate a marketable surplus, and to extend the modified agrarian system, including new agents and technologies, into the ever-wider areas permitted by resource availabilities. In almost all these cases, the nature of the investment was to break particular constraints on the expansion of cultivation. Only in the 1950s did factor substitution and some new cultural practices become relevant.

The second factor to be noted here concerns the manner in which the traditional system was able to address ecological problems, particularly on irrigated land in the centre and south of the country. Prior to land redistribution under the reforms, small individual landholdings in those irrigated parts were largely confined to older agricultural areas along the banks of the rivers, and were frequently served by gravity irrigation. Other exceptions were the few government project areas, but most agricultural land was held in large holdings, frequently by former tribal shaikhs with a continuing relationship to tribal social organisation.

In other words, although individual land title was settled, and agricultural activities were commercial and frequently dependent upon urban-based entrepreneurs, important elements of tribal organisation remained. The most significant of these was the persistence of customary rights of tribesmen to be tenants, and the rotation of the plots to be cultivated and strips to be assigned to tenants, who tended to have a share of the land rather than rights to a specific plot.

The organisation of production was left to the landlord's agent, the *sirkal*, who tended to lead members of a tribal lineage in agricultural activities. In this way, the traditional agrarian system guaranteed the operation of an effective biennial cereal–fallow rotation that was essential for extensive cultivation over large agricultural areas. This rotation made it possible to control the effects of high soil salinity in the short- and medium-terms without the provision of modern drainage.

The system operated by allowing the water table to fall during the fallow year, and for downward leaching of the topsoil to take effect in readiness for cultivation the following year. Whereas this process was accentuating salinisation in the longer term, and whereas it constrained any improvement in land yields, cultivation would otherwise have been restricted to areas of better natural drainage. Any other alternative would have required the installation of costly modern irrigation and drainage systems and extensive operations in land reclamation. The maintenance of the traditional system made it possible to expand the cultivated area of the irrigation zone from a mere 1.5 million *donums* in 1918 to 6.5 million *donums* in 1942.[6] By 1958, the Agricultural and Livestock Census was reporting a probably somewhat exaggerated figure of 13.5 million *donums*.[7]

The intensity of cultivation and land productivity was very low, and the entire growth in output was due to a rapid expansion in the cultivated area associated with a rapid growth of the rural agricultural population. The investment in, and the introduction of, modern irrigation pump technology in the centre and south of the country was land augmentation of a kind that permitted the expansion of cultivation rather than of land yield. Initially, any effect of increasing intensity of cultivation was also precluded by the apparently rapidly decreasing returns due to soil salinity. Further stages of technological innovation that would have resulted in higher intensity and greater yields did not take place under the traditional agrarian system. The system itself militated against this. Landowners with extensive holdings, who were heavily reliant upon cheap labour and traditional social organisation, had no incentive to carry out investments that might have had unfavourable consequences for factor pricing and social organisation.

As noted above, to the extent that they had spread in rural Iraq, entrepreneurship and technical innovation were not a threat to the agrarian system but an aid to its expansion. This was also true of the northern part of the country, particularly in the north-western Jazira plain where cultivation expanded into areas of marginal rainfall towards the 200mm *isohyet*. Given the variability in quantity and the incidence of rainfall, the effect, in many cases, was to turn over good pastureland to a highly risky cereal crop. Low and variable yields were again prevalent, and the expansion was only possible with the introduction of tractors for timeliness of operations, albeit with serious consequences for soil fertility.

Hence, the most rapid expansion of cultivation in these northern and north-western rainfed areas took place during the 1950s, with the widespread introduction of tractors into the region,[8] and concurrently with developments that were taking place in similar environmental and social conditions across the border in Syria.[9]

The biennial cereal–fallow system practised in the north-western Jazira area had a different purpose, which was to replenish nutrients and combat weeds, pests and diseases. All of these functions could feasibly be performed in the areas of more secure rainfall by means of more appropriate crop rotations, better selection of varieties, improved land preparation, cultural practices, and by the use of chemical inputs.

In many, though not in all cases, such activities are scale-neutral, and machinery, unlike irrigation, is not location-specific. The roles of markets and of the entrepreneur were therefore very important for outcomes in the rainfed zone, and the agrarian system served no purpose in maintaining an ecological balance. In any case, even without appropriate practices, a more intensive crop rotation in the rainfed zone would tend to have led to a gradual erosion of the land's productive capacity, rather than to the more rapid detrimental effect of salinisation in the irrigation zone. Absentee landlords had little incentive to follow better management practices, especially where they took a share of gross output without contributing to costs.

Applied in both the irrigation and the rainfed plains of Iraq, the same traditional system had diverging consequences upon the interrelationships between the different agents of the production process; that is, landlords, tenants, entrepreneurs, and the government. The ecological merits in the two areas were also different. In the irrigation zone, the traditional system offered medium-term ecological respite, while in the rainfed zone, it had no such function. As a result, the collapse of the traditional agrarian system had more wide-ranging consequences upon agriculture and land utilisation in the irrigated areas than it did in the main rainfed areas.

The relevance of this past experience to the present situation in Iraq will be considered below, but we note here that the agricultural policy framework, under which the traditional system operated, was compatible with the institutional environment described above. This applies to all the three major aspects of policy at that time: settlement of land title, expansion of irrigation, and flood prevention and water storage. Other agricultural policies and the provision of production services were of minor significance. The government was also reducing the burden of agricultural taxes even before oil revenues began to be a major source of revenue. As a result, sufficient resources for an active agricultural policy, including provision of a range of services, were not available until the 1950s.[10]

3. The agrarian system after the reforms

Not only did the agrarian reform of 1958 bring about a redistribution of ownership of land and of income from agriculture, but it also led to a radical transformation in many aspects of the agrarian system. In this chapter, we are concerned with the implications for policy and for performance, and we would want also to consider what broad conclusions might be drawn for current conditions and future policy.

The main policy of expropriation of the large estates and redistribution of the land to former tenants was implemented over two decades until the late 1970s, during which time the government's strategy, commitment and policy on agrarian issues were all erratic. This was also a period during which the economy and the society underwent radical transformations, the most pertinent aspects of which were the relative decline of agriculture and of rural activities, dependence upon oil for foreign exchange and government revenues, and the emergence of a substantial deficit in agricultural products. All those aspects have been discussed in various writings.[11]

Here, we will consider the radical changes in the agrarian system in response both to policy measures and to economic and socio-political changes.

Land reform has involved a major transfer of landholding, initially to the state, and later to reform beneficiaries. Between 1958 and 1977, the total of confiscated private land and requisitioned state land designated as *Miri Sirf* (pure state domain) was about 15 million *donums*. This amounted to 47 per cent of all land reportedly held in agricultural holdings in 1958/9, including state land utilised without settled title.

By 1971, some 13 to 14 million *donums* were subject to tenure reform, of which only 9 million *donums* were reported to be still in agricultural holdings through having been either distributed or leased by the reform authorities. Those 9 million *donums* comprised just under 40 per cent of the total land in agricultural holdings reported in the 1971 Agricultural Census.[12]

Despite considerable government investment in irrigation and reclamation of reform land, the decline in the reported agricultural area between 1958/9 and 1971 had been more rapid on agrarian reform land than in the rest of the agricultural area. A number of factors related to the reform process itself may help to explain this. First, the process was prolonged and uncertain, thereby contributing to neglect. Some landlords withdrew machinery and services and the state was unable to compensate for those services. Second, the break-up of the estates was not carried out with a view to a proper management of all the land affected. Furthermore, the quality of reform land was lower than average, since landlords had been permitted to retain the areas of their choice. Some irrigated land became uncultivable due to the disruption of the traditional irrigation-management system and the failure to institute an alternative.

Problems and performance of a peasant economy

From the first decade of the reform, it was evident that agrarian transformation towards a peasant economy was more an aspiration than a reality. Some of the poorer reform land was abandoned under both ecological and economic pressures. A total of 335,000 farmers were reported to have received land from the agrarian reform authorities by 1971, but in that year's Agricultural Census, only 231,000 were reported to have retained their holdings. It is possible that some may have consolidated their holdings into combined plots rather than abandoned the land altogether, but, as noted above, the total areas concerned confirm the decline of agriculture on much of the reform land.

During that period, the reform authorities adhered to the ideal of individual family farm ownership, but those same authorities deferred to the reality of communal land use in the irrigation zone. Prior to distribution, land was leased jointly to groups of peasants, in order both to simplify procedures and to make it possible to carry out joint organisation of production.

Nevertheless, the disturbance associated with the reform, coupled with the growth of employment in the urban sectors, attracted labour away from agriculture. Substitution of capital for labour proceeded rapidly under conditions of a rentier economy. Labour-intensive tillage, harvesting and threshing were mechanised, thereby easing peak-period labour shortages. Other practices of sowing, crop care and irrigation changed more slowly, and low levels of productivity continued to be prevalent.

While the cultivated area was contracting, a simultaneous process of intensification of cultivation was under way in the irrigation zone, where cropping intensity rose from under 50 per cent to at least 68 per cent between the early 1950s and 1971.[13] This process raised the water table, and precluded downward leaching of salts accumulated in the topsoil. Without a basin-wide drainage system, it would appear that the breakdown of the traditional system and the abandonment of irrigated agricultural land had, in turn, made it possible to raise cropping intensity by leaching salts on to neighbouring land.

Agricultural output recovered from the disruption of the early reform, and, by the end of the 1960s, agricultural value added stood twenty per cent above the pre-reform peak. With terms of trade having also moved in favour of agriculture, per capita agricultural income had risen by 50 per cent in a decade,[14] despite environmental degradation.[15] Although the growth of agricultural output had been broad-based in its commodity composition, there was no general rise in productivity, and output growth tended to rely upon heavy utilisation of the most scarce agricultural resources, namely non-saline agricultural land and summer water supplies. Those resources were in particular demand for the growth of fruit and vegetable production.

The change in the agrarian system stimulated the productive potential of agricultural resources, but it also accelerated environmental degradation. Agricultural potential was invigorated by removing restrictions upon the utilisation of resources, by making agriculture more responsive to market stimuli and incentives, by releasing more labour and capital resources into agricultural production, and by exploiting existing agricultural resources more fully. The reform also did away with restrictions upon the movement of labour out of farming estates. Vegetables and livestock production increased as landlord controls were relaxed.

There was in addition a substantial growth of private investment in agriculture. After a two-year decline during 1958 to 1960, the percentage share of agriculture in private Gross Fixed Capital Formation recovered. The relatively high levels of the second half of the 1960s need to be viewed against the background of the relative decline in the share of agriculture in non-oil GDP from 35.5 per cent in 1957 to 24.6 per cent in 1970. Private agricultural investment was largely channelled into machinery purchases, despite growing competition from the public sector in this area. It is also notable that mechanisation increased more rapidly in the irrigated areas where tractor ploughing came to replace animal draft. By the late 1960s, 85 per cent of all cultivated land in Iraq was ploughed using tractors, and 90 per cent of the winter-crop area was harvested by combine.[16] This is in stark contrast to the situation of a decade earlier.

An important aspect of post-reform mechanisation is the fact that it had been undertaken by a large number of entrepreneurs. The number of tractors reported in 1971 totalled 8,752, and these were owned in 8,103 holdings (1.08 machines/holding). The pattern was similar for combines and for irrigation pumps. Machinery ownership had become much more widespread than earlier. From among machinery owners, there emerged a class of entrepreneurs having a small base in agriculture, but mainly providing machinery services with low overhead costs.[17]

Between the constraints of the traditional system and the heavy state controls of the 1970s, the decade of the 1960s was one of relative laissez-faire in the agricultural sector, notwithstanding the implementation of land reform and the growing intervention elsewhere in the economy. There was uncertainty about the direction of policy, and a shift of state-sector investment away from agriculture. The growth of agricultural extension services, and more broadly of an infrastructure of state and private production services, was significant for the transformation of agriculture from traditional systems of tenure and relations. Policy misconception and administrative failure may have contributed to the apparent flight of over 100,000 land-reform beneficiaries from land assigned to them either on temporary lease or permanent title.

Nevertheless, the number of agricultural landholdings increased during the post-reform period to 500,000 by 1971 and to over 800,000 by 1979, in a total rural population of 4 million. Iraq's new agrarian system did not have a large reserve army of landless labourers like the Egyptian *tarahil* workers,[18] but instead acquired a large number of small absentee landowners, a fragmented land-tenure situation, and a propensity to shed both labour and land.

The break-up of the large estates was a direct, as well as an indirect, result of the reform, and well over 100,000 new landholders were created by means other than land redistribution during the reform's first decade. Inheritance alone was probably not responsible for the additional expansion in ownership, but it is likely that the break-up of the estates continued in response to economic pressures and social change, as well in expectation of more radical future reform measures. Many owners might have come to regard the parts of landlords' estates that had not been relinquished in the break-up of large holdings more as a reserve asset, and less as a source of authority and of monopoly power as they had been earlier. In this way, the role of land as a productive resource continued to be uncomfortably combined with its other roles.

Iraq's agricultural sector was at the same time transformed in other ways. Farmers were no longer constrained in allocating their time and resources between livestock-raising and cultivation. Value added in livestock products increased after the reform, in part as a response to the vigour of peasant mixed farming. This feature did not continue in later periods, and components of the production of the livestock sector other than poultry and intensive dairy were already stagnant or in decline by the 1970s. Natural pasture, being sparse and spread over wide open ranges, placed feed production in competition with the production of other agricultural commodities.

The general stagnation and subsequent decline of agriculture in the 1970s brought with it a reversal in the trend of rising on-farm livestock output. The index number of the output of beef and lamb declined from an average of 115 for 1973 and 1974 to 91 in 1981 and 1982. By 1988–9, it had fallen further to 75, taking 1974–6 as base.[19]

A slightly rising trend in overall livestock output over the decades of the 1970s and 1980s was due entirely to rising production of poultry, eggs and, to a lesser extent, dairy and fish farming products. All these activities had the advantage of high state investment, and concessionary finance and subsidies, at least initially. Subsidies were also implicit in the pricing of imported feed, vaccines and equipment, so long as these activities had enjoyed an official foreign-exchange allocation. In contrast, other products, especially cereals, were constrained by official pricing based upon comparisons with competitive imports, prices of which were compared on the basis of the overvalued official exchange rate.[20]

The reform generated a measure of economic dynamism in the agricultural sector. It broadened asset ownership, increased labour mobility, varied the output mix and encouraged private investment. Above all, it began a process of modernising the agricultural sector and of integrating it into the rest of the economy.

The relative stagnation of agriculture meant that the sector had developed its backward linkages, particularly with imported goods, but not its forward linkages into industry and exports to any great degree. As was the case with other non-oil sectors, Iraq's agriculture was the recipient of considerable financial investments generated in the oil sector. In one crucial respect, the reform was slow in bringing about necessary changes. It did not succeed in reversing the trend of falling land productivity, and failed to act as an effective catalyst to the development of an integrated system of policy and implementation, research, extension, production and distribution.

4. The agrarian system in the 1970s and the 1980s

We have argued elsewhere[21] that the policies of the 1970s had all operated with negative consequences for agriculture, especially in their emphasis upon the state sector and upon directives and detailed state intervention in production and pricing. In its effort to strengthen its authority in rural areas, the state weakened the thriving peasant economy of the 1960s, and set out to impose direct bureaucratic institutions. This experience to some degree repeated (in mirror-image) the earlier attempt to force individual peasant holdings upon what was essentially a communal system.

The new policy offered heavy subsidies to both state-sector and capital-intensive activities, while at the same time constraining other elements of the agricultural economy. Rural incomes were kept down by agricultural price controls. For a time, farmers found that the value of chaff exceeded their return on grain, and it was acknowledged that official wheat pricing gave rise to serious inefficiencies during the 1970s.[22]

The policies of the 1970s reinforced the pressures of a rentier economy. Incomes and prices rose rapidly in other sectors, and agriculture was placed at a disadvantage. In the 1960s the vigour of the agricultural sector had been due partly to its ability to mobilise family labour for a range of activities on the farm. The transformed labour market that followed the increases in oil revenues precipitated a decline in farm employment, and this continued into the 1980s under the pressure of military conscription.

The failure of the experiment in state control led to a further reversal in policy, this time in favour of privatisation and private capital-intensive enterprises. Based on the provisions of Law No. 35 for 1983, the government leased 6 million *donums* of agricultural land to the private capitalist sector. This was a radical break with agrarian reform principles, since the lessees were entrepreneurs rather than farming families.

Most of this land was in the irrigation zone, and there appears to have been a very substantial reversal of the reform in these areas. In 1979, there were about 4.5 million *donums* of reform land in the centre and south of the country.[23] Furthermore, 4 million out of the 6 million *donums* rented since 1983 to new enterprises and capitalist farmers are also situated in the same area. If, as is likely, most newly developed agricultural project land has been upgraded from pre-existing farmland, there would seem to have been an almost total reversal of policy in this part of the country.

5. Some conclusions

The present agrarian structure has evolved from the conflux of a number of processes and from previous policy interventions. Market forces, elite behaviour, political events, inheritance laws and ecological factors have all contributed to the development of a complex tenure structure encompassing forms of ownership and landholding that have their origins in specific historical circumstances.

There has been no appropriate settling of rights to land and to water in order to protect agricultural resources and their use in socially optimal combinations. Therefore agricultural land has continued to be owned and managed in uneconomic units that are frequently badly served by the necessary infrastructure. Rational private decisions tend therefore to result in undesirable social outcomes. Political instability and economic dislocation, together with the failure successfully to manage the environment, have accentuated the conditions of uncertainty. Since agricultural policy is dominated by political decisions and imperatives, agricultural and rural institutions are increasingly characterised by a combination of crony capitalism and crony tribalism. At the same time, fragmentation of agricultural units and absentee ownership remain common. Yet the policy formulation capability of government institutions is now severely damaged, and this paralysis of policymaking and the loss of professional skills is one of the most potent effects of economic sanctions.

The agrarian policy adopted during the 1980s was formulated during the Iraq–Iran war and at a time of labour shortages, and there was an implicit assumption of continued growth in the reliance upon mechanical and chemical inputs. Iraq's present circumstances are radically different, and the country's financial position has now been transformed. Policies appropriate to the decades of the oil boom when farmers abandoned agriculture in their hundreds of thousands should now be reviewed.

Employment generation now has to be one of the main criteria of an agricultural strategy, and the flexibility of labour supply and the social safety-net provision of family farms is now a great advantage. The land-lease policies adopted during the 1980s relied upon a business elite that was not only close to the centres of political power, but that was also assumed to be able to invest heavily in capital equipment and industrial inputs in order to raise agricultural productivity. Under current and foreseeable circumstances, more labour-intensive solutions are called for, on social as well as on economic grounds. The current land-tenure policy may also be one which, since land disputes abound, discourages resettlement of recent migrants back on to the land. It also accentuates the polarisation of income and wealth that is a stark and disturbing feature of present-day Iraq.

Outside the large private estates, the government has not developed any new tenure policy, but has reverted to reliance upon tribal networks for the resolution of local disputes, for its product procurement policy, and for the exercise of power. The impact on local social and economic conditions is not yet clear, especially given the widespread disruption of the 1990s.

As was noted earlier, management of irrigated land in larger blocks is appropriate in Iraqi conditions. However, size of holding can by no means guarantee ecologically sound management. There is a need for investment in irrigation and drainage and proper management of the land. Under an apparently indefinite siege of the country, Iraq's public sector is disintegrating, and the irrigation and agricultural services are not excluded from this process. There is evidence that the severely constrained resources and the diminishing water supplies of recent years are being concentrated in the regions north of Baghdad that provide the government and the bureaucratic-military elite with maximum short-term political advantage.[24]

The policies of leasing agrarian reform land and of regenerating tribal power structures cannot be substitutes for a more general strategy to cover the entire sector. Different parts of the agrarian system cannot operate independently, and it is essential that land and water management policies are addressed for the country as a whole, rather than as a piecemeal response to localised problems. The paralysis of policy under sanctions has led to emphasis being placed upon immediate problems and localised solutions. This has been working to the short-term advantage of politically influential groups that are able to steer limited government efforts towards particular sectional interests in conditions of degenerating physical and administrative infrastructure and of breakdown of regulatory and policy institutions.

The main tool of agricultural policy used in the 1990s has been that of government procurement of staple crops at relatively high prices. Although these prices were less than competitive at international levels (see M. Ahmad in this volume), given the collapse of the value of the Iraqi dinar, their relative valuation in terms of local wage rates has increased dramatically (cf. Gazdar and Hussain in this volume). The use of this policy tool was an essential emergency measure in the face of a famine-inducing UN blockade, and it was partially successful in the early 1990s. However, it has also been inequitable, both within agriculture and between agriculture and other sectors. After the food-for-oil deal, compulsory procurement has been dropped and market prices of grains are now subject to strong competition from imports. A more comprehensive and consistent sector-wide policy is a growing necessity, and it cannot wait indefinitely for an end to sanctions. An agrarian system that addresses the questions of the development of efficient production units and effective utilisation of agricultural resources is an essential part of the design of an agricultural strategy.

NOTES

1 The term is due to S. Haidar, 'Land Problems of Iraq', PhD thesis, University of London, 1942.
2 Cf. Knappen-Tippetts-Abbett-McCarthy, *Report on the Development of the Tigris-Euphrates River Systems*, 1954; abridged version of the company's 1952 report, Baghdad.
3 See Food and Agriculture Organisation (FAO), *Mediterranean Development Project, IRAQ: Country Report*, Rome, 1959 (FAO 59/8/6039).
4 It should be noted that agricultural censuses such as those of 1952-3 and 1958-9 did not report land *ownership* but 'holdings'. There are no sample or other studies to relate *ownership* to *holding* information. It is, however, important to point out that the 1958-9 census was conducted within the context of preparation for land reform, and respondents with large *holdings* would have had no incentive to exaggerate their *ownership*.
5 See Edith Penrose and E. F. Penrose, *Iraq: International Relations and National Development*, London: Croom Helm, 1978; Rony Gabbay, *Communism and Agrarian Reform in Iraq*, London: Croom Helm, 1978.
6 A. Sousa, *Iraq Irrigation Handbook, Part I*, Baghdad, Iraqi State Railways Press. 1944.
7 Central Statistical Organisation, *Nata'ij al-ihsa' al-zira'i wa al-hayawani li-sanat 1958/59* [Results of the Agricultural and Livestock Census 1958/9], Baghdad, 1961.
8 K. A. Mahdi, 'Agricultural Labor and Technological Change in Iraq' in Dennis Tully, (ed.), *Labor and Rainfed Agriculture in West Asia and North Africa*, Dordrecht, Kluwer, 1990.

9 Raymond Hinnebusch, *Peasants and Bureaucracy in Ba'thist Syria: The Political Economy of Rural Development*, Boulder, Westview Press, 1989.

10 Kamil A. Mahdi, *State and Agriculture in Iraq: Modern Development, Stagnation and the Impact of Oil*, Reading: Ithaca Press, 2000.

11 For example, Penrose and Penrose, *Iraq . . .*, 1978, op. cit.; Mahdi, *State and Agriculture . . ., ibid.*

12 Central Statistical Organisation (CSO), *Results of 1971 Census of Agriculture*, Parts 1 and 2, December, Baghdad, 1973.

13 Mahdi, *State and Agriculture . . ., ibid.*

14 The land reform also brought about a less inequitable income distribution, and reduced the share of agricultural income going to urban-based landowners.

15 Mahdi, *ibid.*

16 FAO/UNDP, Institute of Co-operation and Agricultural Extension, Abu-Ghraib, Iraq, *Study of Selected Agrarian Reform Programmes*, Report to the Government of Iraq based on the work by S.A. El-Shishtawi *et al.*, Technical Report No. 1, Rome, 1970 (Reference: ESR: SF/IRQ 8).

17 See Awshalim L. Khammo, 'The Role of Mechanisation in the Development of Agriculture in Iraq', PhD thesis, School of Economic Studies, University of Leeds, 1977.

18 Cf. M. Abdel-Fadil, *Development, Income Distribution and Social Change in Rural Egypt (1952-1970)*, University of Cambridge, Department of Applied Economics, Occasional Paper 45, 1976.

19 This data is derived from various issues of the *Annual Abstract of Statistics* (AAS), published in Baghdad by the Central Statistical Organisation (CSO).

20 No distinction is being made here between the period of balance of payments surpluses until 1980, and the subsequent period of constrained foreign exchange resources.

21 K. A. Mahdi, 'The Political Economy of Iraq's Agriculture' in Eric Watkins, ed., *The Middle Eastern Environment*, Cambridge: St Malo, 1995.

22 See ESCWA/FAO, *Tatweer nudhum al-takhteet al-zira'i fi al-jumhuriyya al-iraqiyya* [Development of Agricultural Planning Systems in the Iraqi Republic], United Nations Economic and Social Commission for Western Asia, 1992 (E/ESCWA/AGR/1992/4).

23 Central Statistical Organisation (CSO), *Annual Abstract of Statistics*, 1990, Table 3/28, p. 145.

24 ESCWA/FAO, *Tatweer nudhum al-takhteet . . .*, 1992, op. cit.

PART IV

REFLECTIONS AND FUTURE PROSPECTS

10

Long-term Consequences
of War and Sanctions

Abbas Alnasrawi

Any attempt to reflect on the future course of the economy of Iraq must, if one is to do justice to the task at hand, address a very large number of variables, both external and internal. However, the following contribution restricts itself to consideration of three variables – namely, foreign debt, oil, and the sanctions regime.

1. External debt

A developing country should at all costs avoid external debt, since debt is a trap from which it will be very difficult to escape. While it may be easy to obtain loans in the world's financial centres, it is not easy at all to generate the necessary balance of trade surplus to service the debt. The history of the past 50 or so years tells us that very few countries have managed to extricate themselves from the debt trap.

A debt-free developing country is a fortunate developing country. Until the early 1980s, Iraq was more fortunate than most in that not only was it debt-free but it also had considerable foreign reserves, estimated at between US$35 and US$40 billion. To be sure, Iraq was carrying a small debt of US$2.5 billion created during the course of trade, but relative to its foreign holdings this was relatively insignificant. In 1980 Iraq's debt was less than 5 per cent of its GDP, slightly more than 9 per cent of its exports and only US$189 per capita. Eight years later, however, Iraq's foreign reserves were exhausted and its foreign debt had climbed to US$82 billion.[1] The new debt constituted 143 per cent of the country's GDP and 607 per cent of its exports. On a per capita basis the debt had jumped to US$4,660 per person. Nor is the story over yet. Although the debt had increased by less than US$12 billion, to US$94 billion by 1995, the collapse of the economy and the embargo increased the ratio of debt to GDP to 558 per cent and of debt to exports to 10,400 per cent – debt indicators the like of which have never obtained anywhere else.

Naturally, the question is why did the debt soar so high? The answer can be found, as is well known, in Iraq's economic and foreign policies – in other words, in the failure of development policies to increase domestic output, in the unprecedented reliance on oil, in the militarisation of the economy and the expanding size of the armed forces, in the war with Iran and the destructive impact of the Gulf War, and in the prolonged siege of the people and the economy that has remained in effect since 1990.

What about the future? How will this problem be managed? The difficulty is that Iraq's huge foreign debt is only a fraction of the country's external obligations. The managers of the UN Compensation Fund have said that more than US$320 billion of claims have been filed against Iraq, but a proportion of these will be rejected The UN has also said that the cost of the damage to Iran's assets was close to US$100 billion. In short, Iraq could have up to US$400 billion worth of claims that it will have somehow to face in the future.

Here, however, we will focus only on the external debt. A combination of the relevant literature and of experience indicates that there are many mechanisms to manage the debt, including rescheduling, refinancing, new financing, debt–equity swap, and debt reduction. The problem with these mechanisms is that, even without new debts, Iraq's existing burden is high and growing (because of the growing interest on the debt), and that lenders are going to place their funds in more viable economies.

One can always wish, of course, that the international community would come to the rescue of Iraq. If the mother of all coalitions could demolish the infrastructure of an entire economy in six weeks, it can most certainly restore the Iraqi economy to some normalcy in a matter of a few years. However, since one must be realistic in these matters, we have to look to other remedies.

There are really two unorthodox or drastic ways to deal with the problem. First, Iraq should attempt to convince its OPEC partners, especially Saudi Arabia, that it should be given a much higher quota to compensate for the oil revenues that it has lost since 1990. Secondly, Iraq could technically default and refuse to pay its debt well into the future.

2. Oil in the future of Iraq

For several reasons, it is of course to be hoped that the role of oil in the Iraqi economy in the future will be far better than its role has been in the past. First, oil – or to be more precise, the pattern of the utilisation of oil revenue – over the last five decades has been nothing but a calamity for the people of Iraq. Successive governments and policymakers failed, or indeed *had* to fail, to use oil to develop the country's agricultural and industrial potential. Rather, oil represented a source of easy money for rulers to realise their ill-conceived schemes.

No less important is the fact that oil enabled the government to corrupt the old principle of 'no taxation without representation' into a new principle of 'no representation without taxation'. Since the oil money gave the government all the revenue it needed, it refrained from imposing taxes and simultaneously denied the people the right to participate and the right to criticise. How else can one explain the fact that the present regime managed in one decade (the 1980s) to squander more treasure than all the oil revenue that Iraq had received in its entire petroleum history?

Third, the heavy reliance on oil as a source of income and foreign exchange exposed the fragility of the whole edifice of the Iraqi economy. No sooner had the revenue declined than the economy plunged in a state of crisis and underdevelopment. One can only look at per capita GDP – which (measured in constant 1980 prices) rose from US$650 in 1950 to US$4,200 in 1979, and then collapsed to US$500 in 1993 – and reflect on the human tragedy behind those cold numbers. What makes the tragedy so severe is the fact that Iraq relied on foreign suppliers for 70 per cent of its population's consumption of major food items. Keeping in mind the principle of the relationship between taxation and representation, one has to conclude that had the people been given the opportunity to determine their destiny, much of the tragic state of affairs that now prevails in Iraq could have been avoided.

What about the future? The writer's view is that there needs to be a whole new formulation of the relationship between oil on the one hand, and the people and their economy on the other. As noted above, it is not really necessary to be an economist to conclude that the policies of the past have been bankrupt, dangerous and catastrophic. Certainly many people, especially the experts, will maintain that Iraq has the second largest oil-reserves endowment after Saudi Arabia (some say even larger than Saudi Arabia), and that Iraq should develop its oil as speedily as it can. They will also claim that desperate conditions call for drastic measures, and that Iraq will benefit immensely from selling its oil reserves to foreign oil companies in order for it to pay its foreign debts.

The above arguments are good as far as they go, but they do not go particularly far, or even far enough. More specifically, they do not answer the fundamental question regarding the role of oil in the future of the Iraqi people and their economy. In other words, and in the writer's opinion, these arguments will have the unwelcome effect of reproducing the same pattern of relationships that has prevailed in Iraq since the emergence of oil as an important sector of the economy.

What then might one advocate? Very briefly one might propose a return to the basics – to the original planning and development objectives, which are to raise the country's sustainable non-oil national output. If this were to become the national goal again, then oil would become an instrument of economic policy, to be used to raise national output rather than to displace it as the primary source of income and employment. An important benefit arising from this policy would, I believe, be the adoption of the principle of taxation with representation.

To be sure, Iraq's GDP will grow at lower rates under this policy, and it will therefore take the country longer to restore the earlier GDP levels. The question as to whether this will cause further problems is really very simple and can be answered by a further question: how much of that oil rent was actually spent to develop the country's human and natural resources and how much of that wealth was squandered on wasteful and suicidal projects? In other words, the high GDP figures of the 1970s and early 1980s were deceptive: they did not reflect high living standards because most of that GDP was diverted to expand the military and security structures.

3. Economic sanctions

When Iraq invaded Kuwait in 1990, the UN Security Council (UNSC) passed Resolution 661 imposing sanctions that were, in the words of the resolution, intended to bring the invasion and occupation by Iraq to an end. Some ten years later, and after well over a million Iraqi deaths, the sanctions are still in place, performing their grisly task of what has been described as silent slaughter. As long as the sanctions are effective, there is obviously little the people of Iraq can do. In addition to death, the sanctions regime has led to poverty, underdevelopment, stunted growth, social disintegration and the unprecedented emigration of huge numbers of professionals and skilled workers.

Legal studies have shown that what the Security Council has been doing to the people of Iraq is against international law, as evidenced in the articles of such documents as the Geneva Convention, the Convention on the Rights of the Child and other agreements and declarations.

It is true, of course, that the government of Iraq has exhibited callous indifference to the plight of its people, but this indifference cannot justify the UNSC's own violation of the human rights of an entire population. In other words, the Iraqi government's disregard of the human rights of its own people, or its failure to comply with UNSC resolutions, should not give the UNSC a licence to abrogate its own independent obligations to respect the human rights of the Iraqi people. In effect, the UNSC has given itself the liberty to impose collective punishment on an entire population simply for the decisions of its leader.

It is important to keep in mind that, in the case of Iraq, the destructive impact of the sanctions regime has been intensified because of the war damage to the infrastructure, and that it has disproportionately affected children. As is well known, hundreds of thousands of children have died because of the sanctions. Such a tragic loss of life constitutes an outright violation of the most fundamental human right – the right to life, the supreme right from which no derogation is permitted even in time of public emergency.

4. Concluding remarks

The people of Iraq have been going through the agonies of a humanitarian emergency of historical proportions which has resulted from a series of man-made decisions.

After having enjoyed nearly three decades of growth, the economy was subjected to the impact and the consequences of the Iraq–Iran war of 1980–8. This was followed by the effects and the consequences of the invasion of Kuwait, which were symbolised by the nearly total blockade of Iraq's economic interaction with the world economy. The Gulf War superimposed another series of destructive and devastating circumstances on the country, which added to, intensified and amplified the impact of the sanctions.

Contrary to presumption and expectations, the sanctions that had been imposed in August 1990 were retained by the UNSC and are still in effect, long after the aims of the sanctions and the war were secured. It is difficult to see when and how the sanctions will be lifted. In the meantime underdevelopment, starvation and death will plague Iraq and its people for some time to come.

As a concluding observation, one may note that when a new Iraq does begin to emerge, all political, economic and social paradigms, assumptions and policies must be examined: the aim must be to retain only what will be good for the welfare of the people, and to discard the rest. Otherwise history will be unmerciful in its judgement.

NOTE

1 Including an estimated US$40 billion extended by the Arab Gulf states to Iraq during the Iran–Iraq war.

11

The Iraqi Economy: Some Thoughts on a Recovery and Growth Programme

Sinan Al-Shabibi[1]

1. Introduction

Among its neighbours and within the Arab world, Iraq is distinguished by its endowments in both natural and human resources. Not only does it possess the second largest petroleum reserves in the world, as well as enormous agricultural potential, but it also has educated human resources. Needless to say, such potential could have turned Iraq into a relatively advanced country by now. Instead, a mixture of internal conflicts, wars, international sanctions and inappropriate development strategies and economic policies has prevented the Iraqis from achieving the standards that their country's resources would have permitted, had those resources been properly and effectively utilised. While the potential certainly remains, its present and future utilisation will, in addition to political and social stability, require considerable financial resources comparable at least to the level that was available in the 1970s. However, it is highly unlikely that the experience of the 1970s – facilitated at the time by the huge jump in oil prices – will be repeated.

This chapter's main purpose is to examine the prospects for the Iraqi economy, but it will also look at why developments in the 1970s and 1980s failed to capture the opportunities offered by the financial resources then available, and will consider the consequences of that failure.

2. The Iraqi economy: sequence of collapse

Developments in the 1970s: brief review and evaluation

Like the rest of the oil exporters, Iraq benefited from the jump in oil prices in 1973–4. Oil revenues almost quadrupled, increasing from US$1.8 billion in 1973 to US$6.5 billion in 1974. The authorities were encouraged by this to embark on a huge development programme, which was labelled at the time 'the explosive development'. This strategy implicitly assumed that the constraint on development was financial.

However, the relaxation of financial constraints exposed the real nature of the economy's limited ability to spend the oil revenues. There proved to be two main reasons for this.

First, there were the constraints on absorptive capacity, accentuated by the absence of sufficient complementary factors such as labour and managerial skills, and also aggravated by the undeveloped infrastructure and the sluggishness of bureaucratic procedures, especially in the public sector, which was the dominant sector at the time. Second, there were inappropriate economic policies, the design of which was not adapted to take into account the new and large inflow of financial resources. These policies were just as responsible as the lagging infrastructure for the limited absorptive capacity.

Labour policies remained inflexible, in terms of both wage levels and labour mobility. Wage levels, especially in the public sector, continued to be low, adversely affecting that sector's productivity and degree of absorption. The private sector was able to award higher wages but public-sector employees were mostly unable to move there, although a labour transfer of this kind would have narrowed the gap between the level of productivity and wages between the two sectors. Determination of the wage level is basically a labour policy, but it also forms an important part of fiscal policy. The government had a substantial programme of investment and consumption but this was not reflected in higher wages in the public sector. Fiscal policy was thus aimed at increased expenditures but, by maintaining a low wage level, it offered no incentives for any productivity increases.

While labour mobility between the public and the private sector remained difficult, movement from the countryside to cities was on the whole tolerated. More specifically, there was a shift from the agricultural sector to the construction, services and military sectors. Labour mobility favoured the non-tradable sector, and it thereby adversely affected the performance of the agricultural sector, leaving it with limited absorptive capacity for investment because of the shortage of agricultural labour.

Furthermore, exchange-rate policies did not favour agricultural expansion and thus did not increase the absorptive capacity for investment. Because the petroleum-trade-dominated exchange rate was overvalued for the non-petroleum tradable sector, this had deleterious effects on the expansion of agricultural output. Nevertheless, it encouraged importation of food to meet the demands of the urban population which, because of the movement to the cities from the countryside, had risen appreciably.

The interest-rate policy did not encourage people to save in domestic financial institutions. Domestic absorption took the form of domestic consumption, but goods and services were in short supply, leading to inflationary pressures. Private investment, normally encouraged by low interest rates, did not increase to a degree that would have made any difference to absorptive capacity. Additionally, ideological considerations prevented the country from investing abroad, further limiting the outlets for funds. Finally, despite the fact that the country was awash with money, there was frequently a ban on foreign travel.

All this relates to civilian spending. By contrast, military spending experienced a huge increase during the period 1975 to 1980. Military expenditures grew from US$3.1 billion in 1975 to around US$20 billion in 1980, which in terms of GDP represented 22.5 per cent and 39 per cent respectively. This high absorptive capacity was made possible in part by the availability of labour; the ratio of labour in the military sector to the total labour force increased from 2.9 per cent in 1975 to 13.4 per cent in 1980. As for military imports, they increased fivefold during the same period, whereas non-military imports increased only threefold. Therefore, as far as policy was concerned, military expenditures had priority and did not appear to suffer from any absorptive capacity constraints.

In short, with regard to civilian spending (especially capital spending and the limited spending outlets due to inadequate policies), a combination of factors relating to the restricted absorptive capacity had, on the eve of the war with Iran, led the economy to acquire a number of important characteristics. These included large financial reserves, amounting in 1980 to about US$36.5 billion, and a high degree of militarisation. There was also relatively rapid development in infrastructure, especially in the construction and transport sectors.

As far as the evaluation of this development is concerned, the fact that the economy possessed such characteristics generated a perception among Iraqi policy-makers that it was strong and resilient. To what extent was that thinking justified?

In the first instance, as already noted, the accumulation of financial reserves was the result of the limited domestic absorption that went hand in hand with the low implementation ratio of development projects. That accumulation was not the result of productive activity; rather, it was a financial phenomenon resulting from successive increases in the price of oil in the 1970s. In other words, the financial reserves were obtained and accumulated even before the process of development had begun.

This means that, as far as resources are concerned, the state in Iraq played an allocating role rather than a mobilising one, which, in turn, meant that the nature and orientation of that allocation would have important implications for development performance. This was due to the fact that a significant portion of the resources available could be allocated to an unproductive activity – which, more or less, was the case with Iraq because of the heavy emphasis on the military sector. This, too, is why Iraq and the rest of the oil producers should be differentiated from the East Asian countries where the mobilisation of resources was the result of successful development performance.

Second, Iraq had built a large military industry, but it remained one of the biggest importers of arms in the world up to 1990. Apart from generating a demand for labour that was probably needed much more in the civilian sector, the military sector did not have substantial linkages with the rest of the economy, remaining labour-intensive, but making a minimal technological contribution to the economy. The establishment of a military industry has to be based on a developed manufacturing and electronics sector within the civilian economy. Admittedly the performance of Iraq's military sector needs further detailed investigation and evaluation, but in the final analysis, the Iraqi military sector turned out to be a labour-intensive importing sector in terms of both its intermediate and its final goods.

The above developments do not indicate a strong and resilient economy, and on the eve of the war with Iran the economy possessed a number of other characteristics that also reflected serious weaknesses and vulnerability. These characteristics included:

(a) *Heavy import dependence* This was the case for the major sectors in the economy. The agricultural sector imported about three-quarters of Iraq's food requirements. The manufacturing sector also imported most of its inputs and, because of its poor performance, Iraq became a major importer of manufactured products, especially consumption goods. The intensive import content of the military sector aggravated this import dependence.

(b) *The disappointing performance of the state enterprises* These mostly operated at a loss and increasingly depended on the state budget to generate positive value added.

(c) *The satisfactory performance, by and large, of the construction sector* Even so, this sector remained indirectly dependent on oil revenues, which determined the contracts awarded to local and foreign companies.

(d) *A consequent and increasing dependence on oil exports* This was due to the fact that no other sector was able to generate its own finance.

(e) *In the case of oil exports, Iraq's heavy dependence on neighbouring countries* since the pipelines that passed through other countries carried almost 3.2 million barrels per day.[2]

One must ask whether in these circumstances Iraq should ever have gone to war, since a country possessing such characteristics is very vulnerable to any external shock, let alone war. Such vulnerability will determine very clearly the direction an economy will need to take in the event of war breaking out. In Iraq's case, the country was not in a position to compensate for any loss of oil revenues, and the oil sector was, of course, the first to be hit by the war with Iran.

The war with Iran – the genesis of the new realities

Apart from the destruction caused by the war with Iran,[3] the fact that the oil sector was its main casualty meant that Iraq lost its war-financing capacity. Oil exports declined from US$26 billion in 1980 to US$10 billion in 1980/1, while imports (civilian and military) jumped in the same period from US$8.7 billion to around US$21 billion, since Iraq chose not to reduce its civilian imports. The question of war finance therefore assumed great importance, and the following sequence of events was observed:

1. The government resorted to its stock of financial reserves, estimated in 1980 to be about US$36.5 billion. In the first two years of the war, and because of successive current account deficits, the reserves declined to almost US$14 billion.[4] While war finance was

an important reason for the decline of reserves, the government continued during the first two years to increase civilian imports, in the hope of cushioning the war's effect on the population. It could therefore be said that initially the government tried to use its own resources to finance the war.

2. The government's options for financing the war started to narrow. Having seen oil exports decline and financial reserves shrink, the government resorted to external finance. Although interest free, the loans extended by the Gulf states, which amounted to about US$40 billion, more or less constituted the beginning of the debt problem in Iraq. The government also obtained loans from Western and East European governments and financial institutions, amounting to US$42.1 billion.[5]

3. As the war continued, Iraq started to develop a problem of credit-worthiness. Oil exports, which stagnated because of war damage and the closure of the Syrian/Iraqi pipeline in 1982, were spent mostly on military imports; the latter represented 75 per cent of the former. Following the drying up of external funds, the main option left to the government was to turn to the private sector to finance the war. In effect, the government wanted to devote its financial resources exclusively to the war and to make the private sector contribute indirectly to this financing.

The government's recourse to the private sector took two forms. First, since it wished to concentrate exclusively on military spending and imports, the government initially allowed the private sector to import civilian goods using its own externally-held resources. This policy amounted to encouraging the repatriation of Iraqi capital in the form of goods. However, because of the shortage of goods and the fact that Iraqi capital was reluctant to come back home, the Iraqi dinar was being smuggled to neighbouring countries to be exchanged for US dollars. The value of the Iraqi dinar thus began to depreciate, but ironically the depreciated dinar was increasingly being used for imports! Because of this mechanism for civilian imports and the growing demand for those imports, the parallel exchange rate of the Iraqi dinar continued to decline. Meanwhile the government continued to import military goods using the official exchange rate, which effectively meant that it was subsidising these imported goods.

Second, the government started a privatisation drive in the latter part of the 1980s. This was undertaken basically for fiscal reasons since, in the absence of an effective tax system in Iraq, the government wanted to obtain additional resources from the private sector. Because of the pressing need for resources, the privatisation programme was not implemented on the basis of proper evaluations of the enterprises sold, and the sale itself was not made to an authentic private sector. As Chaudhry comments, 'The new importers, as well as the investors that bought and leased the large government holdings in agriculture and industry, were not traditional private sector elites; rather, most of them made their fortunes in contracting.'[6]

The two ways in which the government resorted to the private sector were an indication of its desire to relinquish some of its responsibilities in the civilian sector of the economy. First, it refrained to a large extent from civilian imports (thereby saving resources for military imports), and second, it pursued the selling of public enterprises (thereby obtaining additional finance from the private sector through such sales and from the cessation of subsidies).

The war with Iran ended with Iraq facing new economic realities. It was now heavily indebted, its currency was significantly depreciated, and it suffered from high inflation. In addition it had to face all the human and social problems associated with large-scale death and disability arising from a major war.

The second Gulf War

By the end of the war with Iran, Iraq's need for resources had proved to be very substantial, with its petroleum exports barely paying for its imports, and with its debt servicing running at more than three billion dollars – about one-third of total exports. Given these new realities, there were three ways in which Iraq could acquire the resources it required.

First, the price of oil needed to increase. Second, Iraq needed to obtain some debt relief. Third, it needed fresh external finance. It is not difficult to see the relationship between these ways of obtaining additional finance and the demands made by Iraq in mid-July 1990 in a letter submitted to the Arab League.[7] Iraq started by accusing Kuwait and the United Arab Emirates of flooding the oil market to lower the price of oil, thereby preventing Iraq from using the price instrument to obtain additional resources. In Iraq's opinion, the price of oil could be increased to US$25 per barrel from an average level of US$16 per barrel for the period 1988–9. Next, Iraq sought debt relief on the loan which had been made to it by Kuwait during the war with Iran by pressing for the loan to be considered as a grant. Furthermore, Iraq accused Kuwait of stealing oil from its Rumaila oil field, and on these grounds requested additional finance.

In addition to the historic border dispute, it is thought that these demands provided the economic rationale for the invasion of Kuwait, even though the price of oil had increased by that time to US$21 per barrel.[8] However, the invasion resulted in the imposition by the United Nations of comprehensive sanctions[9] on Iraq, thereby depriving it of oil revenues that had represented around 98 per cent of total exports. Furthermore, the UN imposed war reparations in the form of claims that amounted to more than US$320 billion; most of this is in large corporate claims that have not yet been dealt with by the relevant UN commission.[10]

Any country that suddenly loses almost all its export revenue un-doubtedly faces a daunting situation. The sanctions caused considerable shortages in medical and humanitarian goods and in food supplies, and created a highly unstable macroeconomic environment in which inflation became rampant.[11] It might therefore have been expected that, given such circumstances, government policy would try to mitigate the effects of sanction and the loss of resources.

A closer examination of the situation points to the contrary, for the government responded to the sanctions – which effectively involved a problem of deficit in the balance of payments – by adopting a rationing system under which around one-half of the basic foodstuff requirements were met. At the same time, however, the government resorted to printing money to finance its expenditures. Money printing produced substantial pressures on the national currency, because the resource deficit was essentially a foreign exchange one. This policy led to a vicious circle of inflation that in turn caused further deficits financed by more printing, and the currency weakened further. This is indicative of the extent to which monetary policy failed to harmonise money printing with the economy's ability to produce and supply goods.

Thus, while the rationing system moderated increases in price levels, this monetary expansion contributed significantly to hyperinflation and to the soaring price levels of non-rationed goods. The policy that was being followed in response to the embargo therefore served to aggravate the adverse effects of the embargo.

It is thought that the handling of the embargo in this manner enabled the government to raise some of the resources that it needed. Printing of money, and trading in it, provided the means for the government to collect any foreign currency that the public had possessed. This policy benefited the few who possessed dollars, since they were protected from the consequences of the hyperinflation to which that policy contributed. The rest of the population was deprived of any such protection, because it had to live on the depreciated dinars. The effect of this policy was thus to shift the burden of the sanctions to the Iraqi population.

What might have been considered a more appropriate policy under these conditions? The Iraqi government should, to a large extent, have followed an adjustment policy. The supply of goods would have revolved around the rationed goods instead of the government having to resort to money printing which did not contribute much to the real supply. Needless to say, such a policy would not have ensured that the needs of the population were met satisfactorily, but it would have been less inflationary. Printing money made goods so expensive that people were not able to buy them. In effect, such goods were non-existent for the majority of the population. From the national economy point of view it is certainly undesirable to create an environment where goods are made available but at exorbitant prices. In this case, the majority of the Iraqis would be encouraged to sell their assets (jewellery, cars and even houses); that is, they had to 'dis-save' in order to acquire those goods.

Through this 'dis-saving', there was a transfer of resources from the majority to the few who had the purchasing power and the foreign currency to buy the assets, and this transfer actually distorted the distribution of national income and wealth. One outcome of this distortion was to be the virtual disappearance of the once powerful middle class in Iraq.

The evolution of the Iraqi economy since the 1970s has therefore gone through a sequence during which it collapsed in almost every respect. The details in Table 11.1 illustrate this self-explanatory sequence.

3. The Iraqi economy: the road to recovery and growth

An earlier study, published in 1997,[12] concluded that, as far as the external sector of the economy was concerned, the growth prospects of the Iraqi

TABLE 11.1
Sequence of collapse in the Iraqi economy

Growth

Growth in some years prior to 1980 exceeded ten per cent → growth in the 1980s was about -9 per cent → afterwards it further declined to around -11 per cent.

Financial position

Current account surplus and huge reserves → depletion of reserves → deficit and dependence on external finance → aggravation of deficit due to sanctions → potential deficit because of reparations and debt service.

Domestic and external sectors

External sector was financing the domestic sector → domestic sector will have to remedy the deficit in the external sector.

Inflation

Inflation was being caused by development pressures (economic fundamentals) → weakness of currency and military spending → sanctions, money printing, collapse of currency and hyperinflation.

Financing consumption

Saving and wealth used, in the main, to finance investment → now they are used to finance consumption and current living.

Social state

A measure of social stability → high casualties in war with Iran undermined family unit and caused widespread disability → with the second Gulf War, breakdown in social norms, widespread violent crime and the middle class virtually disappeared → collapse of institutions, social services, and law and order.

---

economy would depend on the extent to which the country was relieved of the financial constraints resulting from wars and their aftermath. This would be particularly true when sanctions were lifted, because Iraq would then be deprived of the benefits it could obtain from a full return to the oil market. The study also decided that the most generous relief would be forgiveness from international claims. However, in the event that this did not materialise, the study showed that the claims were not to be paid unless a trade surplus was generated; that is, imports would be given priority over payment of reparations and debt service.

This indicates the importance of mobilising resources for imports. However, while the simulation in the 1997 study points to the general direction in which the money should go, a development strategy needs to be elaborated.

Such a strategy necessitates taking action in the short run – in other words, if Iraq wants to resume growth, it must first restore economic stability and create the conditions under which it maintains such stability. The strategy envisages the reintegration of Iraq into the world economy and if domestic conditions improve – in the sense that necessary institutions are built – Iraq will be able to maximise the benefits of globalisation and minimise its costs. The rest of this chapter will deal with these issues in detail.

Priority areas in the short run

In the short run, the immediate priority must be given to raising the external value of the dinar, and to controlling high inflation because of its adverse economic, social and political consequences. In other words, the immediate priority is to restore macroeconomic stability.

With high inflation, produced basically by the collapse of the currency, it is almost impossible to implement macroeconomic and development policies effectively. The private non-oil sector will see the resources that it wishes to use for imports continuously declining in real terms, thereby ruining development prospects. Under high inflation it is always difficult to maintain a budget surplus (something important in the first stages) because government expenditures are determined by the high rate of increase in price levels, whereas revenues keep on losing their real value.

Also, high inflation more or less cripples certain important tools of monetary policy since the cost of capital will be high. This is detrimental to investment in the private sector, since the purchasing power of its cash flows will continuously be eroded by inflation. Such inflation will also make it difficult to establish guidelines for wages and other incomes because of the indexation problem. Wage levels will contribute to the high cost of production.

Needless to say, inflation hurts the poor and fixed-income earners. High inflation in Iraq is associated with enormous social problems, such as increased crime, and is creating a new social situation under which savings and real assets are liquidated to meet immediate needs, while the middle class has virtually disappeared. There is now a worsening of the distribution of both income and wealth.

If inflation is not reduced, political protest is likely to erupt, and there is a danger that repressive measures will be used to quell it. This would be damaging to economic stability. In other words, in order to help stabilise the situation, a resolute effort should be made to address the question of how to reduce inflation, and in this respect the mobilisation of a substantial volume of financial resources, internationally and regionally as well as at the domestic level, would seem to be necessary.

On the international and regional level, Iraq should export the volume of oil commensurate with its oil-export capacity, and in order for this re-entry not to affect the market, maximum cooperation by other OPEC members and oil producers is essential, even though many of them have budgetary difficulties. These countries are certainly aware of the suffering that Iraqis have experienced, and they should also be aware that economic and political stability in Iraq will have favourable repercussions on regional security. Agreement on a new oil-production level for Iraq will involve a process of dialogue and negotiation with other OPEC members, and through this process Iraq can be reintegrated into the region and into the international community.

A standstill[13] on the payment of debt and reparations should be granted to Iraq. If this is not done then Iraq may remain in deficit. (While sanctions are currently causing the resource deficit, payments of debt and reparations may later become the deficit's principal cause.) Experience shows that high debt-service payments engender economic instability and fuel inflationary pressures.

The regional and international community should extend substantial financial assistance to Iraq. This assistance, which should be on concessional terms or preferably in the form of grants, assumes particular importance in the event that the full re-entry of Iraq into the oil market is only partially accommodated.

If undertaken, these measures would indicate an element of goodwill on the part of the international community towards the new regime in Iraq, and would contribute significantly to its stability. Success in the mobilisation of resources depends on Iraq's creditors in the region and/or outside the region, on the United Nations, and on other oil-exporting countries.

On the domestic front, Iraq should complement the actions of the international community by refraining from money printing to finance its expenditures, since at this stage (the short run) it does not have the productive capacity to back this additional money supply. Money printing can, however, be tolerated if foreign exchange flows into the country, although it should be carefully synchronised with the growth in domestic production and foreign exchange. Government expenditures should, in any case, be tightly controlled, but government support for agriculture and other similar essential services will remain important during this period. Agriculture, for example, is a labour-intensive sector, and employment will be a serious problem facing Iraq, especially when a large part of the army will be demobilised.

The resources thus accumulated should be used immediately for the following purposes:

(a) *For imports, especially of consumer goods and food as a matter of priority* This is not inconsistent with the aforementioned policy of supporting agriculture. The latent demand for agricultural products and food in Iraq is almost certainly so huge that supply from imports and domestic production will be needed in the short run.

(b) *For the provision of social services, especially in the fields of health and education* UN reports stress the poor state of hospitals, and the shortages of medicine and medical equipment as well as of school materials.

(c) *For reconstruction and rehabilitation,* especially of power and water plants, sanitation, sewage facilities and telecommunications.

(d) *For a new programme of human development and technological rehabilitation,* so that Iraq can bridge the technological gap so far as access to information technology is concerned. This gap is due to sanctions, as well as to the authoritarian nature of government which prevents access to information in general. Information technology will be an essential prerequisite for growth in the next decade.

These policies and measures, particularly the importation of essential goods, should, if undertaken, help to stabilise the economic, social and political situation, and to put the country back on the path of development and growth through implementation of macroeconomic policies and development strategies. An approach that gives priority to imports over payment of external claims can, therefore, be expected to encourage growth as well as help to achieve economic stability.

From the foregoing, it is clear that the main concern will be to regain stability through economic measures that will need to be implemented as quickly as possible. Certainly the package suggested above can be initiated rapidly since, apart from oil exports, it does not depend on production or exports. When correcting their external payments position, many countries resort initially to external financing, which provides them with a breathing space in which to start adjustment policies.

Measures to maintain economic stability

Once the situation is relatively stable, the country should exert every effort to maintain this stability. Thus it is essential that the country does not slide back into a situation of a large resources deficit. If the deficit recurs, it will again lead to inflationary pressures and a lower value of the dinar. Maintaining stability will depend more on domestic effort and macroeconomic policies than on financial resources, as in the period of restoring stability. However, the macroeconomic policies will not need to be excessively restrictive since they can be complemented by an infusion of resources from the crude-oil sector. Here, once again, the importance of Iraq's regional and international environment should be emphasised. In other words, if policies to maintain macroeconomic stability are to work smoothly, Iraq needs to avoid external shocks (such as an oil-price collapse), and it must maintain the purchasing power of its exports. Therefore a friendly and favourable regional and international environment remains essential for Iraq. In this regard the following are important:

Iraq should cooperate with other oil producers to bring about a price level that will guarantee it the necessary level of revenues. At the same time, this should not have an unfavourable effect on the economic prospects of oil-importing countries, whether developed or developing, since Iraq, like other oil exporters, will have every interest in maintaining the demand of these countries for oil.

Having agreed a temporary standstill on the payment of its obligations with its creditors and with the United Nations, Iraq should propose the renegotiation of those obligations. Many proposals and schemes, including forgiveness, swaps and rescheduling, have been put forward for restructuring the external debt of developing countries, though specific details will not be discussed here. Without a clear idea of the structure of the debt and the development of the macroeconomic situation of the country concerned, it is not easy to suggest any specific solution or initiative. However, what is certain is the need for debt relief, so that high debt-service payments will not contribute to the widening of the financing gap, bring back instability and impede the growth prospects.[14]

As far as war reparations are concerned there are also historical precedents. Lessons can be drawn from the German experience after the

First and Second World Wars. Although the claims on Iraq submitted to the United Nations Compensation Commission amount to more than US$320 billion, the vast majority of those claims are those of governments and corporations. According to UN Security Council Resolution 687, the payment of reparations should take into account the needs of the Iraqi economy. Germany actually received financial assistance after the Second World War, in contrast with the burden of reparations imposed after the First World War.

The last thing Iraqis wish to see happening is the return of a resource deficit. Though the present deficit is the result of sanctions, it might also, when sanctions are lifted, result from payment of debt service and reparations. It should also be clear that Iraq's economic problems, including inflation and shortages, are directly related to the deficit. In the final analysis, it is the adverse effect on the importing capacity of the country that needs to be addressed. This import capacity can be negatively affected by a shortfall in, or collapse of, export earnings due to sanctions, or by claims on those export earnings through payment of debt and reparations.

If a situation of resource deficit returns, then inflation, together with macroeconomic instability, will also return. History provides many examples of deficits accompanied by inflation. Many Latin American countries experienced hyperinflationary situations because of their debt crisis in the 1980s, and Germany's hyperinflationary period in the 1920s was partly the result of that country having had to lower its exchange rate to generate the huge trade surplus needed to meet reparations payments.

The above analysis shows that a solution to the problem of Iraq's external obligations must be found. Such a solution, however, can be reached through negotiations and agreements with the international community and the United Nations. It is not in Iraq's interests to take unilateral action.

In the short run the immediate objective is to restore and maintain economic stability, but this period is by no means confined solely to this aim, since a lot of work needs to be done to design policies for implementation at a later period. There should also be an initiative to draft a new constitution as well as laws – including electoral and press-freedom laws – to ensure democratic rule. Work should also be done to prepare economic and social programmes that will guarantee the resumption of economic growth and provide solutions to the huge social problems caused by wars and misguided policies such as displacement and deportations. Unemployment will be a serious problem, so that retraining and re-employing the released military workforce will have to occupy an important place in the development strategy.

While these programmes must be vigorously implemented once economic stability is regained and maintained, this does not, at this stage, preclude the provision of financial resources to agriculture and the social sector, since they make an obvious contribution to stability. However, the real solution to the displaced population and the released military workforce will rest mainly with the satisfactory growth of the civilian economy, since this will determine its capacity to absorb additional labour.

In the short run, therefore, while stability is a priority, it will also be important to design policies and programmes, and pave the way for building institutions.

Iraq's enabling environment for the resumption of growth

When its economy has attained, and is able to maintain, economic stability, and when the country has been reintegrated into the world economy and community, Iraq can proceed to implement an orderly development strategy that will ensure faster growth. Experience of developing countries shows that growth under unstable conditions and fluctuating financial flows cannot be sustained. At this stage it is only possible to outline overall policy prescriptions and the orientation of future development strategy. Since the recent experience of the Iraqi economy has been one of mismanagement, wars and comprehensive economic sanctions,[15] its development strategy should therefore be based on a new orientation that will emphasise effective economic management.

Development policy in Iraq has always stressed the importance of diversifying the economy by lessening dependence on oil, although achievements in this regard were far from impressive. However, this objective is assuming greater relevance and importance since there is now some doubt as to whether the external sector will still be able to finance the domestic sector, especially if relief from external obligations fails to come about. There is now a pressing need to increase the efficiency of the non-oil sector in order to increase its contribution to overall output and growth.

Based on this, the new orientation should emphasise (a) the intensive use of economic policies to mobilise resources for development and not merely to attain and maintain economic stability (ample oil revenues, historically, led to the adoption of lax economic policies); and (b) a redefinition of sectoral priorities, in terms both of production structure and of ownership.

The use of macroeconomic policies in the economic management of the non-oil sector will involve a different relationship between the state and the private sector. In the past, state intervention in Iraq was driven by ideological motivations that attributed the leading role in the development process to the public sector. This led the state to establish many enterprises, including both small and medium-sized projects, in the productive sector. However, because of their inefficient and loss-making operations, the state ended up financing these enterprises, therefore intervening through microeconomic as much as macro-economic management. Moreover, because of the ideological nature of the intervention, the state's relation with the private sector was often confrontational. Finally, the state was unable to realise the advantages that could have been reaped through combining ample resources with good macroeconomic policies. As a result, the potential contribution of the available financial resources to effective and development-friendly macroeconomic policies was not achieved.

Direction of macroeconomic policies

The objective of *fiscal policy*, in the long run, is to diversify the structure of government revenues in order to reduce dependence on the oil sector. On the revenue side the government will have to establish an efficient tax system. Without going into too much detail, it is clear that government expenditures will have to be rationalised and the ratio of military to total expenditures substantially reduced.

The development dimension of fiscal policy includes, *inter alia*, the granting of incentives to support development in the non-oil sector, especially the private sector. This will include policies for regional development.

It is, of course, impossible to have a budget surplus or even a balanced budget with oil revenues excluded. But the budget deficit must be minimised: because the oil sector's financing ability will be weaker from now on, it will also have to finance debt service and reparations claims. Under these conditions the government will have to resort to deficit financing, which will reduce the latitude for monetary policy.

Monetary policy needs the right infrastructure. In particular, this must include reform of the financial and banking system and, more importantly, independence for the Central Bank, otherwise, the Bank will end up financing the budget deficit. However, Central Bank independence does not mean that it should adopt a neutral stance vis-à-vis the development process. Its policies should aim at easing credit to the non-oil productive sector, especially the private sector, and resources for this purpose can be generated from reducing military expenditures and from the relief from external obligations, should this materialise. The channelling of resources into society will be through the specialised development banks, which means that these banks should be assisted for a period of time by the government.

Exchange-rate policy has never been used in Iraq before – historically, the exchange rate was determined by the surplus created by the oil sector. However, the real economy in Iraq is the non-oil sector, and an oil-dominated overvalued exchange rate could prove detrimental to this sector's prospects. It is very difficult at this stage to form an opinion on this issue, but a realistic exchange rate for the non-oil tradable sector may improve its competitiveness, given that the diversification of foreign-exchange sources will assume a high priority from now on. A realistic exchange rate will not affect oil exports because these are dollar-denominated. Nor will the government's import capacity be affected by a change in the exchange-rate policy, as long as those imports are paid directly or indirectly through oil revenues. However, the government should use its oil revenues to mitigate the harmful effect of a more realistic exchange rate on the private sector's import capacity.

Sectoral priorities

Sectoral policies have two dimensions – ownership, and productive structure. As far as ownership is concerned, the private sector should play a leading role in the development process. This is one way of developing the non-oil sector. The state should also play a significant role in providing an enabling environment for the private sector so that it can assume a development role through its macroeconomic policies. Apart from efficiency considerations, emphasis on the private sector will be a matter of necessity. The government has limited choices in this respect since it will be burdened by the payments of debt (which is public in Iraq's case) and of reparations, if relief does not materialise. Privatisation is one way of relying on the private sector. However, the privatisations undertaken in Iraq in the second half of the 1980s were driven by the need to finance the war with Iran, and efficiency considerations were secondary. In addition, many of the state enterprises were sold at book value to the government-linked private sector. The strategy governing reliance on the private sector will have to reconsidered and re-evaluated with a view to putting efficiency considerations first.

With respect to the productive structure, priority should be on human development and agriculture, especially after the economy has regained stability. As noted earlier, since the economy suffered from isolation due to reasons of ideology and international sanctions, it was unable to reap the benefits of new technology, especially in the communication and information sectors. Therefore Iraq is at present poorly placed to reap the benefits or to meet the challenges of globalisation unless it makes serious efforts to enhance its technological base in these areas.

Another productive sector that should receive considerable attention is the agricultural sector. The development of this sector should save the country ample foreign exchange, which will be a precious resource in Iraq in the future. Through agriculture, Iraq can mobilise foreign exchange by way of export earnings and/or import savings. In addition, reviving this sector will not require substantial foreign exchange.

In the industrial sector, light industry (especially food processing) should receive priority because of existing domestic demand and its export potential. Development of other sectors should be based on careful market studies that take into account domestic and foreign demand conditions and existing supply capacities in the region.

Because domestic resources may not be sufficient for consumption and development purposes, development of these and any other sectors that can rapidly generate the necessary foreign exchange can also be financed through foreign direct investment (FDI) by Arab and multi-national corporations. FDI is a source of finance and technology that can be undertaken within the country's development strategy. Iraqis who are resident abroad represent another source of finance.

NOTES

1 The author is a staff member at the UNCTAD secretariat. The views expressed in this contribution are his own and do not necessarily reflect the opinion of UNCTAD.

2 This excludes the pipeline that passess through Syria with a capacity of about 1.4 mbd. The figure in the text relates to the Iraqi–Turkish pipeline and the IPSA pipeline system through Saudi Arabia, each of which has a capacity of 1.6 mbpd. While it is a good strategy to have such a pipeline network, it does require the maintaining of good relations with neighbouring countries. As an example of Iraq's dependence on this network, Iraq exported 2.7 mbd in January 1990, of which 87 per cent passed through Turkey (1.5 mbd) and Saudi Arabia (0.85 mbd); see *Middle East Economic Survey (MEES)*, 15 January 1990.

3 For details on this, see A. Alnasrawi, *The Economy of Iraq, Oil, Wars, Destruction of Development and Prospects, (1950–2010)*, Westport, Conn.: Greenwood Press, 1994.

4 Data on financial reserves up to 1977 was obtained from the International Monetary Fund's *International Financial Statistics (IFS)*, Washington, DC, and updated from current account balance data available in the OPEC Secretariat's *OPEC Statistical Bulletin 1995*.

5 See Alnasrawi, op. cit., p. 109.

6 See Kiren Aziz Chaudhry, 'Economic Liberalization in Oil-exporting Countries: Iraq and Saudi Arabia', in I. Harik and D. Sullivan (eds), *Privatisation and Liberalisation in the Middle East*, Indiana University Press, 1992.

7 See the letter sent from Iraq's Deputy Prime Minister to the Secretary-General of the Arab League in P. Salinger and E. Laurent, *Secret Dossier: The Hidden Agenda behind the Gulf War*, London: Penguin Books, 1991, Annex.

8 See Alnasrawi, op. cit., p. 117.

9 These sanctions were imposed on 6 August 1990 under United Nations Security Council Resolution 661.

10 The Security Council resolutions (SCR) under which reparations are imposed and institutionalised are 687 (1991), 692 (1991) and 705 (1991). SCR 687 dealt with the reparation issue in section E. SCR 692 creates a fund for reparations to be financed by a certain level of Iraqi oil revenues and establishes a Commission to administer that fund. As regards SCR 705, it specified that the compensations to be paid by Iraq shall not exceed 30 per cent of exports of petroleum and petroleum products, and that the ratio could be reviewed.

11 A nutrition survey carried out by UNICEF, with the participation of the World Food Programme (WFP) and the Iraqi Ministry of Health, indicates that 27 per cent of Iraq's 3 million children were at risk of acute malnutrition; see WFP, *Emergency Report*, no. 23, June 6, 1997.

12 See Sinan Al-Shabibi, 'Prospects of Iraq's Economy, Facing the New Reality' in J.Calabrese (ed.), *The Future of Iraq*, Middle East Institute, Washington 1997.

13 A standstill is an agreement between the debtor and its creditors temporarily to defer payment of long-term debt and freeze short-term debt. A moratorium may take the form of a unilateral action. In any attempts to reintegrate into the international community, Iraq should opt for a standstill.

14 Iraqi debt is mostly bilateral. It is owed to governments or guaranteed by governments. In general, bilateral debt is more easily restructured than multilateral debt.
15 See Sinan Al-Shabibi, op. cit.

12

The Iraqi Economy:
Reflections and Prospects

Abdul Munim Al-Sayyid Ali

1. Introduction

Four factors determine the future course of an economy: these are the availability of economic resources, the soundness and rationality of economic policies, a suitable internal political framework, and a permissive and cooperative external environment. Iraq is no exception to this notion.

It would not be possible to outline the future course of the Iraqi economy without reviewing its past performance and assessing the present state of affairs. With Iraq, it has always been the case that much-needed critical reviews and evaluations of past and present economic performance have been based more on political than on economic grounds.

Iraqi economists of different ideological persuasions have often applied harsh criticisms to the country's economic development performance and to past policies, so much so that very little seems ever to have escaped their unfavourable assessments. As clearly shown by many past and present examples, they rarely attribute positive outcomes to development policies of any political regimes. Certainly many failures were registered under the various regimes in Iraq, but it is also true that one way or another, and notwithstanding diverse and difficult economic and political conditions, there were many positive achievements, domestically as well as internationally. Even so, actual achievements have often fallen short of the aspirations of the citizenry, and especially of the expanding number of highly motivated intellectuals who are constantly driven to seek greater achievements and more efficient performances.

In predicting the future of the Iraqi economy under the harsh material, practical and psychological constraints currently prevailing, it is necessary to be as objective as possible, and to focus on the essential factors that will shape the economy at the beginning of the twenty-first century. In doing this, the impact of political factors in defining the pattern, course, processes and trends of Iraq's economic and social development will not be ignored. However, greater emphasis will be placed on the economic factors, while the possible effects of the political circumstances through which Iraq is currently passing will be put to one side. This approach is adopted mainly because of the difficulty of predicting the political intentions and tendencies of the hostile Western powers, particularly the United States and the United Kingdom.

The savage attack on Iraq in January 1991 by the major Western powers and their Arab and non-Arab allies resulted in a massive demolition of the Iraqi economy in its human as well as its physical aspects. By destroying the country's social, economic and scientific facilities and institutions, the allies effectively pushed Iraq back to the pre-industrial stage, thereby achieving their original basic goal.

However, this destruction was not the only indicator of Iraq's economic future to have emerged from these recent events. The fact is that during much of the decade following the 1991 attack, and notwithstanding its own limited capabilities and the inadequate resources at its disposal, Iraq was able to rebuild most of its ruined infrastructure. In spite of the severe shortages stemming from sanctions, boycotts, and a state of comprehensive political and economic siege, reconstruction activities included roads, bridges, postal systems, harbours, airports, railways, schools, electricity supplies, water and communications facilities, oil refineries and oil-production establishments. It is true that Iraq paid dearly for this effort, and hyperinflation and profound human suffering reflect the high cost borne by the country. Yet Iraq succeeded in reconstructing its physical infrastructure with unusual speed and astonishing effectiveness. Would it not therefore be reasonable to consider this successful performance, immediately after the 30 allied states had suspended their acts of war, as indicative of an optimistic future for the Iraqi economy?

In order to answer this question, it is useful briefly to evaluate the outcome of the Iraqi development experience in the 1970s and 1980s, and to assess the state of the Iraqi economy in the second half of the 1990s. We will then utilise certain assumptions and elaborate on specific expectations as a means of exploring the future of the Iraqi economy as it enters a new millennium.

2. Economic policies and their outcomes

The 1970s and the 1980s

The 1970s were characterised by substantial financial surpluses, which were generated by successive swift and sharp oil price rises in 1973 and 1979, and by the remarkable expansion in oil production and export capacities associated with rising prices. These developments resulted in considerable current-account surpluses in Iraq's balance of payments. The assured availability of an unlimited source of funding to meet the extensive financial requirements of economic and social development led to a relaxed financial discipline, which, in turn, resulted in escalating development expenditures, industrial growth (sometimes injudicious), and agricultural decline (compensated for by vast agricultural imports, particularly of food). Financial indiscipline was also responsible for extravagant public as well as private consumption expenditures (especially on imports of consumer goods and services), disguised unemployment (due to inappropriate employment policies), and poor managerial performance (resulting from a lack of the managerial skills required to sustain efficient and effective growth).

The 1980s saw the deterioration of international economic conditions, exacerbated by rising unemployment and inflation rates and a consequent decline in world demand for oil. This resulted in lower oil prices and in turn led to a reduction in oil revenues and to erosion of financial surpluses. The real value of the accumulated foreign assets of the Arab oil-exporting countries, including Iraq, deteriorated further, due to inflation and also to the weakened value of the US dollar; the currency in which oil is priced on the international markets.

The first Gulf War aggravated Iraq's economic difficulties still further. All the country's oil exports from the Mediterranean and Gulf ports were either seriously reduced or completely stopped – a situation that continued until new oil pipelines through Saudi Arabia and Turkey had been constructed and made fully operational. The collapse in oil revenues, coupled with huge expenditures on the war with Iran, compelled Iraq to seek external borrowing and, domestically, to rely heavily on deficit financing through the banking system. The accumulated budget deficit exceeded national income by many degrees, and by the end of the decade the high rates of growth in the money supply had resulted in accelerated inflation rates. In addition, there was a lack of coherence, consistency and coordination among development, financial, monetary and price policies.

Even so, there was substantial growth in the industrial sector throughout the 1970s, although this more or less stopped during the 1980s. Industrial development was based largely on the principle of import-substitution, but because most industry requirements depended on imports, this proved to be ineffective and simply increased the country's reliance on foreign sources and weakened its independent development policies. Emphasis was given to light consumer industries at the expense of productive industries, although substantial oil-based industrial growth (petrochemicals, oil products and fertilisers) was achieved. Even though manufacturing industry achieved respectable annual growth rates, varying between 10 per cent in 1977 and 13.5 per cent in 1980, its contribution to Iraq's economic expansion was limited. Crude-oil production remained the major contributor to growth in the 1970s and 1980s.

The agricultural sector registered a significant decline just at a time when demand for the commodities that it produced was escalating. Since supplies did not match demand, the shortfall had to be met through imports, and the increasing sums earmarked for subsidies and for the balance of payments constituted a heavy burden on government finances. The deterioration of agriculture led to further neglect, and the lack of good management, combined with inappropriate use of allocated funds, had a deleterious effect on overall policy. Low and declining productivity caused by problems of soil salinisation was also worsened by rural–urban migration and by a movement of labour to other sectors of the economy.

Investment planning during the 1970s encountered many problems, including misuse of resources and a consequent failure to utilise these resources effectively because of poor planning. The highest implementation rates were attained in the construction and service sectors, and the lowest in the agricultural sector, due to misplaced developmental priorities, while frequent changes in planning strategies, priorities and allocations slowed down the investment process. The absorptive capacity of the national economy was unable to cope with the level of investment, and the failure to meet rising demand generated inflationary pressures that in turn led to escalation in prices. This was particularly true in the 1980s, when war conditions and related military expenditures were associated with shortages in civilian services and commodities.

The situation in the 1990s

Under the prevailing conditions of total siege and economic sanctions, the Iraqi economy remains afflicted with serious problems; namely, a severe shortage of goods, services, and medical supplies; a critical shortage in foreign currency reserves; and at the time of writing (1997), an almost complete stoppage of oil exports. Furthermore, the economy suffers from a highly depreciated dinar in the foreign exchange markets; a very high monetary growth rate and very large and extensive liquidity; a hyperinflation rate; and a large budget deficit and a near collapse of internal revenues.

This has led to bank financing of the deficit, to a huge internal public debt and money supply, and large liquidity at the disposal of the public. Fiscal policy has taken the lead, with the result that monetary policy has become totally accommodating and has therefore lost its independence. The development process, too, has become practically non-operative, under widespread conditions of excess idle productive capacity, and the price policy has fluctuated from free pricing to one of hesitant controls.

An extensive maldistribution of wealth and income has created a terrible divergence between the small and rich elite group at the top, and the large and utterly crushed classes of the brutally impoverished masses at the bottom. The overall result has been extensive migration abroad of Iraq's professional classes, academics, and skilled labour, culminating in a huge loss of highly-qualified human resources and human capital.

The Iraqi economy is thus afflicted with numerous and widespread distortions. These include structural distortions, and distortions in financial, monetary, foreign exchange, pricing and interest-rate systems, as well as divergent and uncoordinated economic policies.

Under these conditions, where is the Iraqi economy heading, and what kind of future can be expected?

3. Prospects for the Iraqi economy

Under the prevailing conditions of economic imbalances, severe external political limitations, lack of sufficient financial resources, and very limited oil exports and revenues, it could be concluded that, at the onset of the twenty-first century, prospects for the future of the Iraqi economy are indeed bleak.

However, if one looks at the four basic elements that determine the course of an economy's future development – that is, availability of resources, soundness of economic policies, a suitable and liberal internal political system, and a permissive and cooperative external environment – it will be noted that the nature of the first two factors is economic, while that of the other two is political. One of these latter is internal and is thus under the control of the ruling political elite, who could easily implement it and abide by its democratic dictates (popular participation, freedom of opinion, plurality, and so on). The other is external, and though outside the domain of the government is nevertheless subject to the government's foreign policy and to its ability to reflect on and to adjust policy in one way or another, whilst taking into consideration the legitimate interests of the country and its people.

Yet the attitude of the foreign powers – the West and the United States in particular – is absolutely crucial in this respect, and, in the interest of all parties concerned, we will assume it to be positive in the long run. For this reason we will presuppose the neutrality and magnanimity of both the internal and external factors. We will also assume that these factors will culminate internally in the building up of a democratic political system, and externally in the ending of sanctions and in the removal of the economic barriers that are preventing the free flow of oil exports as well as normal trade with the rest of the world.

Such conditions have to be brought about, because if they are not met, the future of Iraq and its economy will a priori be doomed. On the other hand, if sanctions are totally or even partially removed, and if the internal and external political stipulations are met, then available resources and sound economic policies, together with rational planning, will enable Iraq to achieve economic progress. It would then be able to meet the development requirements of the new millennium, both internally and externally, and to occupy the position it deserves among world nations. It would be at peace with itself as well as with others, and would also enjoy internal security as well as security beyond its borders – locally, nationally and regionally.

The future course of the Iraqi economy cannot be predicted without the presence of two key elements. First, a radical change is required in the political, economic and social conditions prevailing within the country; and secondly, a programme of comprehensive and total economic reform must be established.

The first of these will depend basically on the removal of the UN sanctions; this will permit a free flow of oil and a consequent expansion in oil exports and revenues and in foreign earnings. The second demands an effective economic reform programme, which will be exemplified by sound economic policies, economic restructuring, and efficient social policies, and which will place greater emphasis on the managerial side of these policies. However, Iraq has first to decide on the fundamental nature of its economic and political system: will it be centrally directed as before or will it be decentralised and follow the current trend of economic liberalisation, in common with many other countries around the world?

When these questions have been answered, and when the necessary political and economic decisions have been taken and are ready to be implemented, then the first step will be in the direction of a wide-ranging economic structural reform and adjustment. Greater emphasis should be put on agricultural development, especially in the area of basic foodstuffs and industrial crops. Industrial development basically needs to be internally oriented, in the sense that it should utilise more raw materials from within the country and depend less on foreign requirements. A clear demarcation line should be drawn between the public and the private sectors, and the emphasis should be on cooperation rather than on competition between the two of them. A rational oil policy should also be devised that will preserve Iraq's proper share in the world oil market, with stable and fair oil prices that are aimed at optimising oil revenues with the least volume of oil production.

At the same time, it is imperative that all distortions should be properly and speedily removed from the tax and financial systems. The foreign exchange system should be reformed, as should also the interest-rate structure, bank credit policies, and monetary policy in general. The budget deficit, including deficit financing through the banking system, should be curtailed and reduced to a minimum. With these factors in mind, and with the high inflation rate and the excessive liquidity in the hands of the public, a careful, extensive, and well-planned monetary reform should be undertaken. Development policy should be reconsidered as well, with more emphasis placed on independence; in other words, development should be based more on local and national factors than on external incentives.

4. Conclusion: limits, constraints and future prospects

From the foregoing, we can safely conclude that, following the removal of sanctions, any future Iraqi movement to meet the anticipated requirements of economic and social development and to promote national independence will be entirely limited by the following constraints: a huge internal financial deficit; a large external deficit; an excessive internal public debt; a heavy external debt; a heavy reparations burden; a profound maldistribution of income and wealth; excessive liquidity in the hands of the public; large military expenditures; and a severe lack of financial resources to meet the requirements of both growth and reconstruction. In addition there is the dependent nature of the Iraqi economy on oil production and exports; an extensive loss of Iraqi professionals and their expertise as a result of extensive migration, especially in the 1990s; and lastly, constrained external relations with the outside world, including the Arab world, the wider region, and internationally.

To transcend these obstacles, Iraq has to organise and institutionalise its economic policies in relation to growth, money, finance, trade, foreign exchange, and so on. It has to initiate comprehensive economic, financial and monetary reforms. Rational, systematic, and objective analysis should replace rigid and subjective decision making that is based upon moribund ideological positions.

The recent experience of Iraq in rebuilding its ruined infrastructure, including its demolished factories and its destroyed oil refineries and oil production and export facilities, attests to the following highly significant facts:

(a)   The ability of the Iraqi individual to innovate, achieve and develop has been proved. The various economic sectors that were destroyed by the war were effectively and successfully rebuilt, entirely by Iraqi hands, so as to function satisfactorily again.

(b)   In motivating people to work and to participate actively in the reconstruction efforts, the importance and effectiveness of material incentives were also confirmed.

(c)   The importance of the agricultural sector in the development and reconstruction processes was confirmed, as well as its significance in mitigating the restrictive effects of the severe and extensive sanctions.

(d)   Iraqis have proved to be capable and astute and to have excellent organisational abilities, as exemplified by the highly successful rationing system based on family food coupons that was set up and implemented in spite of the widespread shortages resulting from the imposition of sanctions.

Finally, there are two pertinent questions that might justifiably be asked. The first of these concerns inflation.

Inflation in Iraq was, and still is, an epidemic that has destroyed the majority of the populace while enriching the parasitic groups such as the middlemen, the money and commodity speculators, and the big merchants who are always hungry for overly excessive profits. The writer of this chapter is constantly preoccupied with the question as to whether it was possible for the government to avoid inflation, given the very large financial requirements of reconstruction, and the dearth of financial resources at its disposal. Adequate sovereign financial sources, such as taxes, were simply not available under conditions of extensive war destruction, as well as deteriorating production facilities, widespread idle productive power and general unemployment. Therefore, in order to cover its basic financial needs and its large and increasing budget deficit, Iraq had no other source of finance to turn to except the banking system; nor had it any other alternative but to resort to deficit financing and bank borrowing. But it exaggerated its financial needs and over-exploited this source by much irrational and non-urgent expenditure, and thus pushed inflation to unnecessarily high levels by creating excessive money supply and widespread liquidity.

Second, there remains the question of the sanctions and reparations imposed on Iraq hard on the heels of the second Gulf War. Due to this factor most, if not all, writers tend to be pessimistic about the future of Iraq's economy. This, of course, is understandable: because of the large size of the reparations, much time and vast funds will be required to cover them, and extensive efforts will be needed to meet the costs, even under highly favourable assumptions concerning the volume of future oil exports, prices and revenues. Reparations thus constitute a heavy burden on the Iraqi economy and its future development, and hence represent a severe constraint on it. This is an undeniable fact.

However, it should not be forgotten that sanctions, reparations and boycotts are all political variables with an economic content. They are thus unstable and subject to change. It is unwarranted to assume that they will last forever; for politics, whether national or international, is like shifting sands. For this reason I do not expect these burdens on Iraq to stay put. The Middle East is in ferment and it is witnessing events and undergoing changes of all kinds. As the international community, including the Americans and the British, knows very well, a divided, economically degraded and politically unstable Iraq will harm the whole region, and perhaps the whole world. It is therefore in the interest of all parties concerned, including the major world powers, to take a pragmatic view and look towards a more stable future by supporting a free, independent, unified and prosperous Iraq.

Contributors

Mahmood Ahmad is Senior Policy Analyst with the FAO's Regional Near East Programme, based in Cairo. He has led research teams in many countries including Iraq.

Abbas Alnasrawi is Professor of Economics at the University of Vermont. His numerous publications include *Financing Economic Development in Iraq*, New York: Praeger, 1967; and *The Economy of Iraq*, Westport, Conn.: Greenwood Press, 1994.

Fadhil Chalabi is Executive Director of the Centre of Global Energy Studies. Between 1983 and 1988 he was Acting Secretary-General of OPEC and he had earlier served as Under-Secretary for Oil at the Iraqi Ministry of Petroleum. He has published widely on the economics and political economy of oil.

Kiren Aziz Chaudhry is Associate Professor in Politics at the University of California at Berkeley. She has written on Iraq's privatisation programme, and more widely on states and economies in a number of developing countries. Her 1997 book, *The Price of Wealth*, received wide acclaim.

Haris Gazdar is political economist associated with the Social Science Research Collective, Karachi. His research interests include poverty, public goods, problems of collective action, and conflict. He was co-author of the economy section of the Harvard Study Team's 1991 post-war report on Iraq, and his present contribution, which updates the earlier report, was written when he was based at the London School of Economics.

Sarah Graham-Brown has wide research experience of Middle East issues and organised the Gulf Information Project after 1990. Her book, *Sanctioning Saddam: The Politics of Intervention in Iraq*, was published by I. B. Tauris in 1999.

Athar Hussain is Deputy Director of Research at the Asia Research Centre, London School of Economics. Among his publications is *The*

Political Economy of Hunger with Jean Drèze and Amyarta Sen, Oxford: Oxford University Press, 1995.

Ahmed M. Jiyad has published on debt, poverty and development issues. He has a long professional experience as a senior economist in Iraq, with the Ministry of Oil and with the External Economic Relations Commission. Subsequently he worked at the University of Bergen.

Tariq Al-Khudayri is an expert on Arab industry who worked at UNIDO for 15 years, having previously been with the Iraqi Ministry of Industry and Baghdad University.

Michiel Leezenberg teaches at the University of Amsterdam, and has a wide research interest and publications on the Kurdish question.

Kamil Mahdi teaches economics of the Middle East at the University of Exeter. He has written widely on the economy and agriculture of Iraq.

Abdul Munim Al-Sayyid Ali is Professor of Economics at Al Al-Bayt University, Jordan and Emeritus Professor at Al-Mustansiriyyah University, Baghdad. He has numerous publications, particularly in the field of money and monetary policy in Arab countries.

Sinan Al-Shabibi is a Senior Economist with UNCTAD, Geneva. He has worked on debt management, balance of payments, the development implication of globalization, and accession to WTO. Before joining UNCTAD in 1980 he worked with the Iraqi Ministry of Planning. His publications are in areas of finance, economics of disarmament, and prospects for the Iraqi economy.

Index